COMPREHENSIVE GLOSSARY OF
TELECOM
ABBREVIATIONS
AND ACRONYMS

COMPREHENSIVE GLOSSARY OF
TELECOM
ABBREVIATIONS
AND ACRONYMS

Ali Akbar Arabi

Auerbach Publications
Taylor & Francis Group
Boca Raton   New York

Auerbach Publications is an imprint of the
Taylor & Francis Group, an **Informa** business

Auerbach Publications
Taylor & Francis Group
6000 Broken Sound Parkway NW, Suite 300
Boca Raton, FL 33487-2742

Library of Congress Cataloging-in-Publication Data

Arabi, Ali Akbar.
    Comprehensive glossary of telecom abbreviations and acronyms / Ali Akbar
Arabi.
        p. cm.
    Includes bibliographical references and index.
    ISBN-13: 978-1-4200-5866-6 (alk. paper)
    ISBN-10: 1-4200-5866-5 (alk. paper)
    1. Telecommunication--Abbreviations--Dictionaries. 2.
Telecommunication--Dictionaries. I. Title.

TK5102.A72 2011
384.01'48--dc22   2007012820

Visit the Taylor & Francis Web site at
http://www.taylorandfrancis.com

and the Auerbach Web site at
http://www.auerbach-publications.com

Table of Contents

Preface

Abbreviations are the important part of any scientific language of modern life. In recent decades, all branches of science have been widely expanded and many words and terms have been utilized to express their concept. Many of them are compound phrases and high-frequency terms, and if we use them as they are, our speech and writing becomes very lengthy and boring. That is why people prefer the short forms (abbreviations) for frequently used compound phrases.

The main objective of technical abbreviations is to save time and decrease speech and writing to facilitate engineers' communication. The other intention is the explicitness, fluency, and beauty of our speech and writing. Any method, which can help us to achieve these goals, is acceptable. Abbreviations contribute to faster writing, reading, speaking, and better understanding of technical and scientific articles, reports, and lectures. They can save time and money by reducing the time and cost of writing, typing, editing, and composition resetting costs. They also reduce the space required for texts, tables, and diagrams. Therefore, they are very good tools to enhance any live and dynamic language.

The words of the English and other European languages comprise a few special characters implying distinct technical meanings, such as λ for wavelength or φ for phase. In these languages, abbreviations play a similar role, but each of them implies the meaning of a "phrase" instead of a "word." Thus, they can be considered as the elements of that language and are treated like the original words. Sometimes, the short form of a phrase is more common than that of its original form because they are very proper and effective tools for communication. In most languages, especially in English, coining abbreviations is a common practice and most of the high-frequency compound terms are used in short forms. For this reason, the abbreviations of a language are considered as "a language within that language", and from the socio-linguistic point of view, they are comparable with a "dialect." Therefore, utilization of abbreviations is inevitable and it is a wise decision to take action for their standardization.

Without standard rules and principles, there will be many debates and discussions about writing style, when engineers are writing their manuscripts. To avoid these debates, it is recommended that writing principles and rules are formulated and standardized. With this in mind, I have followed the principles and rules adopted by John Markus, the composer of "Electronics Dictionary" as described in the fourth edition and recommend them to be standardized for international application.

This book is entitled as "Comprehensive Glossary of Telecom Abbreviations", but since each branch of science listed below are somehow interrelated and intermingled with telecommunications engineering, I have also included the abbreviations of those branches. The abbreviations listed in this work have been compiled over years from intensive study of various resources. A glossary of abbreviations is a growing work and will never be perfect and fully comprehensive. Despite this, I have tried all my best to include a collection of over 16,200 entries from almost all the fields of telecommunications, satellite communications, electronics, computer, Internet, broadcasting, fiber optics, information technology (IT), ICT, live Internet chat, e-learning, e-commerce, remote sensing, cellular networks, avionics, Ham radio, radar, and military communications terms to satisfy the needs of engineers, technical writers, technicians, and university students.

Particular attention is paid to the abbreviations used in ITU, IMO, Intelsat, Inmarsat, and APSCC documents. Telecommunication magazines are not forgotten and many titles and volumes have been searched. You may also find many abbreviations used in contracts and financial documents. The most famous telecom companies and operators are included. A larger number of abbreviations made and used by the specialized institutions and standard organizations such as IEEE, ETSI, IETF, and ISO are also included. New technologies and systems such as ISDN, ATM, TMN, SONET, SDH, DECT, and Bluetooth are of special importance in this glossary.

All entries are listed in alphabetical order. Where two or more entries differ only in capitalization or punctuation, the forms comprised of uppercased (capital) letters will follow the lowercased forms, and those having dots, slashes, or spaces precede the capital forms. Entries commencing with numbers are collected before all others. Where more than one phrase stand for a given abbreviation, both or all phrases are mentioned and marked by numbers. Entries of Greek symbols or letters are also alphabetized as if they were spelt out. Following the phrases, the relevant field of each entry has also been shown in parentheses or brackets to clarify or limit the usage field. If an entry does not fall into a specific category, the name of the company which coined that abbreviation is mentioned in parentheses.

Ali Akbar Arabi

Acknowledgments

My sincere gratitude goes to dear friends Abbas Pourkhesalian and Peyman Rezaee for their constructive help and valuable consultations. They contributed so much to solve and cope with the problems, which I faced while compiling this glossary.

Introduction

STRUCTURE OF ABBREVIATIONS

Abbreviation, in its simplest definition, is a shortened form of a word or phrase used mostly in writing to imply and represent the complete form of that word or phrase. There are many ways to build abbreviations for words or phrases which fall into eight main categories as follows.

1.   Acronyms: The pronounceable abbreviations coined from the combination of the first letters of each successive words of a compound term or phrase, as in **radar**, **laser**, **maser**, and **sonar**. The most common acronyms are normally written in lowercase letters and pronounced in the same way as the ordinary words. They may be either initially capitalized or set in all capitals in titles and headings.

2.   Initials: A category of abbreviations in which the first letters of a series of words are selected and spelt out letter by letter but are not separated by dots, as in **ISDN**, **CDMA**, and **DWDM**.

3.   Blendings: Another category of abbreviations in which the first syllables of two or three words are combined and pronounced as an ordinary word, such as **codec** for **co**der–**dec**oder, **modem** for **mod**ulator–**dem**odulator, and **Inmarsat** for **In**ternational **mar**itime **sat**ellite Organization. Other forms of blendings are made of the first syllable of the first word and the last syllable of the second word, such as **netiquette**, which stands for **net**work-et**iquette** and **webinar** standing for **Web**-based sem**inar**.

4.   Clippings: A category of abbreviations in which some of the letters or sounds of a word are omitted and the key letters are combined, e.g., **Bldg** for **b**ui**ld**in**g** and **MUX** for **mu**ltiple**x**er.

5.   Truncations: A category of abbreviations in which a word is simply shortened by cutting off the first or last syllable, such as **phone** for tele**phone**, **amp** for **amp**lifier, and **Fig** for **fig**ure. For building such short forms, it is recommended to select at least three letters of the word.

6. Contractions: A category of abbreviations in which the first and last letters of a word is selected, such as **Mr** for **M**iste**r** and **Dr** for **D**octo**r**.

7. Symbols: Single letters generally used to represent an element, a quality, a physical or numerical quantity, often appended with subscripts or superscripts to qualify their meaning, such as **C** for **C**apacitor and **R** for **R**esistor. Symbols may not be related to the words that they represent. Examples are **C** for speed of light in a vacuum, **G** for electrical conductance, **H** for magnetic field strength, and **L** for inductance or an inductor.

8. Signs: A category of abbreviations which are printed or written figures (mostly Greek letters) each one conventionally stands for a word, phrase, or mathematical operation, e.g., Ω for **ohm**, Φ for **phase**, μ for **micro**, λ for **wavelength**, and ° for **degree**.

RULES FOLLOWED IN THIS GLOSSARY
(Courtesy of John Markus and McGrawHill)
Part A: Rules for abbreviations

1. All the abbreviations formed from first letters of key words in a phrase are uppercased (capital) letters, except the units of measures which conform to the French SI units of measure adopted by the American National Standards Institute. There is another exception for the soundalike words, such as laser, maser, and radar, which are lowercased.

2. Abbreviations of phrases are generally pronounced letter by letter, but this is not a requirement for capitalizing. Thus, FET can be pronounced either letter by letter or as a word. Long abbreviations such as JFET and MOSFET are usually pronounced partially or entirely as words.

3. A Hyphen (-) is seldom used in abbreviations. When used, it will generally have one of the meanings represented by the slash. Thus, in A/D-D/A the hyphen stands for *or*, but in P-P it stands for *-to-*.

4. A hyphen should be used after an abbreviation when it is combined with another word to form a compound adjective, such as "JPEG-formatted picture."

5. A hyphen should be used before an abbreviation when it is combined with another word to form a compound adjective, such as "Broadband-ISDN."

6. For plural forms of capital-letter abbreviations, especially the acronyms and blendings, regardless of whether in text or headings, same as the ordinary words, simply add a lowercase "s", e.g., ICs, VSATs, FETs, RAMs, LEDs, SCRs, MICs, and PCBs. Same rule applies to plurals of numbers, e.g., 0s and 1s.

7. Use an apostrophe with the plurals of lowercase letters, as in "programming has two m's."

8. Use an apostrophe for the possessive form of a capital-letter abbreviation, as in "the SCR's heat sink."

9. Use a space between a number and its abbreviated unit of measure in noun phrases, such as "5 kΩ and 100 VAR."

10. Use a numerical exponent to indicate repetition of a letter in abbreviations for logic terms, such as T^2L which stands for TTL.

11. A slash (/) generally means -to-, such as S/N (signal-to-noise). It will also have the meaning of *per* when it is used with letter symbols, such as kHz/s which stands for "kilo Hertz per second."

12. Although the names of programming languages are acronyms and therefore are pronounced as words, it is now a common usage to type or print them in capital letters, e.g., ALGOL, COBOL, BASIC, and FORTRAN.

13. Do not italicize R, L, and C where they represent resistance or resistor, inductance or coil, and capacitance or capacitor, respectively. When R, L, and C are used in combination, run the letters together without hyphens, as in RC coupling and RLC circuit. Italicize these letters only in mathematical equations where they represent values of resistance, inductance, and capacitance, respectively, and in expressions like LC product, L/C ratio, and RC constant.

Part B: Rules for SI Units of measure

1. When a letter symbol in SI system of units is derived from the name of a pioneer in the field, capitalize the letter that represents the first letter of the surname, such as "μF" which stands for microfarad.

2. Standard prefixes are used in lowercase letters as multipliers with the SI letter symbols. Note that only six largest prefixes have capital letters for their symbols, e.g., **M** for **m**ega-, **G** for **g**iga-, **T** for **t**era-, **P** for **p**eta-, **Y** for **y**otta-, and **Z** for **z**etta-.

3. Letter symbols always represent both singular and plural forms of units of measure. Never add a lowercase "s" for plural because it represents seconds in the SI system of units.

4. Use an exponent after a length symbol to change it to an area or volume symbol. (The abbreviations **sq** for square and **cu** for cubic are used with SI units.) Pronunciation is unchanged; thus, **in**^{2} is pronounced "square inches" and **in**^{3} is pronounced "cubic inches."

5. Use a space between a number and its abbreviated unit of measure in noun phrases.

6. Use a hyphen (-) between a number and its abbreviated unit of measure in compound adjectives, such as "a 20-kHz signal."

7. Use a hyphen between two numbers that are the upper and lower limits of a range. In this case, the meaning of hyphen is "up to and including."

8. Do not use a letter symbol without a numerical value. Spell out units of measure whenever they are used with words that represent approximations.

The Author

Ali Akbar Arabi is a senior technical engineer with Inmarsat Mobile Satellite Communication Systems, handling the affairs of the Inmarsat network plus the IMSO and APSCC organizations in Iran. He holds degrees in Telecommunications Engineering and Marine Communications Engineering.

He served for five years as "Radio Officer" on board the Iranian Shipping lines, two years in Railway Communications, six years in Inmarsat satellites new systems, many years as the Inmarsat network operations coordinator of Iran, and now serves as the senior expert of satellites affairs for the Iranian Space Agency (ISA).

As an author of five books, Mr. Arabi is a well-known name in the Iranian publisher's society. In any Iranian public library you may find one of his books. He holds membership in the Iranian Association of Electrical and Electronics Engineers (IAEEE), the Iranian branch of IEEE. He is also a member of the Academy of Persian Language and Literature on standardization of Persian scientific words. He has cooperated with the Institute of Standards and Industrial Research of Iran (ISIRI) in compiling the standards for Iranian telecom companies. He has invented a "home-based automatic earthquake alarm" and may be contacted at *aarabi@dci.ir* .

Acronyms and Abbreviations

Numbers (0–9)

0	The standard number dialed for accessing a local telephone company operator
00	The standard number dialed for accessing an international telephone company operator
0RL	**Zero R**elative **L**evel (transmission)
0TLP	**Zero T**ransmission **L**evel **P**oint (transmission)
1×EV-DO	**1×E**volution-**D**ata **O**ptimized
1.44M	The capacity of high-density 3.5-inch floppy disks
10/100	An Ethernet supporting **10** Mbps **and 100** Mbps at same port
10BaseF	A **10**-Mbps **Base**band Ethernet using **F**iber-optic cabling (IEEE)
10BaseFP	A **10**-Mbps **F**iber-**P**assive **Base**band Ethernet (IEEE)
10BaseT	A **10**-Mbps **Base**band LANs using **T**wisted-pair cabling (IEEE)
10GBE	**10 G**iga**b**it **E**thernet (IEEE)
10GEA	**10 G**igabit **E**thernet **A**lliance (IEEE)
100BaseFX	A **100**-Mbps **Base**band **F**ast Ethernet using multimode Fiber-optic cabling (IEEE)
100BaseT	**100** Mbps, **Base**band, **T**wisted pair (IEEE)
100BaseT2	**100** Mbps, **Base**band, with **two T**wisted pair (IEEE)
14.4	Having maximum data transfer rate of **14.4** kbps (modems)
16-QAM	**16-state Q**uadrature **A**mplitude **M**odulation
1BL	**One B**usiness **L**ine (class of service)
1F4	**One F**our-user party line (class of service)
1FB	**One F**lat-rate **B**usiness phone line (class of service)
1FL	**One F**amily phone **L**ine (class of service)

1FR	**One F**lat-rate **R**esidential phone line (class of service)
1G	**First G**eneration (mobile phones)
1MB	**One M**easured-rate **B**usiness phone line (class of service)
1MR	**One M**easured-rate, **R**esidential phone line (class of service)
1NF	**First N**ormal **F**orm (computer database)
2-FSK	**Two**-level **FSK** (modulation)
2/4-QPSK	**Two Q**uadrature **PSK** (modulation)
21CN	**21**st **C**entury **N**etwork
23B+D	Having **23 B**earer channels **and** one **D**ata channel (ISDN)
24×7	**24** Hours of a day, **Seven** days of a week (technical support)
2B1Q	**Two B**inary, **One Q**uaternary (ISDN)
2B+D	Having **two B**earer channel **and** one **D**ata channel (ISDN)
2F	**Two-F**iber (line or cable)
2FR	**F**lat-**R**ate party line having **two** subscribers (service code)
2G	**Second G**eneration (mobile phones)
2.5G	**Second-and-half G**eneration (mobile phones)
286	A simple expression for Intel's 80**286** microprocessor
287	A simple expression for Intel's 80**287** microprocessor
28.8	Having a maximum transfer rate of **28.8** kbps (modems)
2NF	**Second N**ormal **F**orm (computer database)
2W	**Two-W**ire (line or cable)
3+	An operating system developed by **3Com** Company (networking)
386	A simple expression for Intel's 80**386** microprocessor
387	A simple expression for Intel's 80**387** microprocessor
3B/2T	**Three B**inary bits encoded into **two T**ernary symbols
3D	**Three-D**imensional
3D API	**3D A**pplication **P**rogramming **I**nterface (computer)
3DES	**Triple D**ata **E**ncryption **S**tandard
3DGF	**3D G**eometry **F**ile
3FR	**F**lat-**R**ate party line having **Three** subscribers (service code)
3G	**Third G**eneration (mobile phones)
3GL	**Third G**eneration Programming **L**anguage
3GPP	**Third G**eneration **P**artnership **P**roject (GSM)
3GPP2	**Counterpart** of **3GPP** responsible for CDMA-2000 (GSM)
3GSM	**Third**-Generation **GSM**
3GTS	**Third-G**eneration **T**est **S**ystem (Agilient Company)
486	A simple expression for Intel's 80**486** microprocessor
486DX	A simple expression for Intel's 80**486DX** microprocessor
486SL	A simple expression for Intel's 80**486SL** microprocessor
476SX	A simple expression for Intel's 80**476SX** microprocessor
4B/3T	**4 B**inary bits encoded into **3 T**ernary (data encoding)
4B/5B	**4 B**inary bits encoded into **5 B**inary bits (data encoding)
4-DEPSK	**4-phase D**ifferentially-**E**ncoded **PSK** (modulation)
4F	**Four-F**iber (cable)

4FR	Flat-**R**ate party line having **four** subscribers (service code)
4FSK	**Four**-level **FSK** (modulation)
4G	**Fourth**-**G**eneration (mobile phones)
4GL	The **Fourth**-**G**eneration Programming **L**anguage
4NF	**Fourth** **N**ormal **F**orm (computer database)
4PDT	**Four**-**P**ole **D**ouble-**T**hrow (switches)
4PST	**Four**-**P**ole **S**ingle-**T**hrow (switches)
4W	**Four**-**W**ire (line or cable)
4WL-WDM	**Four**-**W**avelength **W**avelength-**D**ivision **M**ultiplexing
5×5	**Five by Five** (teletype communication)
5×B	No. **5 CrossB**ar Circuit Switch
56K	Having a maximum data transfer rate of **56 k**bps (modems)
586	An unofficial name for Intel's **Pentium** microprocessor
56 kbps	**56,000 B**its **P**er **S**econd
5B6B	**5 B**inary bits encoded into **6 B**inary bits (data encoding)
5ESS	Number **5 E**lectronic **S**witching **S**ystem (AT&T)
5NF	**Fifth** **N**ormal **F**orm (computer database)
64 kbps	**64,000 B**its **P**er **S**econd
64QAM	**64**-state **Q**uadrature **A**mplitude **M**odulation
6800	An 8-bit microprocessor developed by Motorola
73	Best Regards (Morse code transmissions)
80286	A 16-bit microprocessor developed by Intel
80287	A coprocessor for use with the **80286** microprocessors (Intel)
80386	See **80386**DX
80386DX	A 32-bit microprocessor developed by Intel
80386SL	A microprocessor developed by Intel for laptop PCs
80386SX	A low-cost version of **80386DX** microprocessor (Intel)
80387	A coprocessor for use with the **80386** microprocessors (Intel)
80387SX	A coprocessor for use with the **80386SX** microprocessors (Intel)
80486	See i**486**DX
80486SL	See i**486SL**
80486SX	See i**486SX**
8080	One of the first chipsets served as the basis of PCs
8086	The original microprocessor in the **80×86** family (Intel)
8087	A coprocessor developed by Intel for use with the **8086/8088** and microprocessors (Intel)
8088	The original microprocessor of the first IBM PCs
88	Love and Kisses (Morse code transmissions)
8B/6T	**8 B**inary bits encoded into **6 T**ri-state symbols (data encoding)
8B/10B	**8 B**inary bits encoded into **10 B**inary bits (data encoding)
8FR	Flat-**R**ate party line having **eight** subscribers (service code)
8PSK	**O**ctant **P**hase **S**hift **K**eying (modulation)
9600	Having a maximum data transfer rate of **9600** bps (modems)

A a

a	Symbol for prefix **a**tto-, denoting one-quintillionth or 10^{-18}
A	1. Symbol for **A**mpere (electronics)
	2. Symbol for **A**node (electronics)
	3. Symbol for **A**cceleration
	4. Symbol for **A**rea
Å	Symbol for **A**ngstrom (unit of wavelength of light)
A²DP	**A**dvanced **A**udio **D**istribution **P**rofile
A-BPSK	**A**viation **B**inary **P**hase **S**hift **K**eying (modulation)
A-QPSK	**A**viation **Q**uadrature **P**hase **S**hit **K**eying (modulation)
A-TDMA	**A**dvanced **TDMA** (access)
A/D	**A**nalog-**to**-**D**igital (converters)
A/D-D/A	**A**nalog-**to**-**D**igital **or** **D**igital-**to**-**A**nalog (converters)
A/m	**A**mpere **per m**eter
aA	**a**tto**a**mpere
AA	1. **A**ccounting **A**uthority (Inmarsat)
	2. **A**utomated **A**ttendant
	3. **A**uto **A**nswer (modems)
	4. **A**ccess **A**dapter
AAA	1. **A**uthentication, **A**uthorization, and **A**ccounting (LAN server)
	2. **ATM A**ccess **A**dapter
AAAC	**A**ll **A**luminum **A**lloy **C**able
AAAI	**A**merican **A**ssociation for **A**rtificial **I**ntelligence

AAAS	American Association for the Advancement of Science
AABS	Automated Attendant Billing System
AAC	1. Airline Administrative Communications
	2. Aeronautical Administrative Communications
	3. Augmentative and Alternative Communications (ICT)
	4. Automatic Amplitude Control
	5. Abbreviated Address Calling
	6. Advanced Audio Codec
	7. Access Authorization Certificate (Inmarsat)
AACR	All America Cables and Radio, Inc.
AACS	Attitude and Articulation Control Subsystem (spacecraft)
AAIC	1. Accounting Authority Identification Code (Inmarsat)
	2. Active Addressing Integrated Circuit
AAL	ATM Adaptation Layer
AAL-1	ATM Adaptation Layer 1
AAL-2	ATM Adaptation Layer 2
AAL-3/4	ATM Adaptation Layers 3 and 4
AAL-5	ATM Adaptation Layer 5
AAN	Associated Account Number (wireless)
AAP	1. Application Access Point
	2. Address Allocation Protocol
	3. Alternative Access Provider
AAPI	Audio Application Programming Interface (computer)
AAR	Automatic Alternate Routing (network feature)
AARP	AppleTalk Address Resolution Protocol
AARS	Asian Association of Remote Sensing
AARTS	Automatic Audio Remote Test Set
AAS	1. Authorized Application Specialist
	2. Aeronautical Advisory Station
AATS	Access Approval Test System
AATSR	Advanced Along-track Scanning Radiometer (remote sensing)
AAV	Automated Address Verification
AB	1. Access Burst (GSM)
	2. Additional Bits (MUX)
	3. Asynchronous Balanced mode (OSI model)
ABARIS	Advanced Broadband Architecture for Internet Services
ABC	1. Arbitration Bass Controller (electronics)
	2. Automatic Bill Calling (payphone)
	3. Automatic Bias Compensation (circuit)
	4. Activity-Based Costing (ICT finance)
	5. Automated Brightness Control (television)
	6. Australian Broadcasting Corporation (TV broadcaster)

	7. **A**rmored **B**rush **C**able
	8. **A**utomated **B**usiness **C**onnection
ABCS	**A**dvanced **B**usiness **C**ommunications via **S**atellite
ABD	**A**verage **B**usiness **D**ay
ABDN	**A**ttendant **B**locking of **D**irectory **N**umber
ABEC	**A**lternate **B**illing **E**ntity **C**odes
ABEL	**A**dvanced **B**oolean **E**quation **L**anguage
ABEND	1. **Ab**normal **END** (computer)
	2. **Ab**ortive **END**
ABF	**A**ir-**B**lown **F**iber
ABI	**A**pplication **B**inary **I**nterface
ABIOS	**A**dvanced **BIOS** (computer)
ABIST	**A**utonomous **B**uilt-**i**n **S**elf-**T**est
ABM	1. **A**synchronous **B**alanced **M**ode (LANs)
	2. **A**pogee **B**ooster **M**otor
ABME	**A**synchronous **B**alanced **M**ode **E**xtended
ABN	**abn**ormal (alarm status)
ABR	1. **A**vailable **B**it **R**ate (ATM)
	2. **A**rea **B**order **R**outer
	3. **A**uto **B**aud **R**ate detect (data communications)
abs	**abs**olute value (mathematics)
ABS	1. **A**lternate **B**illing **S**ervices (INs)
	2. **A**verage **B**usy **S**eason
	3. **A**eronautical **B**roadcast **S**tation
ABSBH	**A**verage **B**usy **S**eason **B**usy **H**our
ABT	**A**dvanced **B**roadcast **T**elevision
	about (Morse code transmissions)
ABU	**A**sia-Pacific **B**roadcasting **U**nion
ABUI	**A**ssociation of **B**anyan **U**sers **I**nternational
ABX	**A**dvanced **B**ranch **Ex**change
AC	1. **A**lternating **C**urrent (electronics)
	2. **A**ssignment **C**hannel
	3. **A**bsorption **C**oefficient (fiber optics)
	4. **A**ccess **C**ontrol (LANs)
	5. **A**ccess **C**ode
	6. **A**cquisition **C**ycle
	7. **A**uthentication **C**enter
AC/AC	**A**lternating **C**urrent **to A**lternating **C**urrent (power)
AC/DC	1. **A**lternating **C**urrent **to D**irect **C**urrent (power)
	2. Operating both with **AC and DC** power lines (equipment)
ACA	1. **A**merican **C**ommunication **A**ssociation
	2. **A**ustralian **C**ommunications **A**uthority
	3. **A**utomatic **C**ircuit **A**ssurance (PBX feature)

AcademNet	**Academ**ic Research **Net**work (Russia)
ACAR	**A**luminum **C**onductor **A**lloy-**R**einforced
ACARD	1. **A**dvisory **C**ouncil for **A**pplied **R**esearch and **D**evelopment
	2. **A**cquisition **Card** program
ACARS	**A**RINC **C**ommunications **A**ddressing and **R**eporting **S**ystem
ACAT	**A**dditional **C**ooperative **A**cceptance **T**esting (test method)
ACATS	**A**dvisory **C**ommittee on **A**dvanced **T**elevision **S**ervice (standard)
ACB	1. **A**nnoyance **C**all **B**ureau
	2. **A**rchitecture **C**ontrol **B**oard
	3. **A**TM **C**ell **B**us
	4. **A**utomatic **C**all**b**ack
ACC	1. **A**nalog **C**ontrol **C**hannel (wireless)
	2. **A**rea **C**ontrol **C**enter (COMSAR)
	3. **A**uthorization **C**ontrol **C**enter
	4. **A**daptive **C**ruise **C**ontrol (car radar)
	5. **A**utomatic **C**allback **C**alling
	6. **A**rthur **C**. **C**larke Institute for Modern Technologies
ACCS	1. **A**utomatic **C**alling **C**ard **S**ervice
	2. **A**ssociated **C**ommon-**C**hannel **S**ignaling
ACCOLC	**Acc**ess **O**ver**l**oad **C**lass
ACCU	**A**ssociation of **C** and **C**++ **U**sers
ACCH	**A**ssociated **C**ontrol **Ch**annel (GSM)
ACD	1. **A**utomatic **C**all **D**istributor (PBX)
	2. **A**verage **C**all **D**istance
ACE	**A**utomatic **C**ross-connection **E**quipment
ACEC	**A**dvisory **C**ommittee for **E**lectronics and **C**ommunications
ACELP	**A**lgebraic **C**ode **E**xcited **L**inear **P**rediction (voice coding)
ACeS	**A**sian **Ce**llular **S**atellite
ACF	**A**dvanced **C**ommunication **F**unction (software)
ACF/NCP	**ACF** **N**etwork **C**ontrol **P**rogram (software)
ACF/VTAM	**ACF** **V**irtual **T**erminal **A**ccess **M**ethod (software)
ACFG	**a**uto**c**on**fig**uration (BIOS extensions)
ACG	**A**utomatic **C**all **G**apping
ACH	**A**ttempts per **C**ircuit per **H**our (call centers)
ACI	**A**djacent **C**hannel **I**nterference
ACIA	**A**synchronous **C**ommunications **I**nterface **A**dapter (computer)

A

ACID	Atomic, Consistent, Isolation, Durable (database transaction)
ACIF	Australian Communications Industry Forum
ACIR	Adjacent Channel Interference Ratio (microwave)
ACIS	1. Advanced Cargo Information System
	2. Automatic Caller/Customer Identification Service
	3. Andy, Charles, Ian's System (computer modeling)
ACITS	Advisory Committee on Information Technology Standardization
ACK	acknowledge (control character)
ACL	1. Access Control List (networking)
	2. Advanced CMOS Logic (digital electronics)
	3. Application Connectivity Link (Siemens' protocol)
	4. Association for Computational Linguistics
ACLC	Adaptive Communication Line Converter
ACLR	Adjacent Channel Leakage Ratio (microwave)
ACM	1. Association for Computing Machinery
	2. Automatic Call Manager
	3. Address Control Memory
	4. Address Complete Message (ATM)
	5. Advanced Communications Function
ACMPA	Aperture-Coupled Microstrip Patch Antenna (microwave)
ACNA	Access Customer Name Abbreviation
ACO	1. Additional Call Offering (ISDN)
	2. Alarm Cut Off (MUX)
ACON	Administrative Operating Company Number
ACONet	Academic Computer Network (Australia)
ACOST	Advisory Council on Science and Technology
ACP	1. Activity Concentration Point
	2. Adjacent Channel Power (microwave)
	3. Association of Computing Professionals
	4. Azimuth Change Pulse
ACPI	Advanced Configuration and Power Interface (computer)
ACPR	Adjacent Channel Power Ratio (microwave)
ACPW	Asymmetric Coplanar Waveguide (microwave)
ACR	1. Allowed Cell Rate (ATM)
	2. Abandon Call and Retry
	3. Attenuation to Crosstalk Ratio (transmission)
ACRFNET	Academic Computing Research Facility Network
ACS	1. Automatic Call Sequencer
	2. Advanced Communication System (AT&T)
	3. ATM Circuit Steering

	4. **A**ustralian **C**omputer **S**ociety
ACSB	**A**mplitute **C**ompandored **S**ingle Side**B**and
ACSE	1. **A**ccess **C**ontrol and **S**ignaling **E**quipment (Inmarsat)
	2. **A**ssociation **C**ontrol **S**ervice **E**lement (OSI model)
ACSL	**A**dvanced **C**ontinuous **S**imulation **L**anguage (programming)
ACSnet	**A**ustralian **C**omputer **S**ociety national computer **net**work
ACSS	**A**ssociation of **C**omputer **S**upport **S**pecialists
ACT	1. **A**pplied **C**omputer **T**elephony
	2. **A**uthorization **C**ode **T**able
	3. **A**ssociation of **C**ommunications **T**echnicians
ACTA	1. **A**merica's **C**arriers **T**elecommunications **A**ssociation
	2. **A**dministrative **C**ouncil for **T**erminal **A**ttachments (FCC)
ACTAS	**A**lliance of **C**omputer-based **T**elephony **A**pplication **S**uppliers
ACTE	**A**pprovals **C**ommittee for **T**erminal **E**quipment (GSM)
ACTGA	**A**ttendant **C**ontrol of **T**runk **G**roup **A**ccess
ACTIUS	**A**ssociation of **C**omputer **T**elephone **I**ntegration and **U**sers and **S**uppliers (U.K.)
ACTL	**A**ccess **C**ustomer **T**erminal **L**ocation (code)
ACTOM	**A**dvisory **C**ommittee on **T**echnical and **O**perational **M**atters (ITU)
ACTRIS	**A**ssociation for **C**ooperation in **T**elecommunications **R**esearch **I**n **S**witzerland
ACTS	1. **A**dvanced **C**ommunications **T**echnologies and **S**ervices
	2. **A**dvanced **C**ommunications **T**echnology **S**atellite (NASA)
	3. **A**ssociation of **C**ompetitive **T**elecommunications **S**uppliers
	4. **A**utomatic **C**oin **T**oll **S**ervice
ACU	1. **A**ntenna **C**ontrol **U**nit (GSM)
	2. **A**utomatic **C**alling **U**nit (IBM)
ACUTA	**A**ssociation of **C**ollege and **U**niversity **T**elecommunications **A**dministrators (standards organization)
ACV	**A**dvanced **C**ommon **V**iew
ACWG	**A**symmetrical **C**oplanar **W**ave-**G**uide (microwave)
ACWPBX	**A**dvanced **C**ordless **W**ireless **P**rivate **B**ranch **Ex**change
AD	**A**dministrative **D**omain
ADA	1. **A**verage **D**elay to **A**bandon
	2. **ADA** Lovelace Language (programming)
ADACC	**A**utomatic **D**irectory **A**ssistance **C**all **C**ompletion

A

ADAD	**A**utomatic **D**ialing and **A**nnouncing **D**evice
ADAR	**A**dvanced **D**esign **A**rray **R**adar
ADAS	**A**utomated **D**irectory **A**ssistance **S**ervice (Northern Telecom)
ADB	**A**pple **D**esktop **B**us (Macintosh)
ADC	1. **A**nalog-to-**D**igital **C**onverter (electronics)
	2. **A**utomated **D**ata **C**ollection
	3. **A**sia **DAB C**ommittee
ADCA	1. **A**utomatic **D**ata **C**ollection **A**ssociation
	2. **A**utomatic **D**ata **C**apture **A**ssociation
ADCCP	**A**dvanced **D**ata **C**ommunications **C**ontrol **P**rotocol (ANSI)
ADCU	**A**ssociation of **D**ata **C**ommunications **U**sers
ADDMD	**Ad**ministrative **D**irectory **M**anagement **D**omain (X.500)
ADE	**A**bove-**D**ecks **E**quipment
ADEOS	**Ad**vanced **E**arth **O**bserving **S**atellite (remote sensing)
ADF	1. **A**utomatic **D**irection **F**inder (equipment)
	2. **A**utomatic **D**ocument **F**eeder
ADFOC	**A**ll **D**ielectric **F**iber-**O**ptic **C**able
ADH	1. **A**verage **D**elay to **H**andle
	2. **A**utomatic **D**ata **H**andling
ADI	**A**lternate **D**igit **I**nversion (data encoding)
ADIO	**A**nalog/**D**igital **I**nput/**O**utput
ADK	**A**pplication-**D**efinable **K**eys
ADL	**A**dvanced **D**istributed **L**earning (e-learning)
ADLNet	**A**dvanced **D**istributed **L**earning **Net**work (e-learning)
ADM	1. **A**daptive **D**elta **M**odulation
	2. **A**dd and **D**rop **M**ultiplexer (MUX)
ADMD	**Ad**ministration **M**anagement **D**omain (X.400)
ADML	**A**symmetric **D**igital **M**icrocell **L**ink (Telcordia standard)
ADN	**A**dvanced **D**igital **N**etwork (Bell)
ADNT	**A**dvanced **D**igital **N**etwork **T**runking
ADO	**A**uxiliary **D**isconnect **O**utlet
ADONIS	**A**rticle **D**elivery **O**ver **N**etwork **I**nformation **S**ystems (ICT)
ADP	**A**utomatic **D**ata **P**rocessor (computer)
ADPCM	1. **A**daptive **D**ifferential **PCM** (modulation)
	2. **A**ssociation for **D**ata **P**rocessing and **C**omputer **M**anagement
ADPCOD	**Ad**aptive **P**rediction **En**co**d**er
ADPDEC	**Ad**aptive **P**rediction **Dec**o**d**er
ADPE	**A**utomatic **D**ata **P**rocessing **E**quipment
ADPS	**A**utomatic **D**ata **P**rocessing **S**ystem
ADPSSO	**A**utomatic **D**ata **P**rocessing **S**ystem **S**ecurity **O**fficer

ADQ	Average Delay in Queue
ADR	1. Achievable Data Rate
	2. Aggregate Data Rate
	3. Analog to Digital Recording
	4. Alternate Destination Routing (AT&T)
ADRMP	Auto-Dialing Recorded Message Player
ADRT	Approximate Discrete Radon Transform
ADS	1. Automatic Dependent Surveillance (aviation)
	2. Automatic Data System
	3. Advanced Design System (Agilient's microwave)
	4. Angular Displacement Sensor (remote sensing)
	5. AudioGram Delivery Services (Nortel)
ADSI	Analog Display Services Interface (Telcordia standard)
ADSL	Asymmetric DSL (access)
ADSP	AppleTalk Data Stream Protocol (networking)
ADSR	Automatic Data Speed Recognition
ADSTAR	Automated Document Storage And Retrieval
ADSU	1. ATM Digital Service Unit (ATM hardware)
	2. Automatic Dependent Surveillance Unit (aviation)
ADT	1. Abstract Data Type
	2. Audio, Data, and Teletext
	3. Automatic Detection and Tracking (radar)
ADTF	ACR Decrease Time Factor
ADTV	Advanced Definition Television
ADU	1. Asynchronous Data Unit
	2. Above-Deck Unit (satellite terminals)
	3. Area Decision Unit (MUX)
	4. Average Daily Use
	5. Automatic Dialing Unit
ADX	Automatic Data Exchange
AE	1. Acoustic Emission
	2. Application Entity (OSI model)
	3. Account Executive
AEA	1. American Electronics Association
	2. American Engineering Association
	3. Aerial Experimental Association (Bell)
AEB	Analog Expansion Bus (computer)
AEC	1. Acoustic Echo Canceller
	2. Alternate Exchange Carrier
AECS	Aeronautical Emergency Communications System (avionics)
AECT	Association for Educational Communications and Technology
AEEC	Airlines Electronic Engineering Committee (ARINC)

AEEM	Aerospace Engineering and Engineering Mechanics
AEF	Aircraft Emergency Frequency
AEGIS	1. Advanced Electronic Guidance and Instrumentation System
	2. Airborne Early-warning Ground Integration System
AEI	Automatic Equipment Identification
AEJMC	Association for Education in Journalism and Mass Communication
AEMIS	Automatic Electronic Management Information System (AT&T)
AEP	AppleTalk Echo Protocol (networking)
AER	Advanced Edge Router
AEROSAT	Aeronautical Satellite Communications System
AEROTHAI	Aeronautical Radio of Thailand
AES	1. Aeronautical Earth Station (Inmarsat)
	2. Audio Engineering Society
	3. Advanced Encryption Standard
	4. Application Environment Standard (or Service)
AESA	Active Electronically Steered Array (microwave antennas)
AESS	Aerospace and Electronics System Society
AET	Application Entity Title (OSI model)
AEU	1. Auxiliary Electronics Unit
	2. Aircraft Early Warning
	3. Airborne Early Warning
	4. Asia Electronic Union
AEW	1. Aircraft Early Warning
	2. Airborne Early Warning
AF	1. Audio Frequency
	2. Assigned Frame (Motorola)
AFACTS	Automatic Facilities Test System
AFAQ	French Quality Assurance Association
AFAST	Advanced Flyaway Satellite Terminal
AFC	1. Automatic Frequency Control (electronics)
	2. Amplitude-to-Frequency Converter
	3. Advanced Fiber Communications
AFCEA	Armed Forces Communications and Electronics Association
AFDW	Active Framework for Data Warehousing
AFE	1. Analog Front End (functions)
	2. Antiferroelectric
AFI	1. Authority and Format Identifier (ATM)
	2. Address and Format Identifier (OSI model)

AFIPS	American Federation of Information Processing Societies
AFIS	1. Airborne Flight Information System
	2. Automated Fingerprint Identification System (Motorola)
AFM	1. Adobe Font Manager
	2. Adobe Font Metrics
	3. Antiferromagnetism
AFMR	Antiferromagnetic Resonance
AFN	Aeronautical Fixed Network (COMSAR)
AFNOR	Association Française de Normalization (France)
AFOSR	Air Force Office of Scientific Research (U.S.A.)
AFP	AppleTalk Filing Protocol (networking)
AFPCON	AFP Configuration (Novell NetWare)
AFRN	Armed Forces Radio Network (broadcaster)
AFRS	Armed Forces Radio Service (broadcaster)
AFRTS	American Forces Radio and Television Services (broadcaster)
AFS	1. Advanced Freephone Service
	2. Andrew File System (networks)
	3. Aeronautical Fixed Service (ITU)
	4. Automation File Server
AFSK	Audio Frequency-Shift Keying (modulation)
AFT	Automatic Fine Tuning (receivers)
AFTN	Aeronautical Fixed Telecommunications Network (COMSAR)
AFTRA	American Federation of Television and Radio Artists
AFV	Audio-Follow-Video
AG	Application Generator (software)
AGC	1. Automatic Gain Control (electronics)
	2. Audio Graphic Conferencing
AGCH	Access Grant Channel (GSM)
AGCOMNET	Agriculture's voice and data Communications Network (U.S.A.)
AGE	Aerospace Ground Equipment
AGN	again (Morse code transmissions)
AGP	Accelerated Graphics Port (Intel motherboards)
AGPS	Assisted GPS
AGRS	Air-Ground Radiotelephone Service (aviation)
AGT	1. Audio Graphics Terminal
	2. Albera Government Telephone
AGTK	Application Generator Toolkit
AGU	1. Address-Generation Unit
	2. Automatic Ground Unit

A

AGW	**A**ccess **G**ate**w**ay (switching)
AGWCS	**A**ir-**G**round **W**orldwide **C**ommunications **S**ystem (aviation)
Ah	**A**mpere-**h**our
AH	**A**uthentication **H**eader (IP datagrams)
AHD	**A**udio **H**igh-**D**ensity (recording)
AHDL	**A**nalog **H**ardware-**D**escription **L**anguage (software)
AHDS	**A**rts and **H**umanities **D**ata **S**ervice (ICT of JISC committee)
AHEN	1. **A**labama **H**ome **E**ducator's **N**etwork (U.S.A.)
	2. **A**lberta **H**igher **E**ducation **N**etwork (e-learning)
	3. **A**lternative **H**igher **E**ducation **N**etwork (e-learning)
AHF	**A**daptive **H**igh-**F**requency (radio)
AHFG	**A**TM-attached **H**ost **F**unctional **G**roup
AHP	**A**uthentication **H**eader **P**rotocol
AHR	**A**mpere **H**ou**r**
AHT	1. **A**verage **H**andle **T**ime
	2. **A**verage **H**olding **T**ime
.aif	Name extension for **A**udio **I**nterchange **F**iles (computer)
AI	1. **A**rtificial **I**ntelligence
	2. **A**irborne **I**nterception (radar)
	3. **A**ction **I**ndicator (ISDN)
AIA	1. **A**erospace **I**ndustries **A**ssociation (U.S.A.)
	2. **A**pplication **I**nterface **A**dapter
	3. **A**utomatic **I**nternet **A**dministration
AIC	1. **A**utomatic **I**ntercept **C**enter (Telcordia Technologies)
	2. **A**nalog **I**nterface **C**ircuit (modem)
AICC	1. **A**utomatic **I**ncoming **C**all **C**onnection
	2. **A**utonomous **I**ntelligent **C**ruise **C**ontrol (car radar)
AICE	**A**ustralian **I**nstitute of **C**omputer **E**thics
AID	**A**ccess **Id**entifier (MUX)
AIDA	**A**ccessible **I**nformation on **D**evelopment **A**ctivities
AIDC	**A**utomatic **I**dentification and **D**ata **C**ollection (Mexico)
AIDS	**A**ccess **Id**entifier **S**ystem
AIEE	**A**merican **I**nstitute of **E**lectrical **E**ngineers (now IEEE)
AIFF	**A**udio **I**nterchange **F**ile **F**ormat
AIIA	**A**ustralian **I**nformation **I**ndustry **A**ssociation
AIIM	**A**ssociation for **I**nformation and **I**mage **M**anagement
AILS	**A**dvanced **I**ntegrated **L**anding **S**ystem (ICT)
AIM	1. **A**mplitude **I**ntensity **M**odulation
	2. **A**scend **I**nverse **M**ultiplexing (protocol)
	3. **A**ssociation for **I**nteractive **M**edia
	4. **A**TM **I**nverse **M**ultiplexer
	5. **A**ccess **I**ntelligent **M**ultiplexer

AIMS	1. **A**eronautical **I**nterim **M**onitoring **S**ystem
	2. **A**uto **I**ndexing **M**ass **S**torage
AIMUX	**A**TM **I**nverse **Mu**ltiplexer
AIN	**A**dvance **I**ntelligent **N**etwork
AINC	**A**rabic **I**nternet **N**ames **C**onsortium
AINTCC	**A**utomated **Int**ercept **C**all **C**ompletion (Northern Telecom)
AIOD	**A**utomatic **I**dentification of **O**utward **D**ialing (PBX)
AIP	**A**TM **I**nterface **P**rocessor
AIR	1. **A**llowed **I**nformation **R**ate (frame relay)
	2. **A**dditive **I**ncrease **R**ate (ATM)
	3. **A**irborne **I**maging **R**adar
	4. **A**ll **I**ndia **R**adio (broadcaster)
	5. **A**ssociation of **I**ndependents in **R**adio
	6. **A**ir-**I**ncident **R**ecording (magnetic storage)
AIRB	**A**ssociation of **R**adio **I**ndustries and **B**usinesses
AIRF	**A**dditive **I**ncrease **R**ate **F**actor (ATM)
AIRS	**A**tmospheric **I**nfra**r**ed **S**ensor (remote sensing)
AIS	1. **A**larm **I**ndication **S**ignal (MUX)
	2. **A**utomatic **I**ntercept **S**ystem
	3. **A**utomated **I**nformation **S**ystem
	4. **A**eronautical **I**nformation **S**ervices (COMSAR)
	5. **A**ssociation for **I**nformation **S**ystems
AISS	**A**utomated **I**nformation **S**ystems **S**ecurity
AIST	**A**gency of **I**ndustrial **S**cience and **T**echnology
AISTel	**A**ssociazione **I**taliana per lo **S**viluppo delle **Tel**ecomunicazioni
AIT	1. **A**ssembly, **I**ntegration, and **T**esting
	2. **A**tomic **I**nternational **T**ime
	3. **A**dvanced **I**ntelligent **T**ape (tape format)
	4. **A**utomatic **I**dentification **T**echnology
AITP	**A**ssociation of **I**nformation **T**echnology **P**rofessionals (U.S.A.)
AITS	1. **A**ustralian **I**nformation **T**echnology **S**ociety
	2. **A**dministrative **I**nformation **T**echnology **S**ervices
	3. **A**dvanced **I**nformation **T**echnology **S**ervices
	4. **A**ssociazione **I**taliana **T**ecnici del **S**uono
	5. **A**cknowledged **I**nformation **T**ransfer **S**ervice
AIW	**A**pplication **I**mplementer's **W**orkshop
AIX	**A**dvanced **I**nteractive **ex**ecutive (IBM-operating system)
AJ	**A**nti-**J**amming (electronics)
AJP	**A**merican **J**ournal of **P**hysics
AKM	**A**pogee **K**ick **M**otor (satellite)

A

AL	1. **A**daptation **L**ayer (ATM)
	2. **A**pplication **L**ayer (OSI model)
ALAP	**A**ppleTalk **L**ink **A**ccess **P**rotocol
ALB	**A**nalog **L**oop-**B**ack (testing)
ALBO	**A**utomatic **L**ine **B**uild-**O**ut (transmission)
ALC	1. **A**utomatic **L**evel **C**ontrol
	2. **A**utomatic **L**oad **C**ontrol (fiber optics)
	3. **A**irline **L**ine **C**ontrol (protocol)
ALDC	**A**daptive **L**ossless **D**ata **C**ompression
ALE	1. **A**pplication **L**ogic **E**lement
	2. **A**utomatic **L**ink **E**stablishment (HF radio station)
	3. **A**tlanta **L**inux **E**nthusiasts
	4. **A**pprovals **L**iaison **E**ngineer (British)
ALF	**A**dvanced **L**ibrary **F**ormat
ALFA	**A**utomatic **L**aser-**F**iber **A**ssembly
ALG	**A**pplication **L**evel **G**ateway
ALGaAs	**Al**uminum–**Ga**llium–**Ar**senide (semiconductors)
ALGOCOM	**Algo**rithm **Com**munications Company (Iran)
ALGOL	**Algo**rithmic **L**anguage (programming)
ALI	1. **A**TM **L**ine **I**nterface
	2. **A**utomatic **L**ocation **I**dentification
	3. **A**utomatic **L**ocation **I**nformation
ALIS	**A**ccess **L**ine **I**n **S**ervice
ALIT	**A**utomatic **L**ine **I**nsulation **T**esting
ALL	**A**nalog **L**eased **L**ine
ALLC	**A**ssociation for **L**iterary and **L**inguistic **C**omputing
ALM	1. **A**ppWare **L**oadable **M**odule
	2. **A**utomated **L**oan **M**achine
ALMRBS	**A**irport **L**and **M**obile **R**adio **B**ase **S**tation (aviation)
ALMRS	**A**irport **L**and **M**obile **R**adio **S**tation (aviation)
ALN	1. **A**synchronous **L**earning **N**etworks (e-learning)
	2. **A**daptive-**L**ogic **N**etwork (neural networks)
AlNiCo	**Al**uminum **Ni**ckel **Co**balt (substance)
ALOS	**A**dvanced **L**and **O**bserving **S**atellite
ALPETH	**Al**uminum/**P**oly**eth**ylene (cable)
ALPS	**A**utomatic **L**oop **P**rotection **S**witching
ALS	1. **A**utomatic **L**aser **S**hutdown
	2. **A**dvanced **L**ow-power **S**chottky (TTL)
	3. **A**ctive **L**ine **S**tate
ALT	1. **A**utomated **L**oop **T**est
	2. **A**ssociation for **L**earning **T**echnology (U.K. e-learning)
	3. **A**lternate **L**ocal **T**ransport
ALTEL	**A**ssociation of **L**ong-distance **Tele**phone companies

ALTS	Association for Local Telecommunications Services (U.S.A.)
ALU	1. Arithmetic and Logic Unit (computer)
	2. Average Line Utilization
AM	1. Amplitude Modulation
	2. Angle Modulation
	3. Access Module
	4. Apogee Motor (satellite)
	5. Active Matrix (displays)
	6. Alarm Management
AM band	Radio-frequency **band** ranging from 535 to 1605 kHz
AM-AM	Amplitude-Modulation **to** Amplitude-Modulation conversion
AM-PM	Amplitude-Modulation **to** Phase-Modulation conversion
AM-TFT	Active Matrix, Thin-Film Transistor
AM/FM	Handling **either AM or FM** signals (radio receivers)
AM/SSB	Amplitude-Modulation **with** Single SideBand operation
AM/VSB	Amplitude-Modulation **with** Vestigial SideBand operation
AMA	Automatic Message Accounting
AMADNS	**AMA** Data Networking System (OSS)
AMANDA	Automated Messaging And Directory Assistance
AMAP	Adaptive Mobile Access Protocol
AMATPS	**AMA** Teleprocessing System (OSS)
AMBA	Advanced Microcontroller Bus Architecture
AMBE	Advanced Multiband Excitation (Inmarsat)
AMC	Allgon Mobile Communications (Sweden)
AMCs	Adds, Moves, and Changes (IBM-speak about computer fan)
AMD	American Micro Devices (company)
AMDM	ATM Multiplexer/Demultiplexer
AME	1. Amplitude Modulation Equivalent (transmission)
	2. Automatic Message Exchange
AMEL	Active Matrix Electroluminescence
AMES	Aeronautical Mobile Earth Station
AMF	1. Apogee Motor Firing (satellite)
	2. Automated Module Fabrication
AMHS	Automated Message-Handling System
AMI	1. Alternate Mark Inversion (data encoding)
	2. Active Microwave Instrument (remote sensing)
AMIS	Audio Messaging Interchange Specification (voice mail)
AML	1. Actual Measured Loss

A

	2. **A**utomatic **M**odulation **L**imiting (electronics)
	3. **A**mplitude **M**odulated **L**ink
	4. **A**nalog **M**icrowave **L**ink
	5. **A**RC **M**acro **L**anguage (programming)
AMLCD	**A**ctive **M**atrix **L**iquid-**C**rystal **D**isplay
AMN	**A**bstract **M**achine **N**otation
Amp	1. **Amp**ere
	2. **amp**lifier
AMP	**A**dvanced **M**etal **P**owder (storage technology)
amplidyne	**ampl**ifier **dyne**
AmprNet	**Am**ateur **p**acket **r**adio **Net**work
AMPS	1. **A**dvanced **M**obile **P**hone **S**ervice (cellular networks)
	2. **A**utomatic **M**essage **P**rocessing **S**ystem
AMPSSO	**A**utomatic **M**essage **P**rocessing **S**ystem **S**ecurity **O**fficer
AMR	1. **A**nisotropic **M**agneto-**R**esistance
	2. **A**utomated **M**eter **R**eading
	3. **A**udio **M**odem **R**iser (Intel motherboards)
	4. **A**daptive **M**ulti**r**ate (codec)
AMRAC	**A**ssociation **M**ondale des **R**adiodiffuseurs **C**ommunautaires (France)
AMS	1. **A**ccount **M**anagement **S**ystem
	2. **A**ttendant **M**anagement **S**ystem (NEC)
	3. **A**eronautical **M**obile **S**ervice (COMSAR)
AMS-IX	**Ams**terdam **I**nternet **Ex**change
Amsat	Radio **Am**ateur **Sat**ellite Corporation
AMSC	**A**merican **M**obile **S**atellite **C**orporation
AMSK	**A**uxiliary **M**anual **S**elect **K**eyboard
AMSS	1. **A**eronautical **M**obile **S**atellite **S**ervice (COMSAR)
	2. **A**irborne **M**ulti**s**pectral **S**canner
AMT	**A**ddress **M**apping **T**able (routers and servers)
AMTA	**A**merican **M**obile **T**elecommunications **A**ssociation
AMTFT	**A**ctive-**M**atrix **T**hin-**F**ilm **T**ransistor (electronics)
AMTI	**A**irborne **M**oving-**T**arget **I**ndicator
AMTOR	**Am**ateur **T**eleprinting **O**ver **R**adio
AMTS	1. **A**utomated **M**aritime **T**elecommunications **S**ystem
	2. **A**sia-**P**acific **M**obile **T**elecommunications **S**atellite Pte. Ltd
AmTV	**Am**ateur **T**ele**v**ision
AMVER	**A**utomated **M**erchant **Ve**ssel **R**eport
AN	**A**ccess **N**etwork
ANA	**A**utomatic **N**etwork **A**nalyzer
ANAC	**A**utomatic **N**umber **A**nnouncement **C**ircuit
Anamux	**Ana**log **mu**ltiple**x**er
ANBP	**A**ppleTalk **N**ame-**B**inding **P**rotocol

ANC	1. **A**ll **N**umber **C**alling
	2. **A**ir **N**avigation **C**ommission (COMSAR)
ANCARA	**A**dvanced **N**etworked **C**ities **A**nd **R**egions **A**ssociation
AND	**A**utomatic **N**etwork **D**ialing (service)
ANI	**A**utomatic **N**umber **I**dentification (feature)
ANL	1. **A**utomatic **N**oise **L**imiter
	2. **A**mbient **N**oise **L**evel
ANM	**An**swer **M**essage
ANN	1. **A**rtificial **N**eural **N**etwork
	2. **A**rab **N**ews **N**etwork
ANOVA	**An**alysis **o**f **Va**riance
ANRT	**A**gence **N**ationale de **R**eglementation des **T**elecommunication (Morocco)
Ans	**ans**wer
ANS	1. **A**dvanced **N**etwork and **S**ervices
	2. **A**merican **N**ational **S**tandards
ANSA	**A**lternate **N**etwork **S**ervice **A**greement (ISDN)
ANSI	**A**merican **N**ational **S**tandards **I**nstitute (U.S.A.)
ANSI-C	A version of **C** programming language standardized by **ANSI**
ANSI/SPARC	**ANSI S**tandards **P**lanning **A**nd **R**equirements **C**ommittee
ANT	1. **ant**enna
	2. **A**ntenna **N**oise **T**emperature
	3. **A**dvanced **N**etwork **T**echnologies (company)
	4. **A**ccess **N**etwork **T**ermination
	5. **A**lternate **N**umber **T**ranslation
ANTC	**A**dvanced **N**etworking **T**est **C**enter
ANTIVOX	**A**nti-**v**oice-**O**perated **Transmission** (electronics)
ANX	**A**utomative **N**etwork **E**xchange
AO	**A**coustic-**O**ptical
AO/DI	**A**lways **O**n **D**ynamic **ISDN**
AoC	**A**dvice **o**f **C**harge (GSM supplementary service)
AOC	**A**eronautical **O**perational **C**ommunications
AoCI	**A**dvice **o**f **C**harge **I**ndication (GSM supplementary service)
AoCC	**A**dvice **o**f **C**harge **C**harging (GSM supplementary service)
AOCN	**A**dministrative **O**perating **C**ompany **N**umber
AOCS	**A**ltitude and **O**rbit **C**ontrol **S**ystem
AoD	**A**udio-**o**n-**D**emand (service)
AOHell	**A**merica **O**nline **Hell** (hacker program)
AOL	**A**merica **On**line (information service)

A

AOM	1. Acousto-optic Modulator
	2. Administration, Operations, and Maintenance
AON	1. Active Optical Network
	2. All-optical Network
AOR	Atlantic Ocean Region (Inmarsat)
AOR-E	Atlantic Ocean Region-East (Inmarsat)
AOR-W	Atlantic Ocean Region-West (Inmarsat)
AOS	1. Alternate Operator Services
	2. Area of Service
AOSP	Alternate Operator Service Provider
AOSS	Auxiliary Operator Services System
AOSSVR	Auxiliary Operator Services System Voice Response
AOTF	Acoustic Optical Tunable Filter (MUX)
AOW	Asia and Oceania Workshop
AP	1. Access Point (wireless LAN)
	2. Access Provider
	3. Anomalous Propagation
	4. Application Program
	5. Applications Processor (AT&T)
	6. Adjunct Processor
APA	All Points Addressable (graphics)
APAD	Asynchronous Packet Assembler/Disassembler
APAN	Asia-Pacific Advanced Network (consortium)
APaRT	Automated Packet Recognition/Translation
APC	1. Adaptive Predictive Coding
	2. Automatic Phase Control (electronics)
	3. Advanced Process Control
	4. Aeronautical Passenger Communications
	5. Association for Progressive Communications
APCC	American Public Communications Council
APCN	Asia-Pacific Cable Network (consortium)
APCO	Association of Public-safety Communications Officials (U.S.A.)
APCS	Aeronautical Public Correspondence Service
APD	Avalanche Photo Diode (electronics)
APDC	Avalanche Photodiode Coupler
APDIP	Asia-Pacific Development Information Program (United Nations)
APDU	Application Protocol Data Unit (OSI model)
APES	Antenna Pointing Element Set (Satcom)
APEX	Application Exchange (protocol)
APFD	Aggregate Power Flux Density
API	1. Application Programming Interface (software)
	2. Advanced Publication of Information (Intelsat)
	3. Air Position Indicator

API/CS	**A**pplication **P**rogramming **I**nterface **and C**ommunications **S**ervice
APK	**A**mplitude **P**hase-Shift **K**eying (modulation)
APL	1. **A**rray **P**rogramming **L**anguage (computer)
	2. **A**utomatic **P**rogram **L**oad (PBX feature)
	3. **A**verage **P**icture **L**evel (video system)
APLT	**A**dvanced **P**rivate **L**ine **T**ermination (PBX)
APM	1. **A**mplitude **P**hase **M**odulation
	2. **A**dvanced **P**ower **M**anagement (computer)
APNIC	**A**sia-**P**acific **N**etwork **I**nformation **C**enter
APO	**A**daptive **P**erformance **O**ptimization (technology)
APOLT	**APON** **L**ine **T**ermination
APON	**A**TM **P**assive **O**ptical **N**etwork
APONT	**APON T**ermination (electronics)
APOT	**A**dditional **P**oint **of T**ermination
app	**app**lication
APP	**A**scend **P**assword **P**rotocol
APPA	**A**merican **P**ublic **P**ower **A**ssociation
APPC	**A**dvanced **P**rogram-to-**P**rogram **C**ommunications (IBM)
APPGEN	**App**lication **Gen**erator
APPN	**A**dvanced **P**eer-to-**P**eer **N**etworking (IBM SNA)
APR	1. **A**merican **P**ublic **R**adio (broadcaster)
	2. **A**nnual **P**ercentage **R**ate (contracts)
APRN	**A**laska **P**ublic **R**adio **N**etwork (broadcaster)
APS	1. **A**dvanced **P**hoto **S**ystem
	2. **A**dvanced **P**lanning and **S**cheduling (ICT)
	3. **A**utomatic **P**rotection **S**witching (MUX)
	4. **A**nalog **P**rotection **S**ystem (broadcasting)
APSC	**A**sia-**P**acific **S**pace **C**enter, Inc. (U.S.A.)
APSCC	**A**sia-**P**acific **S**atellite **C**ommunications **C**ouncil
APSE	**A**utomatic **P**rotection **S**witching **E**xtendable
APSI	**A**sia-**P**acific **S**atellite **I**ndustries Co., Ltd
APSK	**A**mplitude **PSK** (modulation)
APSW	**A**ll **P**urpose **S**oftware
APT	1. **A**sia-**P**acific **T**elecommunity (organization)
	2. **A**utomatically **P**rogrammed **T**ools
	3. **A**utomatic **P**icture **T**ransmission
APTS	**A**ssociation of **P**ublic **T**elevision **S**tations
AQAP	**A**llied **Q**uality **A**ssurance **P**ublications
AQCB	**A**utomated **Q**uote **C**ontract **B**illing
AQL	**A**cceptable **Q**uality **L**evel (of products)
AR	1. **A**xial **R**atio (microwave)
	2. **A**lternate **R**oute
	3. **A**utomatic **R**ecall

ARA	1. **A**ppleTalk **R**emote **A**ccess
	2. **A**ddress **R**esolution **A**dvertisement (routing service)
Arabsat	**Arab Sat**ellite Communications (organization)
ARAM	**A**udio grade **RAM** (memory)
ARB	**A**ll-**R**outes **B**roadcast
.arc	Name extension for compressed **arc**hive files (computer)
ARC	**A**ttached **R**esource **C**omputer (LANs)
ARCH	**A**ccess **R**esponse **Ch**annel (wireless)
ARCI	**A**mateur **R**adio **C**lub **I**nternational
ARCNet	**A**ttached **R**esource **C**omputer **Net**work
ARCS	**A**stra **R**eturn **C**hannel **S**ystem
ARD	1. **A**dvanced **R**esearch and **D**evelopment
	2. **A**utomatic **R**ing **D**own
ARE	**A**ll-**R**outes **E**xplorer (ATM)
ARENA	**A**utomated **R**esource to **E**lectronic **N**avigation **A**rchive
ARES	**A**mateur **R**adio **E**mergency **S**ervice (organization)
ARF	1. **A**lternative **R**egulatory **F**ramework
	2. **A**lternate **R**ecovery **F**acility (Satcom)
ARFA	**A**rrested **R**eceive **F**rame **A**cquisition
ARFCN	**A**bsolute **R**adio **F**requency **C**hannel **N**umber (GSM)
ARI	**A**utomatic **R**oom **I**dentification (telephony service)
ARIB	**A**ssociation of **R**adio **I**ndustries and **B**usinesses (Japan)
ARIES	1. **A**ngle-**R**esolved **I**on and **E**lectron **S**pectroscopy
	2. **A**ustralian **R**esource **I**nformation and **E**nvironment **S**atellite
ARIN	**A**merican **R**egistry for **I**nternet **N**umbers (U.S.A.)
ARINC	**A**eronautical **R**adio **Inc**orporated (airline consortium)
ARISE	1. **A**dvanced **R**adio **I**nterferometry between **S**pace and **E**arth
	2. **A**merican **R**enaissance **i**n **S**cience **E**ducation
ARISS	**A**mateur **R**adio International **S**pace **S**tation
ARISTOTELES	**A**pplications and **R**esearch **I**nvolving **S**pace Technologies **O**bserving **T**he **E**arth's Fields from **L**ow **E**arth-orbiting **S**atellites
.arj	A DOS file extension for **ARJ** compression program
.ARJ	Name extension for **A**chieving program by **R**obert K. **J**ung (computer)
ARL	**A**cceptable **R**eliability **L**evel (of products)
ARLE	**A**dvanced **R**un-length **L**imited **E**ncoding (data storage)
ARM	1. **A**synchronous **R**esponse **M**ode (OSI model)
	2. **A**merican **R**adio **M**useum
	3. **A**nswering and **R**ecording **M**achine
	4. **A**pogee **R**ocket **M**otor (satellite)

A

ARMIS	**A**utomated **M**anagement **R**eporting **I**nformation **S**ystem
ARNS	**A**eronautical **R**adio **N**avigation **S**ervice (ITU)
ARNSS	**A**eronautical **R**adio **N**avigation-**S**atellite **S**ervice (ITU)
ARO	**A**ctive **R**esonant **O**scillator (microwave)
AROS	1. **A**mateur **R**adio **O**bservation **S**ervice
	2. **A**miga **R**esearch **O**perating **S**ystem
ARP	1. **A**ddress **R**esolution **P**rotocol (networking)
	2. **A**zimuth **R**eference **P**ulse
ARPA	1. **A**dvanced **R**esearch **P**rojects **A**gency (U.S.A.)
	2. **A**utomatic **R**adar **P**lot **A**id (shipboard radar)
ARPANet	**ARPA Net**work
ARPM	**A**verage **R**evenue **P**er **M**inute (GSM)
ARPU	**A**verage **R**evenue **P**er **U**ser (GSM)
ARQ	**A**utomatic **R**etransmit **Re**quest (data communication)
ARQ-GB	**A**utomatic **R**etransmit **Re**quest - **G**o **B**ack
ARQ-SR	**A**utomatic **R**etransmit **Re**quest - **S**elective **R**etransmit
ARQ-SW	**A**utomatic **R**etransmit **Re**quest - **S**top and **W**ait
ARR	**A**irborne **R**adio **R**elay
ARRL	**A**merican **R**elay **R**adio **L**eague
ARRN	**A**mateur **R**adio **R**epeater **N**etwork
ARS	**A**utomatic **R**oute **S**election/**S**elector
ARSG	**A**ustralian **R**adiocommunications **S**tudy **G**roup
ARSR	**A**ir **R**oute **S**urveillance **R**adar
ART	**A**rab **R**adio and **T**elevision network (broadcaster)
ARTES	**A**dvanced **R**esearch in **Te**lecommunications **S**ystems
Artron	**Art**ificial Neu**ron**
ARU	**A**udio **R**esponse **U**nit
AS	1. **A**utonomous **S**ystem
	2. **A**mateur **S**atellite
	3. **A**pplicability **S**tatement
	4. **A**dvanced **S**chottky (microchips)
	5. **A**uthorization **S**tream
AS&C	1. **A**larm **S**urveillance **and C**ontrol
	2. **A**dvanced **S**ystems **and C**oncepts (U.S. Defense)
ASA	1. **A**coustical **S**ociety of **A**merica
	2. **A**merican **S**tandards **A**ssociation (now ANSI)
	3. **A**verage **S**peed of **A**nswer
	4. **A**ffiliated **S**ales **A**gency
ASAI	**A**djunct **S**witch **A**pplication **I**nterface (AT&T)
ASAM	**A**dvanced **S**ervices **A**ccess **M**anager (Alcatel)
ASAPI	**A**dvanced **S**peech **API** (computer)
ASAR	**A**dvanced **S**ynthetic **A**perture **R**adar
ASARS	**A**dvanced **S**ynthetic **A**perture **R**adar **S**ystem

A

ASAS	**A**ll-**S**ource **A**nalysis **S**ystem
ASBC	**A**dvanced **S**pace **B**usiness **C**orporation
.asc	Name extension for files containing **ASC**I text (computer)
ASC	1. **A**bnormal **S**tation **C**ode
	2. **A**UTODIN **S**witching **C**enter
	3. **A**utomatic **S**lope **C**ontrol (circuitry)
	4. **A**dvanced **S**witching **C**ommunications
	5. **A**eronautical **S**ystem **C**enter (U.S.A.)
ASCA	**A**dvanced **S**atellite for **C**osmology & **A**strophysics
ASCC	**A**utomatic **S**equence **C**ontrolled **C**alculator
ASCENT	**As**sociation of **C**ommunications **Ent**erprises
ASCII	**A**merican **S**tandard **C**ode for **I**nformation **I**nterchange (computer)
ASCU	**A**gent **S**et **C**ontrol **U**nit (IBM)
ASD	**As**ynchronous **D**ata
ASDE	**A**irport **S**urface **D**etection **E**quipment
ASDS	**A**ccunet **S**pectrum of **D**igital **S**ervices (AT&T)
ASDSP	**A**pplication-**S**pecific **D**igital **S**ignal **P**rocessor
ASE	1. **A**pplication **S**ervice **E**lement (OSI model)
	2. **A**mplified **S**pontaneous **E**mission
	3. **A**utomatic **S**witching **E**quipment
ASFB	**A**pplication-**S**pecific **F**unctional **B**lock
ASG	**A**ccess **S**ervice **G**roup
ASH	1. **A**rdire-**S**tratigakis-**H**ayduk (algorithm)
	2. **A**mplifier-**S**equenced **H**ybrid (microwave)
ASI	1. **A**lternate **S**pace **I**nversion (data encoding)
	2. **A**rtificial **S**ensing **I**nstrument
	3. **A**pplication **S**oftware **I**nterface
	4. **A**stronomical **S**ociety of **I**ndia
	5. **A**genzia **S**paziale **I**taliano (Italian Space Agency)
ASIC	**A**pplication-**S**pecific **I**ntegrated **C**ircuits (electronics)
ASK	**A**mplitude-**S**hift **K**eying (modulation)
ASL	**A**daptive **S**peed **L**eveling (modem)
ASMTP	**A**uthenticated **SMTP** (Internet)
ASN	1. **A**eronautical **S**atellite **N**ews (magazine)
	2. **A**cknowledgement **S**equence **N**umber
	3. **A**utonomous **S**ystem **N**umber
	4. **A**bstract **S**yntax **N**otation (OSI model)
ASOCNet	**A**rmy **S**pecial **O**perations **C**ommand **Net**work
ASON	**A**utomatically-**S**witched **O**ptical **N**etwork
ASP	1. **A**pplication **S**ervice **P**rovider (Internet)
	2. **A**dministrable **S**ervice **P**rovider (SCSA)
	3. **A**lternate **S**ervice **P**rovider

	4. **A**bstract **S**ervice **P**rimitive (ATM)
	5. **A**djunct **S**ervice **P**oint (INs)
	6. **A**nalog **S**ignal **P**rocessing
	7. **A**ppleTalk **S**ession **P**rotocol (networking)
	8. **A**verage **S**elling **P**rice (microchip manufacturing)
	9. **A**TM **S**witch **P**rocessor
	10. **A**ttached **S**upport **P**rocessor (Northern Telecom)
	11. **A**ssociation of **S**hareware **P**rofessionals
	12. **A**ctive **S**erver **P**age (HTML)
ASPI	**A**dvanced **S**CSI **P**rogramming **I**nterface (computer)
ASPJ	**A**dvanced **S**elf-**P**rotection **J**ammer (electronics)
ASQ	1. **A**utomated **S**tatus **Q**uery
	2. **A**dministrative **S**cience **Q**uarterly
	3. **A**merican **S**ociety for **Q**uality
	4. **A**pplication **S**tatus **Q**uery
ASR	1. **A**utomatic **S**end/**R**eceive (telex maker company)
	2. **A**utomatic **S**peech **R**ecognition (telephony service)
	3. **A**utomatic **S**ystem **R**econfiguration (computer)
	4. **A**ccess **S**ervice **R**equest (frame relay)
	5. **A**irport **S**urveillance **R**adar
	6. **A**nswer **S**eizure **R**atio
	7. **A**verage **S**ervice **R**ate (percentage)
ASRM	**A**utomatic **S**end/**R**eceive **M**onitor (ASR company product)
ASRT	**A**utomatic **S**end/**R**eceive **T**erminal (ASR company product)
ASS	**A**mateur **S**atellite **S**ervice (ITU)
ASSP	1. **A**coustic **S**peech and **S**ignal **P**rocessing
	2. **A**pplication-**S**pecific **S**tandard **P**roduct (microchips)
ASSTA	**A**ustralian **S**peech **S**cience and **T**echnology **A**ssociation
AST	**A**utomatic **S**cheduled **T**esting
ASTAP	APT **St**andardization **P**rogram (APT)
ASTC	1. **A**ustralian **S**cience and **T**echnology **C**ouncil
	2. **A**ssociation of **S**cience-**T**echnology **C**enters Incorporated
ASTD	**A**merican **S**ociety of **T**raining and **D**evelopment (e-learning)
ASTE	**A**dvanced **S**ystems and **T**elecommunications **E**quipment
ASTERR	**A**dvanced **S**paceborne **T**hermal **E**mission **R**eflectance **R**adiometer (remote sensing)
ASTM	**A**merican **S**ociety for **T**esting and **M**aterials
ASTP	**A**dvanced **S**ystems and **T**echnology **P**rogram
ASTRAL	**A**lliance for **S**trategic **T**oken-**R**ing **A**dvancement and **L**eadership

A

ASU	1. **A**pplication-**S**pecific **U**nit
	2. **A**TM **S**ervice **U**nit
ASV	**A**ir-to-**S**urface-**V**essel (radar)
ASWC	**A**lternate **S**erving **W**ire **C**enter
.asx	Name extension for Windows Media Player files (computer)
ASYNC	**Async**hronous (data transfer)
AT	1. **A**dvanced **T**echnology (IBM PCs)
	2. **A**ccess **T**andem
	3. **A**cceptance **T**est
	4. **A**udiotex(t)
	5. **A**synchronous **T**ransmission
AT Attachment	**A**dvanced **T**echnology **Attachment** (interface standard)
AT&T	**A**merican **T**elephone **and T**elegraph (company)
ATA	1. **A**ir **T**ransport **A**ssociation
	2. **A**uto-**T**racking **A**ntenna
	3. **A**merican **T**elemarketing **A**ssociation (U.S.A.)
	4. **A**nalog **T**erminal **A**dapter (Northern Telecom)
	5. **AT A**ttachment (computer interface)
ATA2	**Second** generation **AT A**ttachment (interface standard)
ATACS	**A**rmy **Ta**ctical **C**ommunications **S**ystem (U.S.A.)
ATAPI	**AT A**ttachment **P**acket **I**nterface specification (CD-ROM)
ATB	**A**ll **T**runks **B**usy (signal)
ATC	1. **A**ir **T**raffic **C**ontrol (aviation)
	2. **A**daptive **T**ransform **C**oding
ATCA	**A**ntique **T**elephone **C**ollectors **A**ssociation
ATCON	**A**pple**T**alk network **Con**trol (Novell NetWare)
ATCONFIG	**A**pple**T**alk **Config**uration (Novell NetWare NLM)
ATCP	**A**pple**T**alk **C**ontrol **P**rotocol
ATCRBS	**A**ir **T**raffic **C**ontrol **R**adar **B**eacon **S**ystem
ATD	1. **A**synchronous **T**ime-**D**ivision
	2. **A**dvanced **T**echnology **D**emonstration
	3. **At**tention **D**ial the phone (modems)
ATDE	**A**daptive **T**ime **D**omain **E**qualizer
ATDM	**A**synchronous **T**ime-**D**ivision **M**ultiplexing
ATDNet	**A**dvanced **T**echnology **D**emonstration **Net**work
ATDP	**At**tention **D**ial **P**ulse (modem command)
ATDRSS	**A**dvanced **T**racking and **D**ata **R**elay **S**atellite **S**ystem (NASA)
ATDT	**At**tention **D**ial **T**one (modem command)
ATE	1. **A**uthorized **T**elecommunications **E**ntity
	2. **A**utomatic **T**est **E**quipment

ATEL	**A**dvanced **T**elevision **E**valuation **L**aboratory
ATELP	**A**daptive-**T**ransform-**E**xcited-**L**inear **P**rediction
ATF	**A**dvanced **T**actical **F**iber
ATG	1. **A**ddress **T**ranslation **G**ateway (Cisco)
	2. **A**utomatic **T**est **G**eneration
	3. **A**ir-**T**o-**G**round (service)
ATHD	**A**verage **T**en **H**igh-**D**ay (data communication)
ATI	1. **A**dvanced **T**elecommunication **I**nstitute
	2. **A**ccelerated **T**echnology **I**ncorporated
ATIA	**A**ir **T**raffic **I**nterface **A**pplication (Motorola)
ATIS	**A**lliance for **T**elecommunications **I**ndustry **S**olutions (U.S.A.)
ATL	**A**ctive **T**emplate **L**ibrary (Microsoft)
ATM	1. **A**synchronous **T**ransfer **M**ode (network technology)
	2. **A**ir **T**raffic **M**anagement (aviation)
	3. **A**utomated **T**ransaction **M**achine
	4. **A**utomatic **T**racking **M**echanism
ATM DSU	**ATM D**igital **S**ervice **U**nit
ATM DXI	**ATM D**ata **Ex**change **I**nterface
ATM-25	**ATM** running at **25 Mbps**
ATMARP	**ATM A**ddress **R**esolution **P**rotocol
ATMP	**A**scend **T**unnel **M**anagement **P**rotocol
ATN	**A**eronautical **T**elecommunications **N**etwork
ATNS	**A**dvanced **T**runking **N**etworking **S**ystem
ATOF	**A**dvanced **T**actical **O**ptical **F**iber
ATOP	**A**utomatic **T**raffic **O**verload **P**rotection (feature)
ATOW	**A**cquisition and **T**racking **O**rder-**W**ire
ATP	1. **A**ppleTalk **T**ransaction **P**rotocol (networking)
	2. **A**cceptance **T**est **P**rocedure
	3. **A**dvanced **T**echnology **P**rovider
	4. **A**utomatic **T**elephone **P**ayment
ATPC	**A**utomatic **T**ransmitter **P**ower **C**ontrol
ATPG	**A**utomatic **T**est **P**attern **G**enerator (broadcasting)
ATPS	**A**ppleTalk **P**rint **S**ervices (Novell NetWare)
ATPSCON	**A**ppleTalk **P**rint **S**ervices **Con**figuration (Novell NetWare)
ATRAC	**A**daptive **Tr**ansform **A**coustic **C**oding (minidisk)
ATRAN	**A**utomatic **T**errain **R**ecognition **A**nd **N**avigation (military)
ATS	1. **A**ir **T**raffic **S**ervices (aviation)
	2. **A**bstract **T**est **S**uite (testing a protocol)
	3. **A**pplications **T**echnology **S**atellite program
ATSC	**A**dvanced **T**elevision **S**ystems **C**ommittee (U.S.A.)
ATSE	**A**cademy of **T**echnological **S**ciences and **E**ngineering

ATSR	1. **A**long-**T**rack **S**canning **R**adiometer (remote sensing)
	2. **A**long-**T**rack **S**canning **R**adar
ATSR/M	**A**long-**T**rack **S**canning **R**adiometer **M**icrowave (sounder)
ATT	1. **A**utomatic **T**oll **T**racking
	2. **A**verage **T**alk **T**ime
ATTC	**A**dvanced **T**elevision **T**est **C**enter
ATTND	**att**end**a**nt
ATU	1. **A**DSL **T**ransceiver **U**nit (ITU)
	2. **A**DSL **T**erminal **U**nit
	3. **A**uxiliary **T**est **U**nit
	4. **A**frican **T**elecommunications **U**nion
ATU-C	**A**DSL **T**ransmission **U**nit-**C**entral office
ATU-R	**A**DSL **T**ransmission **U**nit-**R**emote office
ATUC	**A**DSL **T**ransmission **U**nit-**C**entral office
ATUG	**A**ustralian **T**elecommunication **U**ser **G**roup
ATV	1. **A**dvanced **T**ele**v**ision (HDTV)
	2. **A**mateur **T**ele**v**ision
ATVEF	**A**dvanced **T**elevision **E**nhancement **F**orum
ATX	1. **A**udio**tex**(t)
	2. **A**dvanced **T**echnology E**x**panded (PC motherboards)
AU	1. **A**ccess **U**nit
	2. **A**lternate **U**se
	3. **A**dministrative **U**nit (MUX)
AU PTR	**A**dministrative **U**nit **P**oin**t**e**r** (MUX)
AuC	**Au**thentication **C**enter (GSM)
AUCS	**A**T&T **U**nisource **C**ommunications **S**ervices
AUD	**Aud**io input/output (audio systems)
AUG	**A**dministrative **U**nit **G**roup (MUX)
AUI	1. **A**utonomous **U**nit **I**nterface (Ethernet transceiver)
	2. **A**ttachment **U**nit **I**nterface (Ethernet transceiver)
	3. **A**ccess **U**nit **I**nterface (LANs)
AUP	**A**cceptable **U**se **P**olicy (ISPs)
AUR	**A**ccess **U**sage **R**ecord
AURP	**A**ppleTalk **U**pdate-based **R**outing **P**rotocol
AUS	**A**ccess **U**nit **S**ubrack
AUSREP	**Au**stralian **S**hip **Rep**orting system
AUSSAT	**Aus**tralian domestic **Sat**ellite operator
Auto-SD	**Auto**matic **S**ignal **D**egrade (MUX)
Auto-SF	**Auto**matic **S**ignal **F**ailure (MUX)
AUTODIN	**Auto**matic **Di**gital **N**etwork (U.S. Defense)
AUTOEXEC	**Auto**matically **Exec**uted (computer command)
AUTOSEVOCOM	**Auto**matic **Se**cure **Vo**ice **Com**munication network
AUTOVON	**Auto**matic **Vo**ice **N**etwork (U.S. Defense)
AUU	**A**TM **U**ser-to-**U**ser (bit)

AUUG	Australian Unix User Group
AUX	1. Auxiliary Device (computer serial port or COM1)
	2. Auxiliary signal (audio systems)
AUXBC	Auxiliary Broadcasting
.av	Name extension for Audio–Visual files (computer)
AV	Audio–Visual
AVC	Automatic Volume Control (electronics)
AVD	Alternative Voice/Data (transmission)
AVE	Automatic Volume Expander (electronics)
AVHRR	1. Advanced Very High-Resolution Radiometer (remote sensing)
	2. Advanced Very High-Resolution Radar
.avi	Name extension for audio–visual files (computer)
AVI	1. Audio–Video Interleave (file format)
	2. Automatic Vehicle Identification
	3. Analog VHDL International (standards group)
AVIOS	American Voice Input/Output Society
AVK	Audio–Video Kernel (digital video)
AVL	Automatic Vehicle Location (GSM)
AVNIR	Advanced Visible and Near Infrared Radiometer
AVRS	Automated Voice Response System
AVSG	Advanced Video Systems Group (Canada)
AVSS	Audio–Video Support System (digital video)
AVSSCS	Audio/Visual Service-Specific Convergence Sublayer (ATM)
AWA	Antique Wireless Association
AWACS	Airborne Warning And Control System
AWAN	Analog Wide Area Network
AWC	1. Area-Wide Centrex
	2. Association for Women in Computing
AWCC	Afghan Wireless Communications Company
AWG	1. American Wire Gauge (cable standards)
	2. Arbitrary Waveform Generator
AWGN	Additive White Gaussian Noise
AWM	American Wiring Material
AWNV	All Weather and Night Vision System
AWSI	Association of Wireless System Integrators
AX.25	Amateur-radio implementation of X.25
A×E	Automatic Cross-connection Equipment
AZ	Azimuth (Satcom)
AZ/EL	Azimuth and Elevation (Satcom)
AZON	Azimuth Only (Satcom)
AZRAN	Azimuth and Range (radar)

B b

b	1. Symbol for **b**it (unit of data)
	2. Symbol for **b**aud (unit of character transmission)
	3. Symbol for **b**arn (unit of nuclear cross section)
b/s	bits **per s**econd
B	1. Symbol for **B**yte
	2. Symbol for **B**ase (transistor circuit diagrams)
	3. Symbol for **B**el (unit of measure of signal strength)
	4. Symbol for magnetic flux density
	5. Symbol for **B**eta (Beta test)
B channel	**B**earer **channel** (ISDN)
B Link	**B**ridge **Link** (SS7)
B&S	**B**rawn **and S**harp Gauge
B&W	**B**lack **and W**hite (television set)
B+	**P**ositive terminal of **B**attery or voltage source
B−	**N**egative terminal of **B**attery or voltage source
B-CDMA	**B**roadband **CDMA** (access)
B-DCS	**B**roadband **D**igital **C**ross-connect **S**ystem
B-ICI	**B**-ISDN **I**nter **C**arrier **I**nterface (ATM)
B-ICISAAL	**B-ICI S**ignaling **ATM A**daptation **L**ayer
B-ISDN	**B**roadband **ISDN**
B-ISUP	**B**roadband **ISDN U**ser's **P**art (SS7)
B-LLI	**B**roadband **L**ower **L**ayer **I**nformation
B-LT	**B**roadband **L**ine **T**ermination

B-MAC	**B**road**b**and **M**aster **A**ntenna **C**ontrol
B-NT	**B**road**b**and **N**etwork **T**ermination
B-PCS	**B**road**b**and **P**ersonal **C**ommunications **S**ervices (FCC)
B-Picture	**B**i-directionally predictive-coded **Picture** (MPEG)
B-QSIG	**B**road**b**and ISDN **Q**-interface **Sig**naling
B-Sat	**B**roadcasting **Sat**ellite system Corporation (Japan)
B-TA	**B**road**b**and **T**erminal **A**dapter (ISDN)
B-TE	**B**road**b**and **T**erminal **E**quipment (ISDN)
B2B	**B**usiness-**to**-**B**usiness (e-commerce)
B2C	**B**usiness-**to**-**C**onsumer (e-commerce)
B2E	**B**usiness-**to**-**E**mployee (e-commerce)
B2G	**B**usiness-**to**-**G**overnment (e-commerce)
B3ZS	**B**ipolar with **3 Z**eros **S**ubstitution (data encoding)
B6ZS	**B**ipolar with **6 Z**eros **S**ubstitution (data encoding)
B8ZS	**B**ipolar with **8 Z**eros **S**ubstitution (data encoding)
BA	1. **B**asic **A**ccess
	2. **B**urst **A**cquired
	3. **B**usiness **A**ddress
BAA	**B**lanket **A**uthorization **A**greement
BABT	**B**ritish **A**pprovals **B**oard for **T**elecommunications
BACP	**B**andwidth **A**llocation **C**ontrol **P**rotocol
BACR	**B**illing **A**ccount **C**ross **R**eference (number)
BAF	**B**ellcore **AMA F**ormat (data)
BAFTA	**B**ritish **A**cademy of **F**ilm and **T**elevision **A**rts
BAIC	**B**illing of **A**ll **I**ncoming **C**alls (GSM supplementary service)
.bak	Name extension for **bak**up files (computer)
Balun	**Bal**anced-input to **un**balanced-output converter (microwave)
BAN	1. **B**ase **A**rea **N**etwork
	2. **B**illing **A**ccount **N**umber
BAOC	**B**illing of **A**ll **O**utgoing **C**alls (GSM supplementary service)
BAOL	**B**ritish **A**ssociation for **O**pen **L**earning (e-learning)
BAPCO	**B**ritish **A**ssociation of **P**ublic-safety **C**ommunications **O**fficials
BAPTA	**B**earing **A**nd **P**ower **T**ransfer **A**ssembly
BARITT	**Bar**rier **I**njection **T**ransit **T**ime (semiconductors)
BARS	**B**asic **A**utomatic **R**oute **S**election (Nortel Networks)
BAS	**B**it **A**llocation **S**ignal
BASIC	**B**eginner's **A**ll-purpose **S**ymbolic **I**nstruction **C**ode (computer)
BASR	**B**uffered **A**utomatic **S**end/**R**eceive
.bat	Name extension for **bat**ch program files (computer)

B

BAT	**B**ouquet **A**ssociation **T**able
Batelco	**Ba**hrain **Tel**ecommunications **Co**mpany (operator)
BATT	**batt**ery (diagrams)
BAW	**B**ulk **A**coustic **W**ave (HF device)
BB	1. **B**ase**b**and signal (MUX)
	2. **B**ulletin **B**oard
	3. **B**road**b**and
BBC	1. **B**ritish **B**roadcasting **C**orporation (broadcaster)
	2. **B**roadband **B**earer **C**apability
	3. **B**ack-to-**B**ack **C**onnection
BBD	**B**ucket **B**rigade **D**evice (semiconductors)
BBE	**B**ackground **B**lock **E**rror (MUX)
BBER	1. **B**ulletin **B**oard **E**rror **R**ate
	2. **B**ackground **B**lock **E**rror **R**atio
BBEXP	**B**ase**b**and **Exp**ansion
BBG	**B**asic **B**usiness **G**roup
BBG-I	**I**SDN **B**asic **B**usiness **G**roup
BBIN	**B**road**b**and **I**ntelligent **N**etwork
BBIU	**B**ase**b**and **I**nterface **U**nit
BBL	**B**road**b**and **L**oop
BBLAN	**B**ase**b**and **L**ocal **A**rea **N**etwork
BBM	**B**read**b**oard **M**odel
BBN	**B**olt, **B**eranek, and **N**ewman, Inc. (company)
BBPULB	**B**ig **B**rother **P**rotects **You** from **L**ittle **B**rothers (wireless networks)
BBS	1. **B**ulletin **B**oard **S**ystem (ICT)
	2. **B**ridge-to-**B**ridge **S**tation
BBSS	**B**ase**b**and **S**ub-**S**ystem
BBT	1. **B**road**b**and **T**ransmission (Intelsat)
	2. **B**it-**B**lock **T**ransfer
BBTM	**B**eam-to-**B**eam **T**raffic **M**atrix (Intelsat)
BBU	**B**ase**b**and **U**nit (Intelsat)
BBUL	**B**umpless **B**uild-**U**p **L**ayer packaging
Bc	**C**ommitted **B**urst size
BC	1. **B**ackward **C**ompatible (coding and softwares)
	2. **B**eam **C**oupling
	3. **B**inary **C**ode
	4. **B**road**c**ast
BCC	1. **B**lind **C**arbon **C**opy (e-mail)
	2. **B**ase-station **C**olor **C**ode (GSM)
	3. **B**ellcore **C**lient **C**ompany
	4. **B**lock **C**heck **C**haracter
	5. **B**roadcast **C**ontrol **C**omputer
BCCH	**B**roadcast **C**ontrol **Ch**annel (cellular networks)

BCD	1. **B**inary-**C**oded **D**ecimal (coding)
	2. **B**urst **C**hannel **D**emodulator
BCE	**B**andwidth-**C**ontrol **E**lement (frame relay)
BCH	1. **B**ose, **C**haudhuri, and **H**ocquenghem code (error correction)
	2. **B**roadcast **Ch**annels (GSM)
BCHO	**B**ase-**C**ontrolled **H**and-**O**ff (cellular networks)
BCI	1. **B**road**c**ast **I**nterference
	2. **B**it **C**ount **I**ntegrity
BCL	**B**ase general premises **C**abling **L**icense (Australia)
BCM	1. **B**it-**C**ompression **M**ultiplexer
	2. **B**asic **C**all **M**odel (AINs)
	3. **B**lock-**C**oded **M**odulation
	4. **B**inary **C**oded **M**atrix (Intelsat)
BCN	**B**ea**c**o**n** (navigation)
BCOB	**B**roadband **C**onnection-**O**riented **B**earer (ATM)
BCOB-A	**BCOB** Class **A** (ATM)
BCOB-C	**BCOB** Class **C** (ATM)
BCOB-X	**BCOB** Class **X** (ATM)
BCP	1. **B**atch **C**hange **S**upplement
	2. **B**est **C**urrent **P**ractice
	3. **B**ulk **C**opy **P**rogram (software)
BCR	**B**inary **C**harge **R**egulator
BCRI	**B**usiness **C**ommunications **R**eview **I**nternational (magazine)
BCRS	**B**ell **C**anada **R**elay **S**ervice
BCS	1. **B**asic **C**ontrol **S**ystem (Satcom)
	2. **B**atch **C**hange **S**upplement
	3. **B**oston **C**omputer **S**ociety
	4. **B**-channel **C**ircuit-**S**witched
	5. **B**asic **C**ombined **S**et (OSI model)
	6. **B**ritish **C**omputer **S**ociety
	7. **B**eam **C**ontrol **S**ystem (satellite)
	8. **B**roadcast **C**ontrol **S**ystem
BCSM	**B**asic **C**all **S**tate **M**odel (AINs)
BCVT	**B**asic **C**lass **V**irtual **T**erminal (OSI model)
Bd	**b**au**d** (unit of telex transmission speed)
BD	**B**uilding **D**istributor
BDC	1. **B**ackup **D**omain **C**ontroller (server)
	2. **B**roadband **D**igital **C**ross-connect
BDCS	**B**roadband **D**igital **C**ross-connect **S**ystem (MUX)
BDD	**B**inary **D**ecision **D**iagram
BDE	**B**elow-**D**ecks **E**quipment
BDF	**B**lock **D**ata **F**ormat

B

BDFB	**B**reaker **D**istribution **F**use **B**ay (switch room)
BDI	**B**earing **D**eviation **I**ndicator
BDLC	**B**urroughs **D**ata **L**ink **C**ontrol (protocol)
BDN	**B**ell **D**ata **N**etwork
BDR	**B**attery **D**ischarge **R**egulator
BDSL	**B**roadband **DSL** (access)
BDT	1. **B**illing **D**ata **T**ape
	2. **B**inary **D**ecision **T**ree
	3. Telecommunications **D**evelopment **B**ureau (France)
BDU	**B**elow-**D**ecks **U**nit (satellite terminals)
BE	**B**urst **E**xcess
BEAMOS	**Bea**m-associated **MOS** (semiconductors)
BEARS	**B**illing **E**xchange **A**ccount **R**ecord **S**ystem
BECN	**B**ackward **E**xplicit **C**ongestion **N**otification (frame relay)
BECTA	**B**ritish **E**ducational **C**ommunications and **T**echnology **A**gency of United Kingdom (e-learning)
BEDO RAM	**B**urst **E**xtended **D**ata **O**utput **RAM** (memory)
BEEP	**B**lock **E**xtensible **E**xchange **P**rotocol
BEF	**B**and **E**limination **F**ilter
BEHINCOM	**Behin** Ertebat Mehr **Com**pany (Iran)
BEI	**B**ack-scattered **E**lectron **I**maging
Bellcore	**Bell Co**mmunications **Re**search (company)
BeOS	**Be O**perating **S**ystem (Be, Inc.)
BEP	1. **B**it **E**rror **P**robability
	2. **B**ack **E**nd **P**rocessor
BER	1. **B**it **E**rror **R**ate (transmission)
	2. **B**it **E**rror **R**atio (transmission)
	3. **B**asic **E**ncoding **R**ules
BERT	**B**it **E**rror **R**ate **T**est (transmission)
BES	1. **B**ase **E**arth **S**tation
	2. **B**lackBerry **E**nterprise **S**erver
BETA	**B**usiness **E**quipment **T**rade **A**ssociation
BETRS	**B**asic **E**xchange **T**elecommunications **R**adio **S**ervice
BeV	**B**illion **e**lectron **V**olts (power)
BEX	**B**roadband **Ex**change
BEZS	**B**andwidth **E**fficient **Z**ero **S**uppression
BFI	**B**ad **F**rame **I**ndicator (GSM)
BFICC	**B**ritish **F**acsimile **I**ndustry **C**onsultative **C**ommittee
BFL	**B**uffered **FET** **L**ogic (digital electronics)
BFN	**B**eam-**F**orming **N**etwork (microwave)
BFO	**B**eat-**F**requency **O**scillator (electronics)
BFOC	**B**ayonet **F**iber **O**ptic **C**onnector
BFOG	**B**rillouin **F**iber **O**ptic **G**yroscope

BFSK	**B**inary **FSK** (modulation)
BFSL	**B**est-**F**it **S**traight **L**ine (error measurement)
BFT	1. **B**inary **F**ile **T**ransfer (computer)
	2. **B**atch **F**ile **T**ransmission (computer)
BG	1. **B**asic **G**roup
	2. **B**oard of **G**overnors (Intelsat)
BG/BARC	**BG B**udget and **A**ccount **R**eview **C**ommittee (Intelsat)
BG/P	**BG P**lanning committee (Intelsat)
BG/T	**BG T**echnical committee (Intelsat)
BGA	**B**all-**G**rid **A**rray (microchips)
BGAN	**B**roadband **G**lobal **A**rea **N**etwork (Inmarsat)
BGE-I	ISDN **B**usiness **G**roup **E**lements
BGID	**B**usiness **G**roup **ID** (ISDN)
BGMP	**B**order **G**ateway **M**ulticast **P**rotocol
BGND	**B**ack**g**rou**nd**
BGP	**B**order **G**ateway **P**rotocol (routing)
BGP4	**B**order **G**ateway **P**rotocol version **4** (router)
BH	**B**andwidth **H**og
BHC	**B**ackbone to **H**orizontal **C**ross-connect
BHCA	**B**usy **H**our **C**all **A**ttempts (GSM)
BHCC	**B**usy **H**our **C**all **C**ompletion (GSM)
BHCR	**B**usy **H**our **C**all **R**ate
BHLI	**B**roadband **H**igh-**L**ayer **I**nformation (ATM)
BHM	**B**usy **H**our **M**inutes
BHMC	**B**usy **H**our **M**inutes of **C**apacity
Bi-directional	Radiating or sensitive to **two directions**
BIA	**B**urned **I**n **A**ddress (LANs)
BIB	1. **B**ackward **I**ndicator **B**it (SS7)
	2. **B**ritish **I**nteractive **B**roadcasting
BIBO	**B**ounded **I**nput, **B**ounded **O**utput
BIC	**B**arring of **I**ncoming **C**alls (GSM supplementary service)
BIC-Roam	**B**arring of **I**ncoming **C**alls while **Roam**ing (GSM service)
BICC	**B**earer **I**ndependent **C**all **C**ontrol
BICI	**B**roadband **I**nter-**C**arrier **I**nterface
BiCMOS	**Bi**polar **CMOS** (semiconductors)
BICSI	**B**uilding **I**ndustry **C**onsulting **S**ervice **I**nternational (standards)
BID	**B**ridge **I**dentification **C**ode
BIDDS	**B**ase **I**nformation **D**igital **D**istribution **S**ystem (military)
BiDi	**bi-directional** (transmission)
BIDOPS	**Bi-Dop**pler **S**coring (military)

B

BIDS	**B**roadband **I**nfrastructure for **D**igital TV and multimedia **S**ervices
BIE	**B**ase station **I**nterface **E**quipment
BiFET	**Bi**polar **FET** (semiconductors)
BIFODEL	**B**inary **F**iber-**O**ptic **De**lay **L**ine
BIG	**B**roadband **I**ntegrated **G**ateway
BIGA	**B**us **I**nterface **G**ate **A**rray (microchips)
BIH	**B**ureau **I**nternational del'**H**eure
BII	**B**ase **I**nformation **I**nfrastructure
.bin	Name extension for files encoded with Mac **Bin**ary (computer)
BINAC	**Bin**ary **A**utomatic **C**omputer project
BIND	**B**erkley **I**nternet **N**ame **D**aemon (Novell NetWare)
BIOD	**B**ell **I**ntegrated **O**ptical **D**evice
BIOS	**B**asic **I**nput **O**utput **S**ystem (computer)
BIP	1. **B**it **I**nterleaved **P**arity (data encoding)
	2. **B**illing **I**nterconnection **P**ercentage
	3. **B**and **I**nterleaved by **P**ixel (remote sensing)
BIP-8	**B**it **I**nterleaved **P**arity **8** (data encoding)
BIP-N	**B**it **I**nterleaved **P**arity **N** (data encoding)
BIP-RZ	**Bip**olar-**R**eturn to **Z**ero level (data encoding)
BIΦ-L	**Bi**phase-**L**evel (data encoding)
BIS	1. **B**usiness **I**nformation **S**ystem
	2. **B**order **I**ntermediate **S**ystem
	3. **B**roadband **I**nteractive **S**ystem
BISDN	**B**roadband **ISDN**
BISSI	**B**roadband **I**nter **S**witching **S**ystem **I**nterface
BIST	**B**uilt-**i**n **S**elf-**t**est (operation)
BISYNC	**B**inary **Syn**chronous **C**ontrol (protocol)
bit	**bi**nary digi**t**
bit/s	**bit**s **per s**econd
BIT	**B**uilt-**I**n **T**est
BITBLT	**BIT** **Bl**ock **T**ransfer (Microsoft Windows)
BITE	1. **B**ackward **I**nterworking **T**elephony **E**vent
	2. **B**uilt-**in** **T**est **E**quipment
BITNet	**B**ecause **I**t's **T**ime **Net**work (WANs)
BITNIC	**BIT**Net **N**etwork **I**nformation **C**enter
BITS	1. **B**ase **I**nformation **T**ransport **S**ystem (U.S. Air force)
	2. **B**uilding **I**ntegrated **T**iming **S**upply (clock)
BITT	**BIT** **T**ransparent
BIU	1. **B**asic **I**nformation **U**nit (IBM SNA)
	2. **B**us-**I**nterface **U**nit
BIX	**B**yte **I**nformation **Ex**change (Byte magazine)
BJT	**B**ipolar **J**unction **T**ransistor (semiconductors)

BK	**b**rea**k** (Morse code transmissions)
BL	1. **B**usiness **L**ine
	2. **B**it **L**ine
BLAM	**B**inary **L**ogarithmic **A**ccess **M**ethod
BLAS	**B**asic **L**inear-**A**lgebra **S**ubprogram (software)
BLEC	1. **B**roadband **L**ocal **E**xchange **C**arrier
	2. **B**uilding **L**ocal **E**xchange **C**arrier
BLER	**B**lock **E**rror **R**ate
BLERT	**B**lock **E**rror **R**ate **T**est/Tester (data transmission)
BLES	**B**roadband **L**oop **E**mulation **S**ervices
BLF	**B**usy **L**amp **F**ield
BLIP	**B**luetooth's **L**ocal **I**nformation **P**oints (bluetooth)
BLISS	**B**roadband **L**ightwave **S**ource **S**ystem
BLOB	**B**inary **L**arge **Ob**jects (data type)
BLOS	**B**eyond **L**ine-**of**-**S**ight (microwave)
BLP	**B**lock **L**oss **P**robability (transmission)
BLR	1. **B**ranch **L**evel **R**evenue
	2. **B**lock **L**oss **R**atio (transmission)
BLSR	**B**idirectional **L**ine-**S**witched **R**ing
BLT	**B**lock **L**ine **T**ransfer
BLU	**B**asic **L**ink **U**nit
BM	1. **B**alanced **M**odulator
	2. **B**uffer **M**emory
	3. **B**urst **M**odem
BMC	1. **B**usiness **M**anagement **C**omputer
	2. **B**roadband **M**anagement **C**ontroller
BMEC	**B**atelco **M**iddle **E**ast **C**ompany (Bahrain)
BMEWS	**B**allistic **M**issile **E**arly **W**arning **S**ystem
BMIGRATE	A program used to **migrate** from **B**anyan VINES operating system to Novell Netware 4.1 (networking)
BMLC	**B**asic **M**ode **L**ink **C**ontrol (OSI model)
BMMG	**B**ritish **M**inicomputer **M**anufacturing **G**roup
.bmp	Name extension for **bit**ma**p** format files (graphics)
BMP	**B**asic **M**ultilingual **P**lane
BMR	1. **B**it **M**is-delivery **R**atio (transmission)
	2. **B**lock **M**is-delivery **R**atio (transmission)
BMTI	**B**lock **M**ode **T**erminal **I**nterface
BN	1. **B**ridge **N**umber
	2. **B**order **N**ode
	3. **B**ackground **N**oise
BNA	1. **B**urroughs **N**etwork **A**rchitecture
	2. **B**illing **N**ame and **A**ddress
BNAP	**B**roadband **N**etwork **A**ccess **P**oint (British Telecom)
BNC	1. **B**ritish **N**aval **C**onnector (cables)

B

	2. **B**ayonet-coupling **N**avy **C**onnector
	3. **B**ayonet-coupling **N**ut **C**onnector
BNCC	**B**ase **N**etwork **C**ontrol **C**enter
BNR	**B**ell-**N**orthern **R**esearch (Northern Telecom)
BNSC	**B**ritish **N**ational **S**pace **C**entre (England)
BNT	**B**roadband **N**etwork **T**ermination
BO	**B**ack-**O**ff
BOB	**B**reak-**O**ut **B**ox testing device (transmission)
BOC	1. **B**ell **O**perating **C**ompany
	2. **B**usiness **O**ffice **C**ode
BOE	**B**uffer **O**verflow **E**rror
BOF	**B**usiness **O**perations **F**ramework (wireless)
BOFH	**B**astard **O**perator **F**rom **H**ell (network language)
BOIC	**B**arring of **O**utgoing **I**nternational **C**alls (GSM service)
BOIC-exHC	**BOIC ex**cept the **H**ome **C**ountry (GSM service)
BOL	**B**eginning **o**f **L**ife
BOLT	**B**roadband **O**ptical **L**ine **T**ermination
BOM	1. **B**eginning **o**f **M**essage (ATM)
	2. **B**ill **o**f **M**aterials (contracts)
	3. **B**asic **O**perations **M**onitor
BONAPARTE	**B**roadband **O**ptical **N**etwork using **A**TM **P**ON **A**ccess facilities in **R**ealistic **T**elecommunications **E**nvironment
BONDING	**B**andwidth **On** **D**emand **In**teroperability **G**roup
BONT	**B**roadband **O**ptical **N**etwork **T**ermination
BootP	**Boot**strap **P**rotocol (TCP/IP)
BOP	1. **B**it-**O**riented **P**rotocol
	2. **B**iocomputing **O**ffice **P**rotocol
	3. **B**eginning **o**f **P**acket
BOS	**B**ill **O**utput **S**pecifications
BOSS	**B**illing and **O**rder **S**upport **S**ystem
bot	ro**bot**
BOT	**B**uild, **O**perate, and **T**ransfer (GSM contracts)
BP	1. **B**and-**P**ass
	2. **B**y-**P**ass
	3. **B**ase **P**ointer
	4. **B**eam **P**osition
	5. **B**lock **P**air (telephone wire)
BP14	**B**ody **P**art **14**
BPAD	**B**isynchronous **P**acket **A**ssembler/**D**isassembler
BPCS	**B**roadband **P**ersonal **C**ommunications **S**tandards
BPDU	**B**ridge **P**rotocol **D**ata **U**nit (ATM)
BPF	**B**and-**P**ass **F**ilter
Bpi	**B**ytes **p**er **i**nch (recording surface)
BPM	**B**eam **P**osition **M**onitor

BPON	Broadband Passive Optical Network
bpp	bits per pixel (digital images)
BPP	Brokered Private Peering (industry plan)
BPR	Business Process Reengineering
bps	bits per second
BPS	1. By-Pass
	2. B-Channel Packet-Switched
BPSK	Binary PSK (modulation)
BPU	Boundary Processing Unit
BPV	Bipolar Violation (data encoding)
BQM	Business Quality Messaging
BQOS	Business Quality of Service
BR	1. Beacon Receiver
	2. Bureau of Radiocommunications
	3. Back-Reflection
	4. Bit Rate
BRA	Basic Rate Access (ISDN)
BRADS	Bell Rating Administrative Data System (Telcordia Technologies)
BRC	Bus Repeater Card
BRCS	Business and Residence Customer Service
BRDF	Bidirectional Reflectance Distribution Function (remote sensing)
BREW	Binary Run-time Environment for Wireless (Qualcom)
BRF	Band-Rejection Filter
Brg	bridge (circuits)
BRI	1. Basic Rate Interface (ISDN)
	2. Basic Rate ISDN (service)
BRIDS	Bellcore Rating Input Database System
BRISC	Bell-Northern Research Reduced Instruction Set Computing
BRL	Balance Return Loss
BRM	Bus Repeater Module
BRS	1. Buried Ridge Stripe
	2. Big Red Switch (computer)
BRTC	Bus Repeater Terminating Card
BRTM	Bus Repeater Terminating Module
BS	1. Back-Space (character control code)
	2. Base Station
	3. Bearer Service (ISDN)
	4. Beam Splitter
	5. Back Scatter
	6. Broadcasting Service (ITU-T)
BSA	1. Basic Switching Arrangement

	2. **B**asic **S**erving **A**rrangement (OSI model)
BSAC	**B**it-**S**liced **A**rithmetic **C**oding
BSACE	**B**illing **S**ystem **ACE**
BSAM	**B**asic **S**equential **A**ccess **M**ethod
BSC	1. **B**inary **S**ynchronous **C**ommunications (protocol)
	2. **B**inary **S**ymmetric **C**hannel
	3. **B**ase **S**tation **C**ontroller (GSM)
BSCC	**B**ell **S**outh **C**ellular **C**orporation
BSD UNIX	**B**erkeley **S**oftware **D**istribution **UNIX**
BSDL	**B**oundary-**S**can **D**escription **L**anguage (software)
BSE	1. **B**roadcasting **S**atellite **E**xperiment (Japan)
	2. **B**ack-**S**cattered **E**lectrons
	3. **B**asic **S**witching **E**lement (networking)
	4. **B**asic **S**ervice **E**lement (voice processing)
BSF	**B**and **S**top **F**ilter (electronics)
BSFOCS	**B**lack **S**ea **F**iber **O**ptic **C**able **S**ystem (cable consortium)
BSFT	**B**yte **S**tream **F**ile **T**ransfer
BSGL	**B**ranch **S**ystems **G**eneral **L**icense (British)
BSHR	**B**idirectional **S**elf-**H**ealing **R**ing
BSI	**B**ritish **S**tandards **I**nstitution (U.K.)
BSIC	**B**ase **S**tation **I**dentity **C**ode (GSM)
BSIS	**B**ranch **S**ales **I**nformation **S**ystem
BskyB	**B**ritish **Sky** **B**roadcasting (TV broadcaster)
BSL	**B**ritish **S**ign **L**anguage (programming)
BSMS	**B**roadcast **S**hort **M**essage **S**ervice (GSM)
BSMTP	**B**atch **S**imple **M**ail-**T**ransfer **P**rotocol
BSN	**B**ackward **S**equence **N**umber (GSM)
BSO	**B**it **S**ynchronous **O**peration (SONET)
BSOC	**B**attery **S**tate-**o**f-**C**harge
BSOD	**B**lack **S**creen **o**f **D**eath (windows error message)
BSP	1. **B**ell **S**ystem **P**ractice
	2. **B**yte **S**tream **P**rotocol
BSQ	**B**and **Seq**uential (remote sensing)
BSR	1. **B**it **S**can **R**ate
	2. **B**it **S**can **R**everse
BSRF	**B**ell **S**ystem **R**eference **F**requency
BSS	1. **B**roadcast **S**atellite **S**ervices (ITU)
	2. **B**oeing **S**atellite **S**ystems
	3. **B**ase **S**tation **S**ubsystem (GSM)
	4. **B**roadband **S**witching **S**ystem
	5. **B**usiness **S**upport **S**ystem
	6. **B**asic **S**ynchronized **S**et (OSI model)
BSSAP	**B**ase **S**tation **S**ubsystem **A**pplication **P**art (GSM)
BSSMAP	**B**ase **S**tation **S**ubsystem **M**obile **A**pplication **P**art (GSM)

BSTCE	Base Station Terminal Control Element
BSU	1. Beam Steering Unit
	2. Burst Synchronization Unit
BSVC	Broadcast-Switched Virtual Connections
BT	1. British Telecom (company)
	2. Base Transceiver (GSM)
	3. Burst Tolerance (ATM)
	4. Bit Time
	5. Television Broadcasting service (ITU-T)
BTA	1. Basic Trading Area (cellular networks)
	2. Broadband Telecommunications Architecture
	3. Broadband Terminal Adapter
	4. Business Technology Association (standard organization)
BTag	Beginning Tag (ATM)
BTAM	Basic Telecommunications Access Method (IBM)
BTB	Book-To-Bill (ratio)
BTBT	Band-To-Band Tunneling
BTE	1. Boltzmann Transport Equation
	2. Broadband Terminal Equipment
BTI	British Telecom International
BTL	1. Bell Telephone Laboratories
	2. British Telecommunications Laboratories
	3. Backplane Transceiver Logic (electronics)
	4. Bipolar Transistor Logic
BTM	Broadband Transport Manager
BTMU	Bit Transport Master Unit
BTN	Billing Telephone Number
BTO	1. Built To Order (PC makers)
	2. Bandwidth Trading Organization
	3. Bombing Through Overcast (military)
BT	British Telecom
BTP	Burst Time Plan
BTR	1. Bit Timing Recovery
	2. Block Transfer Rate
	3. British Telecom Requirement
BTRL	British Telecom Research Laboratories
BTS	1. Base Transceiver Station (GSM)
	2. Bit Test and Set
BTSM	Base Transceiver Station Management (wireless)
BTSO	Bit Transport Slave Optical
BTU	1. Basic Transmission Unit
	2. British Thermal Unit
BTV	Business Television

BU	**B**aseband **U**nit
BUFF	**buff**er
BUNI	**B**roadband **U**ser **N**etwork **I**nterface
BURBON	**B**roadband **U**rban **R**ural-**B**ased **O**pen **N**etworks
BUS	**B**roadcast and **U**nknown **S**erver
BVA	**B**illing **V**alidation **A**pplication database (AT&T)
BVR	**B**eyond **V**isual **R**ange
BVS	**B**illing **V**alidation **S**ervice (AT&T)
BW	**b**and**w**idth (parameter)
BWA	1. **B**roadband **W**ireless **A**ccess
	2. **B**ackward-**W**ave **A**mplifier (electronics)
BWF	**B**roadcast **W**ave **F**ormat
BWFN	**B**eam **W**idth, **F**irst **N**ulls (antennas)
BWG	**B**eam **W**ave-**G**uide (antenna)
BWM	**B**and**W**idth **M**anagement System
BWO	**B**ackward-**W**ave **O**scillator (electronics)
Byte	**B**inar**y** **Te**rm
BZT	**B**undesamt für **Z**ulassungen in der **T**elecommunikation (German)

C c

c	1. Symbol for prefix **c**enti-, denoting one-hundredth or 10^{-2}
	2. Symbol for **c**haracter
	3. Symbol for **c**ycle
	4. Symbol for a **c**onstant value (mathematics)
cc	**c**ubic **c**entimeter
C/I Ratio	**C**arrier-**to-I**nterference **Ratio**
c/in	**c**haracters **per in**ch
c/s	1. **c**haracters **per s**econd
	2. **c**ycles **per s**econd
C	1. Symbol for **C**apacitor or Capacitance (electronics)
	2. Symbol for **C**oulomb (unit of electric charge)
	3. Symbol for **C**ollector (transistor circuit diagrams)
	4. Symbol for **C**elsius (temperature unit)
	5. Symbol for speed of light in a vacuum
	6. A programming language
C&C	1. **C**omputer **and C**ommunications
	2. **C**ommand **and C**ontrol
C&D	**C**ontrol **and D**isplay
C&DH	**C**ontrol **and D**ata **H**andling (Intelsat)
C&IT	**C**ommunications **and I**nformation **T**echnology (same as ICT)
C&W	**C**able **and W**ireless

C−	Symbol for **negative** terminal of a **C** battery or voltage source
C+	Symbol for **positive** terminal of a **C** battery or voltage source
C++	An extension of **C** programming language
C-AFM	**C**alibrated **A**tomic **F**orce **M**icroscope
C-band	Radio frequency **band** ranging from 4 to 6 GHz
C-channel	Circuit-mode channel
C-DTE	Character mode **DTE** (computer)
c-HTML	Compact **HTML**
C-Net	**C**oaxial **Net**work
C-To-L	**C**-band **to** **L**-band
C/A	**C**lear **and** **A**cquisition Code (GPS)
C/DTAC	**C**ustomer **D**isability **T**elecommunications **A**dvisory **C**ommittee
C/I	**C**arrier-**to**-**I**nterference power (ratio)
C/IM3	**C**arrier-**to**-**3**rd Order **I**nter **m**odulation Ratio (Intelsat)
C/IMo	**C**arrier-**to**-**I**nter **mo**dulation noise density ratio (Intelsat)
C/Io	**C**arrier-**to**-**I**nterference power spectral density ratio (Intelsat)
C/M	**C**arrier-**to**-**M**ultipath power (ratio)
C/N	**C**arrier-**to**-**N**oise power (ratio)
C/No	**C**arrier-**to**-**N**oise power spectral density (ratio)
C/R	**C**ommand **R**esponse (frame relay)
C/S	**C**all **S**ign (navigation)
C/T	**C**arrier-**to**-noise **T**emperature (ratio)
C^2	**C**ommand and **C**ontrol system (military)
C2B	**C**onsumer-**to**-**B**usiness (e-commerce)
C2C	**C**onsumer-**to**-**C**onsumer (e-commerce)
C2G	**C**onsumer-**to**-**G**overnment (e-commerce)
C^2L	**C**lose **C**omplementary **L**ogic (digital electronics)
C^3	**C**ommand, **C**ontrol, and **C**ommunications system (military)
C^3L	**C**omplementary **C**onstant-**C**urrent **L**ogic (digital electronics)
C3Po	**C**isco **3** **Po**rt switch
C^4	**C**ommand, **C**ontrol, **C**ommunications, and **C**omputers
C^4I	**C**ommand, **C**ontrol, **C**omputer, **C**ommunications, and **I**ntelligence
C7	The European version of SS7
C64	**C**ommodore **64**
CA	1. **C**ommunication **A**gent 2. **C**onditional **A**ccess

	3. **C**all **A**ppearance
	4. **C**ell **A**llocation (GSM)
	5. **C**ertificate **A**uthority (digital certificate)
CAA	1. **C**ivil **A**viation **A**dministration or **A**uthority
	2. **C**omputer-**A**ssisted **A**ssessment (ICT)
	3. **C**ollinear **A**ntenna **A**rray
CAAGR	**C**ompound **A**nnual **A**verage **G**rowth **R**ate
CaAs	**C**admium–**Ars**enide (semiconductors)
.cab	Name extension for **cab**inet files (computer)
CAB	**C**anadian **A**ssociation of **B**roadcasters
CableLabs	**Cable** Television **Lab**oratorie**s**, Inc. (standards organization)
CABS	**C**arrier **A**ccess **B**illing **S**ystem
CABS BOS	**CABS-B**illing **O**utput **S**pecifications
CAC	1. **C**ustomer **A**dministration **C**enter
	2. **C**arrier **A**ccess **C**ode
	3. **C**all **A**dmission **C**ontrol (ATM)
	4. **C**onnection **A**ddress **C**ontrol
CACH	**C**all **A**ppearance **C**all **H**andling (ISDN)
CACM	**C**ommunications of the **A**ssociation for **C**omputing **M**achinery
CACS	**C**entralized **A**larm **C**ontrol **S**ystem
CAD	1. **C**omputer-**A**ided **D**esign (ICT)
	2. **C**omputer-**A**ided **D**rafting (ICT)
	3. **C**omputer-**A**ided **D**ispatching (ICT)
CAD/CAM	**C**omputer-**A**ided **D**esign /**C**omputer-**A**ided **M**anufacturing (ICT)
CADA	**C**omputer-**A**ided **D**esign and **A**nalysis (ICT)
CADB	**C**alling **A**rea **D**atab**a**se (MCI)
CADC	**C**ontrol **A**nd **D**isplay **C**onsole (Intelsat)
CADD	**C**omputer-**A**ided **D**esign and **D**evelopment (ICT)
CADS	1. **C**ode **A**buse **D**etection **S**ystem
	2. **C**omputer **A**buse **D**etection **S**ystem
CAE	1. **C**omputer-**A**ided **E**ngineering (ICT)
	2. **C**ommon **A**pplications **E**nvironment (ICT)
CAFA	**C**omputer-**A**ided **F**inancial **A**nalysis
CAG	1. **C**ustomer **A**ctivation **G**roup (Inmarsat)
	2. **C**onditional **A**ccess **G**ateway
	3. **C**omputer-**A**ided **G**raphics (ICT)
CAGR	**C**umulative **A**nnual **G**rowth **R**ate (finance)
CAI	1. **C**omputer-**A**ided **I**nstruction (ICT)
	2. **C**ommon **A**ir **I**nterface (GSM)
	3. **C**onfiguration and **A**larm **I**nterface

C

CAIS	Conditional Access Interface Server
cal	**cal**orie (unit of heat)
CAL	1. CAN Application Layer
	2. Computer-Aided Learning (e-learning)
	3. Computer-Assisted Learning (e-learning)
CALC	Customer Access Line Charge
CALEA	Communications Assistance to Law Enforcement Act
CALLUM	Combined Analog Locked-Loop Universal Modulator
CALNet	**Cal**ifornia **Net**work
CALS	1. Continuous Acquisition and Life-cycle Support (EDI)
	2. Computer-aided Acquisition and Logistics Support (EDI)
CALSCH	**Cal**endaring and **Sch**eduling working group of the IETF
CAM	1. Computer-Aided Manufacturing (ICT)
	2. Computer-Assisted Makeup (ICT)
	3. Carrier Module
	4. Controlled Attachment Module (LANs)
	5. Content-Addressable Memory
	6. Conditional Access Module (TV viewing card)
	7. Call Accounting Manager
	8. Call Application Manager
CAMA	Centralized Automatic Message Accounting
CAMA-ONI	**CAMA**-Operator Number Identification
CAMA/LAMA	**CAMA** Logical Automatic Message Accounting
CAMAC	Computer-Aided Measurement And Control (ICT)
CAMEL	Customized Application for Mobile network Enhanced Logic
CAMP	Channel **Amp**lifier
CAMC	Conditional Access Management Center (television)
CAN	1. Control Area Network
	2. Controller Area Network
	3. **can**cel (character control code)
CanISDN	**Can**adian **ISDN**
CAO	Computer-Assisted Ordering (ICT)
CAP	1. Competitive-Access Provider
	2. Client Access Protocol
	3. Carrierless Amplitude and Phase modulation
	4. Cellular Array Processor (Fujitsu)
	5. Conditional Access Packet (EMM)
	6. CAMEL Application Part
CAPC	Competitive Access Provider Capacity (transmission speed)
Capex	**Cap**ital **ex**penditures

CAPI	1. **C**ryptography **A**pplication **P**rogram **I**nterface
	2. **C**ommon-ISDN **API**
CAPIMS	**C**onditional **A**ccess **P**rogram **I**nformation **M**anagement **S**ystem
CAPP	**C**omputer-**A**ide **P**rocess **P**lanning (ICT)
Caps	1. **C**all **A**ttempts **P**er **S**econd
	2. **Cap**ital letter**s** (keyboard)
CAPS	1. **C**ode **A**buse **P**revention **S**ystem
	2. **C**ompetitive **A**ccess **P**roviders (telephone network)
CAPTAIN	**C**haracter **A**nd **P**attern **T**elephone **A**ccess **I**nformation **N**etwork
CAPWIRE	**Cap**itol **Wire**less Inc. (company)
CAR	**C**omputer-**A**ided **R**etrieval (ICT)
CARAB	**C**anadian **A**mateur **R**adio **A**dvisory **B**oard
CARE	**C**ustomer **A**ccount **R**ecord **E**xchange
CARE/ISI	**CARE** **I**ndustry **S**tandard **I**nterface
CAROT	**C**entralized **A**utomatic **R**eporting **O**n **T**runks (test & maintenance)
CARP	**C**ache **A**rray **R**outing **P**rotocol
CARS	**C**able **T**elevision **R**elay **S**ervice
CAS	1. **C**hannel **A**ssociated **S**ignaling
	2. **C**entralized **A**ttendant **S**ervice
	3. **C**ommunications **A**pplications **S**pecification
	4. **C**onditional **A**ccess **S**egment (TV broadcasting)
	5. **C**ollision **A**voidance **S**ystem (electronics)
	6. **C**olumn-**A**ddress **S**trobe (memory chip logic signal)
CASBAA	**C**able **A**nd **S**atellite **B**roadcasting **A**ssociation of **A**sia
CASC	**C**ommunications **A**ir **S**upport **C**enter
CASE	1. **C**omputer-**A**ided **S**oftware **E**ngineering (ICT)
	2. **C**omputer-**A**ssisted **S**oftware **E**ngineering (ICT)
	3. **C**omputer-**A**ided **S**ystem **E**ngineering (ICT)
	4. **C**omputer-**A**ssisted **S**ystem **E**ngineering (ICT)
	5. **C**ommon **A**pplication **S**ervice **E**lements
CAST	1. **C**omputer-**A**ided **S**oftware **T**esting (ICT)
	2. **C**omputer-**A**ssisted **S**oftware **T**esting (ICT)
Cat	**cat**egory (cable standard)
CAT	1. **C**omputer-**A**ided **T**eaching (e-learning)
	2. **C**omputer-**A**ided **T**esting (ICT)
	3. **C**omputerized **A**xial **T**omography (medicine)
	4. **C**onditional **A**ccess **T**able
	5. **C**ouncil for **A**ccess **T**echnologies
	6. **C**ommunications **A**uthority of **T**hailand
	7. **C**all **A**ccounting **T**erminal (AT&T)
	8. **C**lear-**A**ir **T**urbulence

CAT3	Category 3 UTP Cable (copper cables)
CAT5	Category 5 UTP Cabling (copper cables)
CATI	Computer-Aided Telephone Interviewing
CATLAS	Centralized Automatic Trouble Locating and Analysis System
CATNIP	Common Architecture for Next generation Internet Protocol
CATS	Consortium for Audiographics Teleconferencing Standards
CATT	1. Controlled Avalanche Transit-time Triode (semiconductors)
	2. China Academy of Telecommunications Technology
CATV	1. Community Antenna Television (broadcasting)
	2. Community Access Television
	3. Cable Television
CAU	1. Controlled Access Unit (LANs)
	2. Connection Arrangement Unit (Northern Telecom)
CAUS	Conditional Access Uplink System (TV broadcasting)
CAV	Constant Angular Velocity (hard disks)
CAVE	1. Cave Automatic Virtual Environment
	2. Cellular Authentication and Voice Encryption
CB	1. Check Bits
	2. Cell Broadcast (GSM)
	3. Compensation Buffer
	4. Citizens' Band (radio service)
CB Radio	Citizen's Band Radio
CBA	1. Could Be Anything (fault type)
	2. Community Broadcasters Association (U.S.A.)
CBAA	Community Broadcasters Association of Australia
CBAC	Context-Based Access Control
CBC	1. Canadian Broadcasting Corporation (company)
	2. Cipher Block Chaining (data encryption)
	3. Cell Broadcast Center (GSM)
	4. China Broadcast Cybercast
CBCH	Cell Broadcast Channel (GSM)
CBCPS	Conductor-Backed Co-Planar Stripline (microwave)
CBCPW	Conductor-Backed Co-Planar Waveguide (microwave)
CBDS	1. Connectionless Broadband Data Service (ATM)
	2. Common Basic Data Set
CBEMA	Computer and Business Equipment Manufacture Association
CBETS	Communications and Broadcasting Engineering Test Satellite

CBF	1. Computer-Based Fax
	2. Community Broadcasters Foundation Ltd. (Australia)
CBGA	Ceramic Ball-Grid Array (microchips)
CBH	Component Busy Hour
CBI	Color Burst Inversion (TV broadcasting)
CBK	Change Back
CBL	1. Computer-Based Learning (e-learning)
	2. Communications By Line
CBM	Commodore Business Machines
CBMS	Computer-Based Managing System (ICT)
CBO	Continuous Bit Oriented
CBP	Coded Block Patterns (TV broadcasting)
CBQ	Class-Based Queuing
CBR	1. Carrier and Bit Timing Recovery
	2. Constant Bit Rate (ATM service)
	3. Committed Bit Rate (ATM service)
	4. Carson Bandwidth Rule
CBS	1. Certified Banyan Specialist (Banyan Systems)
	2. Columbia Broadcasting System
	3. Call Broadcast Services (GSM)
	4. Constant Bit rate Service
CBSC	Canadian Broadcast Standards Council
CBT	1. Computer-Based Training (e-learning)
	2. Core-Based Trees (IP multicast)
	3. Cincinnati Bell Telephone company (U.S.A.)
CBTA	Canadian Business Telecommunication Alliance
CBTR	Carrier and Bit Timing Recovery
CBUD	"Call Before U Dig" (cabling)
CBX	Computerized Branch Exchange (IBM)
CC	1. Call Control (GSM)
	2. Carbon Copy (fax and e-mail)
	3. Courtesy Copy (e-mail)
	4. Cross Connect
	5. Country Code
	6. Company Code
	7. Constant Current
CCA	1. Carrier-Controlled Approach
	2. Circuit Card Assembly
CCAF	Cluster Control Agent Function
CCB	1. Common Carrier Bureau (FCC)
	2. Cluster Control Bus
	3. Customer Care and Billing (GSM)
CCBM	Came Clear By Magic (repair technique)

CCBS	1. Call Completion to Busy Subscriber
	2. Customer Care and Billing System
CCC	1. Clear Channel Capability (data transmission)
	2. Clear Coded Channel
	3. Center for Corporate Communications
	4. Call Control Character (telephony)
	5. Command and Control Center
	6. Customer Care Center (Inmarsat)
	7. Credit Card Call
	8. Communications Competition Coalition
CCCH	Common Control Channel (GSM)
CCCP	Committee on Computer-to-Computer Communications Protocol
CCD	Charge-Coupled Device (semiconductors)
CCDA	Cisco Certified Design Associate (Cisco)
CCDN	Corporate Consolidated Data Network
CCDP	Cisco Certified Design Professional (Cisco)
CCE	Call Carrying Equipment
CCEB	Combined Communications–Electronics Board
CCF	1. Cluster Control Function
	2. Consumer Consultative Forum
CCFL	Cold-Cathode Fluorescent Lamp
CCG	Control Clock Generator
CCH	Connections per Circuit-Hour
CCI	1. Co-Channel Interference
	2. Common Client Interface (Windows)
CCIA	Computer and Communications Industry Association
CCIE	Cisco Certified Internetwork Expert (Cisco)
CCIR	1. Comité Consultatif des International Radiocommunications
	2. Center for Communication Interface Research
CCIRN	Coordinating Committee for International Research Networks
CCIS	1. Common-Channel Interoffice Signaling
	2. Control Command Information System
CCITT	1. Comité Consultatif International Télégraphique et Téléphonique
	2. Consultative Committee on International Telegraphy and Telephony (now ITU)
CCK	Complementary Code Keying (modulation)
CCL	1. Configuration Control Link
	2. Carrier Common Line (charge)
CCM	1. Cross Connection Management
	2. Counter-Counter Measure (electronics)

C³CM	Command, Control, and Communications Countermeasures
CCMA	Call Center Management Association (U.K.)
CCMT	Computer-Controlled Microwave Tuner (microwave)
CCNA	1. Cisco Certified Network Associate (Cisco)
	2. Customer's Carrier Name Abbreviation
CCNE	Cisco-Certified Network Engineer (Cisco)
CCNP	Cisco-Certified Network Professional (Cisco)
CCNR	Call Completion on No Reply
ccNUMA	cache-coherent Non-Uniform Memory Access (computer)
CCP	1. Cluster Control Processor
	2. Compression Control Protocol
	3. Certificate of Computer Programming
CCPCH	Common Control Physical Channel
CCPR	Co-Channel Power Ratio (microwave)
CCR	1. Carrier and Clock Recovery
	2. Current Call Rate (ATM)
	3. Customer Control Routing
CCS	1. Common Channel Signaling (ISDN)
	2. Clear Confirmation Signal (data communications)
	3. Common Communication Support (protocol)
	4. Centi Call Second
CCS7	Common Channel Signaling No. 7
CCSA	1. Common Control Switching Arrangement (AT&T)
	2. China Communication Standards Association
CCSD	Command Communications Service Designator
CCSDS	Consultative Committee for Space Data Systems
CCSL	Compatible Current-Sinking Logic (digital electronics)
CCSN	Common Channel Signaling Network
CCT	1. Continuity Check Tone
	2. Computer Compatible Tape
	3. Consultative Committee of Telecommunications
	4. Calling Card Table
CCTA	Central Computer and Telecommunications Agency
CCTV	Closed-Circuit Television (system)
CCU	1. Cluster Control Unit
	2. Communication Control Unit (minicomputers)
	3. Camera Control Unit
CCV	Credit Card Validation
CCW	1. Cable Cutoff Wavelength (fiber optics)
	2. counterclockwise

cd	1. **c**hange the current **d**irectory (MS DOS Command)
	2. **c**an**d**ela (unit of luminous intensity)
cd/m²	**c**andela **per s**quare **m**eter
CD	1. **C**ompact **D**isk
	2. **C**arrier **D**etect (modems)
	3. **C**ollision **D**etection
	4. **C**ampus **D**istributor (Campus backbone cable)
	5. **C**ount **D**own
	6. **C**all **D**efinition
	7. **C**ell **D**elineation (ATM)
CD-E	**C**ompact **D**isk **E**rasable
CD-I	**C**ompact **D**isk **I**nteractive
CD-R	**C**ompact **D**isk **R**ecordable
CD-R/E	**C**ompact **D**isk **R**ecordable **and E**rasable
CD-ROM	**C**ompact **D**isk **R**ead-Only **M**emory
CD-ROM/XA	**CD-ROM** Extended **A**rchitecture
CD-RW	**C**ompact **D**isk **R**e-**w**ritable (computer)
CD-V	**C**ompact **D**isk **V**ideo
CD-WO	**C**ompact **D**isk **W**rite **O**nce
CDA	1. **C**ommunications **D**ecency **A**ct
	2. **C**all **D**uration **A**dvice
CDAR	**C**ustomer **D**ialed **A**ccount **R**ecording
CDAS	**C**ommanding and **D**ata **A**cquisition **S**tation (remote sensing)
CDB	1. **C**lock **D**istribution **B**oard
	2. **C**ode **D**ata **B**yte (software downloading)
CDC	1. **C**ontrol and **D**elay **C**hannel
	2. **C**ustomer **D**ata **C**hange
	3. **C**hromatic **D**ispersion **C**oefficient (fiber optics)
	4. **C**onnected **D**evice **C**onfiguration (J2ME)
CDCA	**C**ontinuous **D**ynamic **C**hannel **A**llocation
CDCS	**C**ontinuous **D**ynamic **C**hannel **S**election
CDDI	**C**opper **D**istributed-**D**ata **I**nterface (cable)
CDE	1. **C**ommon **D**esktop **E**nvironment (UNIX)
	2. **C**ustom **D**esign **E**ngineering
CDEV	**C**ontrol panel **Dev**ice (Apple Macintosh)
CDF	1. **C**utoff **D**ecrease **F**actor (ATM)
	2. **C**ombined **D**istribution **F**rame
	3. **C**umulative **D**istribution **F**unction
	4. **C**hannel **D**efinition **F**ormat (Internet)
CDFP	**C**entrex **D**ata **F**acility **P**ooling
CDFS	**C**ompact **D**isk **F**ile **S**ystem (CD-ROM drives)
CDG	**CD**MA **D**evelopment **G**roup (U.S. standards organization)

C

CDI	1. **C**lock **D**istribution **I**nterface
	2. **C**ompact **D**isk **I**nteractive
CDL	**C**oded **D**igital **L**ocator
CDLC	**C**ellular **D**ata **L**ink **C**ontrol (protocol)
CDLRD	**C**onfirming **D**esign **L**ayout **R**eport **D**ate
CDM	1. **C**onditioned **D**iphase **M**odulation
	2. **C**hina **D**igital **M**edia (pay TV operator)
	3. **C**ode-**D**ivision **M**ultiplexing
CDMA	**C**ode-**D**ivision **M**ultiple **A**ccess (cellular networks)
CDMA 1X	The **first** generation of **CDMA-2000**
CDMA 1X EV DO	The **E**volved **D**ata **O**nly **first** generation of **CDMA-2000** (GSM)
CDMA-2000 **CDMA2000**	**C**ode-**D**ivision **M**ultiple **A**ccess **2000** (3rd Generation)
1x RTT	**C**ode-**D**ivision **M**ultiple **A**ccess (CDMA) **1x** (single-carrier) **R**adio **T**ransmission **T**echnology
CDMA one	The **First** commercial **CDMA** cellular system
CDMP	**C**ellular **D**igital **M**essaging **P**rotocol
CDN	1. **C**ontrol **D**irectory **N**umber
	2. **C**ontent **D**ata **N**etwork
	3. **C**ontent **D**elivery **N**etwork
CDO	**C**ommunity **D**ial **O**ffice (switching system)
CDP	1. **C**isco **D**iscovery **P**rotocol (Cisco)
	2. **C**ustomized **D**ial **P**lan
	3. **C**ertificate of **D**ata **P**rocessing
CDPA	**C**apture **D**ivision **P**acket **A**ccess (cellular networks)
CDPD	**C**ellular **D**igital **P**acket **D**ata (cellular networks)
CDPDF	**C**ellular **D**igital **P**acket **D**ata **F**orum
CDPSK	**C**oherent **D**ifferential **PSK** (modulation)
.cdr	Name extension for **C**orel**DRAW** format files (graphics)
CDR	1. **C**all **D**ata **R**ecord
	2. **C**all **D**etail **R**ecord (cellular networks)
	3. **C**all **D**rop **R**ate (GSM)
	4. **C**ritical **D**esign **R**eview
	5. **C**lock/**D**ata **R**ecovery (LANs)
	6. **C**omplete **D**ocument **R**ecognition
CDRAM	**C**ache **D**ynamic **RAM** (memory)
CdS	**C**ad**m**ium **S**ulfide (semiconductors)
CDS	1. **C**ircuit **D**ata **S**ervices
	2. **C**hromatic **D**ispersion **S**lope (optical fibers)
CDSA	**C**ommon **D**ata **S**ecurity **A**rchitecture
CDSL	**C**onsumer **DSL** (access)
CDT	1. **C**re**d**it allocation (packet **S**witching)
	2. **C**enter for **D**emocracy and **T**echnology

	3. **C**ontrol **D**ata **T**erminal
	4. **C**ode **D**ownload **T**able (software downloading)
CDTAC	**C**onsumer/**D**isabilities **T**elecommunications **A**dvisory **C**ommittee
CDTV	**C**omodore **D**ynamic **T**otal **V**ision
CDU	**C**entral **D**isplay **U**nit
CDV	1. **C**all **D**elay **V**ariation (ATM)
	2. **C**ompressed **D**igital **V**ideo
	3. **C**ompact **D**isk **V**ideo
CDVCC	**C**oded **D**igital **V**erification **C**olor **C**ode
CDVT	**C**ell **D**elay **V**ariation **T**olerance (ATM)
CE	1. **C**onnection **E**ndpoint (ATM)
	2. **C**ircuit **E**mulation
	3. **C**hip **E**nable (electronics)
C/E Buffer	**C**ompression **and** **E**xpansion **Buffer**
CeBIT	**C**entrum **B**uero **I**nformation **T**elekommunikation (Germany)
CEBus	**C**onsumer **E**lectronics **Bus**
CEC	1. **C**ommission of the **E**uropean **C**ommunities
	2. **C**hina **E**lectric **C**ompany
	3. **C**anadian **E**lectric **C**ode
CECL	**C**ascade **E**mitter-**C**oupled **L**ogic (digital electronics)
CED	**C**apacitance **E**lectric **D**isk
CEDAR	**C**enter of **E**xcellence for **D**ocument **A**nalysis and **R**ecognition
CEDT	**C**alled **E**quipment **I**dentification **T**one
CEG	**C**anterbury **E**lectronics **G**roup (New Zealand)
CEI	1. **C**onnection **E**ndpoint **I**dentification (ATM)
	2. **C**omparable **E**fficient **I**nterface
CEIR	**C**entral **E**quipment **I**dentity **R**egister
CEKS	**C**entrex **E**lectronic **K**ey **S**et
celliquette	Social et**iquette** for using **cell**phones
CELP	**C**ode **E**xcited **L**inear **P**rediction (voice coding)
cellphone	**cell**ular tele**phone**
CEM	**C**entro **E**sercizio de **M**anufenzione (Italy)
CEMA	**C**onsumer **E**lectronics **M**anufacturers **A**ssociation (standards)
CEMF	**C**ounter **E**lectro**m**otive **F**orce
CEMH	**C**ontrolled **E**nvironment **M**an-**H**ole
CEN	**C**omité **E**uropean de **N**ormalisation (Belgium standards)
CENELEC	**C**omité **E**uropean pour la **N**ormalisation en **Elec**trotechnique
CEO	**C**hief **E**xecutive **O**fficer (of companies)

C

CEOS	Committee On Earth Observation Satellite (remote sensing)
CEOS-IDN	CEOS International Directory Network (remote sensing)
CEP	Circular Error Probability (transmission)
CEPS	Color Electronic Prepress System
CEPT	Conference of European Postal and Telecommunications
CER	Cell Error Ratio (ATM)
CERB	Centralized Emergency Reporting Bureau (Canada)
CERDIP	Ceramic Dual Inline Package (microchips)
CERFnet	California Education and Research Federation Network
CERMET	Ceramic Metal (element)
CERN	Centre European des Recherche Nucléaire (Switzerland)
CERPACK	Ceramic Package (microchips)
CERT	Computer Emergency Response Team (organization)
CES	1. Coast Earth Station (Inmarsat)
	2. Circuit Emulation Switching (ATM)
CESA	Cyberspace Electronic Security Act (U.S.A.)
CESID	Caller Emergency Service ID
CEST	Center for the Exploitation of Science and Technology
CETIS	Center for Educational Technology Interoperability Standards
CEU	Commercial End User (DSL)
CEV	Controlled Environmental Vault (room)
CF	1. Call Forwarding (GSM)
	2. Compact Flash (memory)
	3. Center Frequency
	4. Critical Frequency (propagation)
CF+	Compact Flash Plus (memory)
CFA	1. Carrier Failure Alarm (transmission)
	2. Commercial Frame Agreement (Inmarsat)
	3. Carrier Facility Assignment
	4. Connecting Facility Assignment
	5. Connecting Facility Arrangement
	6. Consumer Federation of America
	7. Compact Flash Association
	8. Cross-Field Amplifier (microwave)
CFAC	Call Forward All Calls
CFAMN	Call Forwarding Address Modified Notification
CFAR	Constant False-Alarm Rate (radar)

CFB	Call Forwarding on Mobile Subscriber Busy (GSM service)
CFC	Current-to-Frequency Converter
CFCA	Communications Fraud Control Association (U.S.A.)
CFDA	Call Forward Don't Answer
CFDAMA	Combined Free/Demand Assignment Multiple Access
CFE	Contractor-Furnished Equipment
CFF	1. Coded File Format
	2. Critical Fusion Frequency
CFGDA	Call Forward Group Don't Answer
CFM	1. Companded Frequency Modulation
	2. Carrier Financial Management
	3. Cubic Feet per Minute (PC fans)
CFNRe	Call Forwarding on mobile subscriber Not Reachable (GSM)
CFNRy	Call Forwarding on mobile subscriber No Reply (GSM service)
CFO	Chief Financial Officer
CFP	Channel Frame Processor
CFR	1. Confirmation to Receive frame
	2. Code of Federal Regulations
CFRA	Combined Fixed/Reservation Assignment
CFS	Call Forwarding System
CFSK	Coherent FSK (modulation)
CFU	Call Forwarding Unconditional (GSM supplementary service)
CFV	Call For Votes (Usenet newsgroup)
CFW	Call Forward
CG	Character Generator
CGA	1. Carrier Group Alarm
	2. Color Graphics Adapter (320×200 pixel monitors)
CGI	1. Common Gateway Interface (Internet)
	2. Computer Graphics Interface
	3. Cell Global Identification (GSM)
CGI-bin	Common Gateway Interface-binaries
CGISS	Commercial Government and Industrial Solutions Sector
.cgm	Name extension for Computer Graphics format files
CGM	Computer Graphics Metafile (format)
CGMP	Cisco Group Multicast Protocol
CGS unit	Centimeter-Gram-Second unit (parameter)
CGSA	Cellular Geographic Service Area (cellular networks)
CGW	Customer Gateway
Ch	channel

CHAP	Challenge-Handshake Authentication Protocol (security)
CHEST	Combined Higher Education Software Team (e-learning)
CHI	Concentration Highway Interface (ISDN)
CHINT	Charge-Injected Transistor (semiconductors)
CHKDSK	Check Disk (DOS command)
CHRP	Common Hardware Reference Platform
CHS	Cylinder-Head Sector (PC hard drive)
CHTML	Compact HTML (programming)
Ci	Curie (unit of radioactivity)
CI	1. Clear Indication
	2. Cell Identity (GSM)
	3. Congestion Indicator (ATM)
	4. Certified Integrator
	5. Customer Interface
CIAC	Computer Incident Advisory Capability
CIAJ	Communications Industry Association of Japan
CIAS	Circuit Inventory and Analysis System
CIB	Compliance & Information Bureau (FCC)
CIBER	Cellular Intercarrier Billing Exchange Roamer Record (GSM)
CIC	1. Customer Interface Configuration
	2. Carrier Identification Code
	3. Circuit Identification Code (SS7)
	4. Content Indicator Code
	5. China Institute of Communications
CICS	Customer Information Control System (IBM)
CID	1. Customer Identity (British term)
	2. Caller ID (telephony service)
	3. Circuit Designator
	4. Charge-Injection Device (semiconductors)
	5. Consecutive Identical Digits
CIDB	Calling line Identification Delivery Blocking (feature)
CIDCW	CID on Call Waiting (telephony service)
CIDIN	Common ICAO Data Interchange Network
CIDR	Classless Inter-Domain Routing (method)
CiDSL	Consumer-installable DSL (access)
CIE	1. Commercial Internet Exchange
	2. Computer-Integrated Engineering (ICT)
CIF	1. Cells In Frame (ATM)
	2. Cells In Flight (ATM)
	3. Common Intermediate Format (video compression)
	4. Common Interchange Format

C

CIF-AD	Cells In Frames-Attachment Device (ATM)
CIFAX	Ciphered Facsimile
CIFS	Common Internet File System
CIGOS	Canadian Interest Group on Open Systems
CIGRR	Common Interest Group on Rating and Routing
CIID	Card Issuer Identifier Code
CIIG	Canadian ISDN Interest Group
CIK	Cryptographic Ignition Key
CIM	1. Computer-Integrated Manufacturing (ICT)
	2. Computer-Input Microfilm
	3. Common Information Model (network management)
	4. Customer Information Manager (MCI)
	5. Corporate Information Management
CIO	Chief Information Officer (ICT)
CIP	1. Certified Internet Professional
	2. Carrier Identification Parameter (ATM)
	3. Computer Interface Program
	4. Channel Interface Processor (Cisco)
CIPA	Children's Internet Protection Act
CIPHONY	Ciphered telephony
CIPS	Canadian Information Processing Society
CIR	1. Current Instruction Register (CPU)
	2. Committed Information Rate (frame relay)
	3. Committed Information Range (ATM)
	4. Communications Industry Researchers (company)
	5. Carrier to Interference Ratio
	6. Computer Integrated Railroading (ICT)
CIRA	Canadian Internet Registration Authority
CIRF	Co-channel Interference Reduction Factor
CIS	1. Communications and Information Systems
	2. Card Information Structure (memory card)
CISA	China Information Service Association
CISC	1. Complex-Instruction-Set Computer (processor chips)
	2. Customer Information Control System (software)
CISCC	Collocation Interconnection Service Cross Connection
Cisco	Computer Information Systems Company
CISE	Computer and Information Science and Engineering (U.S.A.)
CISPR	Comité International Special des Perturbations Radioelectriques
CIT	1. Communications and Information Technology
	2. Computer Integrated Telephone

CITA	Canadian Independent Telephone Association
CITC	Communications and Information Technology Commission
CITEL	Comision Inter-Americana de **Tel**ecommunicaciones
CITR	Canadian Institute for Telecommunications Research
CITRIS	Center for Information Technology Research in the Interest of Society
CITU	Central **IT** Unit (United Kingdom)
CIV	Cell Inter-arrival Variation (ATM)
CIVDL	Collaboration for Interactive Visual Distance Learning (e-learning)
CIX	Commercial Internet Exchange
CJC	Canadian Journal of Communication
CK	Ciphering Key
CKL	Circuit Location
CKR	Circuit Reference
CKSN	Ciphering Key Sequence Number (GSM)
CKT	circuit
CL	Connectionless (mode)
CLAMN	Called Line Address Modification Notification
CLAS	Centrex Line-Assignment Service
CLASS	Custom Local Area Signaling Services
CLC	1. Communications Line Controller 2. Carrier Liaison Committee
CLCC	Ceramic Leaded-Chip Carrier (electronics)
CLCI	Common Language Circuit Identification
CLD	Competitive Long-Distance (carrier)
CLDC	Connected Limited Device Configuration (J2ME)
CLDLS	Connectionless Data Link Service
CLE	1. Common Logic Equipment 2. Customer Location Equipment
CLEC	Competitive Local Exchange Carrier
CLEI	Common Language Equipment Identifier
CLEOS	Conference of the Lasers and Electro-Optic Society
CLI	1. Caller Line Identification (ISDN) 2. Command-Line Interface 3. Cumulative Leakage Index (FCC)
CLIB	C Language Interface Library (networking)
CLID	Calling Line Identification (GSM)
CLIP	Calling Line Identification Presentation (GSM)
CLIR	Calling Line Identification Restriction (GSM)
CLIS	Calling Line Identification Signal (GSM)
Clk	clock (pulse)
CLLI	Common Language Location Identifier code (Telcordia)

CLM	Career Limiting Move
CLNP	Connectionless Network Protocol (OSI model)
CLNS	Connectionless-mode Network Service (OSI model)
CLP	1. Cell Loss Priority (ATM)
	2. Certified Linux Professional
CLR	1. Cell Loss Ratio (ATM)
	2. Connection Loudness Rating
	3. Circuit Layout Record (type of service)
CLS	1. Connectionless Server
	2. Control Line Setting
	3. Clear the current Screen (MS DOS Command)
CLSI	Custom Large-Scale Integration (microchips)
CLSP	Competitive Local Service Provider
CLTP	Connectionless Transport Protocol (OSI model)
CLTS	Connectionless Transport Service
CLTU	Command Link Transmission Unit
CLUT	Color Look-Up Table
CLV	Constant Linear Velocity (hard disks)
cm	centimeter
cm³	cubic centimeter
CM	1. Configuration Management (wireless)
	2. Connection Management (GSM)
	3. Connection Matrix (MUX)
	4. Computing Module
	5. Cable Modem
CMA	Communications Managers Association (standards organization)
CMAC	Control Mobile Attenuation Code
CMC	1. Common Messaging Calls
	2. Comsat Mobile Communications
	3. Computer-Mediated Communications (ICT)
	4. Communications Management Configuration
	5. Customer Management Complex
	6. Connection Management Controller (ATM)
CMCI	Cable Modem to CPE Interface
CMD	Command (computer)
CMDR	Command Reject (computer)
CMDS	Centralized Message Distribution System
CMEA	Cellular Message Encryption Algorithm
CMI	1. Computer-Managed Instruction (e-learning)
	2. Coded-Mark Inversion (data encoding)
	3. Control Mode Idle
Cmil	Circular mil (measure of sectional area of a wire)

C

CMIP	Common Management Information Protocol (networking)
CMIS	Common Management Information Service (networking)
CMISE	Common Management Information Service Element (wireless)
CML	1. Current-Mode Logic (digital electronics) 2. Customer Microcircuits Ltd (company)
CMOL	CMIP Over Logical link control
CMOS	Complementary Metal-Oxide Semiconductor (semiconductors)
CMOS RAM	Random Access Memory using CMOS technology (memories)
CMOS SOS	Silicon-On-Sapphire combined with CMOS technology
CMOS SOS RAM	A RAM memory using a combination of CMOS and Silicon-On-Sapphire technologies (memories)
CMOT	CMIP Over TCP/IP
CMP	1. Cluster Management Processor 2. Communications Management Processor 3. Chip-level Multiprocessing
CMR	1. Cellular Mobile Radio 2. Cell Misinsertion Rate (ATM) 3. Common-Mode Rejection (amplifiers)
CMRF	Combined Mode Resonator Filter
CMRFI	Cable Modem to Radio Frequency Interface
CMRR	Common-Mode Rejection Ratio (amplifiers)
CMRS	Commercial Mobile Radio Service (cellular networks)
CMS	1. Cluster Management System 2. Content Management System 3. Changeable Message Sign 4. Call Management Services 5. Color Management System (monitors)
CMS-8800	Cellular Mobile telephone Service (North America)
CMS/MD	Cluster Management System Mediation Device
CMT	Connectionless Mode Transmission
CMTRI	Cable Modem Telco Return Interface (data over cable)
CMTS	1. Cellular Mobile Telephone System 2. Cable Modem Termination System
CMW	Compartmented Mode Workstation (networking)
CMY	Cyan-Magneta-Yellow color model (computer imaging)
CMYK	Cyan-Magneta-Yellow-black color model (computer imaging)

CN	1. Complementary Network
	2. Change Notice
	3. Control Network (ATM)
CNA	1. Certified Novell Administrator (Novell NetWare)
	2. Cooperative Network Architecture
	3. Customer Name and Address bureau
CNAE	Customer Network Access Equipment
CNAM	Caller ID with **Nam**e
CNC	1. Central Node Controller
	2. Computer Numerical Control (ICT)
	3. Computerized Numerical Control
CND	1. Calling Number Display
	2. Calling Number Delivery
CNE	Certified Novell Engineer (Novell NetWare)
CNEPA	Certified Novell Engineer Professional Association (Novell, Inc.)
CNE	Certified Novell Engineer (Novell NetWare)
CNES	Centre National D'Etudes Spatiales (French Space Agency)
CNET	Centre National d'Etudes de Télécommunication (France)
CNF	con**f**erence
CNI	Certified Novell Instructor (Novell NetWare)
CNIP	Calling Number Identification Presentation (wireless)
CNIR	Calling Number Identification Registration (wireless)
CNIS	1. Calling Number Identification Service (wireless)
	2. Channel Navigation Information Service
CNIT	Centro Nazionale Interuniversitario per Telecomunicazioni (Italy)
CNM	1. Customer Network Management (ATM)
	2. Communications Network Management
CNN	Cable News Network (TV broadcaster)
CNO	Corporate Networking Officer
CNR	1. Complex Node Representation (ATM)
	2. Customer is Not Ready
	3. Carrier-to-Noise power Ratio (transmission)
CNRI	Corporation for National Research Initiatives
CNS	1. Complementary Network Service
	2. Communication Navigation Surveillance (aviation)
CNSA	China National Space Administration
CNT	Coaxial Network Termination
CNTL	control
CNX	Certified Network Expert

CO	1. Connection Oriented
	2. Central Office (DSL)
CO-LAN	Central Office Local Area Network
COA	Change Over Acknowledge (GSM message)
COAI	Cellular Operations Association of India
COAM	Customer Owned And Maintained telephone equipments
COAT	Coherent Optical Adaptive Technique (fiber optics)
COB	Close Of Business
COB	Chip On Board
CoBAMs	Consortium of Brick And Mortars (e-commerce)
COBOL	Common Business-Oriented Language (programming)
COBRA	A wrong spelling of **CORBA**
COBRAS	Cosmic Background Radiation Anisotropy Satellite
COC	Central Office Connection
COCOT	Customer-Owned Coin-Operated Telephone
COCUS	Central Office Code Utilization Survey
COD	1. Connection-Oriented Data (ATM)
	2. Catastrophic Optical Damage (fiber optics)
CODAN	Carrier-Operated Device Antinoise
CODAR	Correlation, Detection, And Ranging (submarine detection)
CODASYL	Computer On Data Systems Languages (organization)
Codec	Coder/decoder
CODLS	Connection-Oriented Data Link Service
CODS	Connection-Oriented Data Service
COE	1. Central Office Equipment
	2. Central Office Engineer
COER	1. Central Office Equipment Report (of a telephone company)
	2. Center for Organizational Excellence Research (New Zealand)
COFA	Change Of Frame Alignment
COFDM	Coded Orthogonal FDM (transmission)
COFETEL	Comision Federal de Telecomunicaciones (Mexico)
COG	Centralized Ordering Group
COHO	Coherent Oscillator
COLD	Computer Output to Laser Disk
CoLo	Co-Location (servers)
COLP	Connected Line Identification Presentation (GSM service)
COLR	Connected Line Identification Restriction (GSM service)
COLT	Cellular On Light Truck

.com	The MS DOS name extension for executable program files
COM	1. serial **com**munications port (computer)
	2. **C**omponent **O**bject **M**odel (MS ActiveX platform)
	3. **C**ontinuation **O**f **M**essage (ATM)
	4. **C**ommon **O**bject **M**odel (ICT)
	5. **C**omputer-**O**utput **M**icrofilm
COM port	**Com**munication **port** (computer)
COM1	A **com**munication port used for connecting a serial mouse
COM2	A serial **com**munication port used for connecting a modem
COM3	A **com**munication port used connecting other peripherals
COMB	**comb**iner
Comdex	**Com**puter **De**alers **Ex**position (trade show)
COMINT	**Com**munications **Int**elligence
COMJAM	**Com**munications **Jam**ming
COMLOGNet	**Com**bat **Log**istic **Net**work (military)
comms	**comm**unication**s**
COMP	**Comp**are files or set of files (MS DOS Command)
Compander	**Com**pressor-ex**pander**
COMPSURF	**Comp**rehensive **Surf**ace Analysis (Novell program)
CompTel	**Comp**etitive **Tel**ecommunications Association (U.S.A.)
CompTIA	**Comp**uting **T**echnology **I**ndustry **A**ssociation (U.S.A.)
COMPUSEC	**Compu**ter **Sec**urity
COMS	1. **C**ircuit **O**rder **M**anagement **S**ystem
	2. **C**onsortium for **M**ultipurpose **S**atellite (South Korea)
COMSAR	**Com**munications for **S**earch **A**nd **R**escue (IMO sub-committee)
Comsat	**Com**munications **Sat**ellite Corporation (U.S. operator)
COMSEC	**Com**munications **Sec**urity
CON	**C**ircuit **O**rder **N**umber
CONNEX	**Conn**ectivity **Ex**change
CONF	**conf**iguration
CONFIG	**config**uration information (Novell NetWare utility)
CONP	**C**onnection-**O**riented **N**etwork **P**rotocol
compander	**com**pressor and ex**pander**
CoNS	**C**onsortium for **S**chool **N**etworking
CONS	**C**onnection-**O**riented **N**etwork **S**ervice (networking)
CONTEL	**Con**tinental **Tel**ecom Inc. (company)
CONV	**conv**erter
COO	**C**hange **O**ver **O**rder (GSM message)

C

Co-Pol	**Copol**ar (antennas)
COP	1. **C**able **O**rganizer **P**anel
	2. **C**ertificate **O**f **P**roficiency (ship radio operator)
COPOUS	**C**ommittee **O**n **P**eaceful **U**se of **O**uter **S**pace (United Nations)
COPS	1. **C**ommon **O**pen **P**olicy **S**ervice (IETF protocol)
	2. **C**ommon **Op**erations **S**oftware (Bell Labs)
COPT	**C**oin-**O**perated **P**ay **T**elephone
COPW	**C**ustomer **O**wned **P**remises **W**ire
COR	**C**ircuit **O**rder **R**ecord
CORBA	**C**ommon **O**bject **R**equest **B**roker **A**rchitecture
CORE	**C**ouncil **O**f **R**egistrars
CoS	**C**lass **o**f **S**ervice
COS	1. **C**onnection **O**riented **S**ervice
	2. **C**ompatible for **O**pen **S**ystems
	3. **C**orporation for **O**pen **S**ystems (organization)
COS/MOS	**C**omplementary **S**ymmetry **M**etal-**o**xide **S**emiconductor
COSC	**C**ustomer **O**perations **S**upport **C**enter (Boeing)
COSE	**C**ommon **O**pen **S**oftware **E**nvironment (Unix)
COSETI	**C**olumbus **O**ptical **SETI**
COSINE	**C**ooperation for **O**pen **S**ystems **I**nterconnection **N**etworking in **E**urope
CoSN	**Co**nsortium for **S**chool **N**etworking
COSName	**C**lass **O**f **S**ervice **Name**
COSPAS	**Co**smicheskaya **S**istyema **P**oiska **A**variynich **S**udov (Russia)
	Space System for the Search of Vessels in Distress (Russia)
COT	1. **C**entral **O**ffice **T**erminal/Termination
	2. **C**ustomer **O**riginated **T**race (feature)
	3. **C**ustomer **O**ffice **T**erminal
COTAR	**C**orrelated **O**rientation **T**racking **A**nd **R**ange (vehicle tracking)
COTAT	1. **Co**rrelation **T**racking **A**nd **T**riangulation (trajectories)
	2. **C**ommittee **O**n **T**ransportation **A**nd **T**elecommunications
COTP	**C**onnection-**O**riented **T**ransaction **P**rotocol
COTS	1. **C**onnection-**O**riented **T**ransport **S**ervice
	2. **C**ommercial **O**ff-**T**he-**S**helf (military)
	3. **C**ompact, **O**utdoor, **T**ransportable, **S**ingle-site (Marconi)
COV	**C**ontrol **O**ver **V**oice (Mitel's protocol)
COW	1. **C**ell site **O**n **W**heels
	2. **C**haracter-**O**riented **W**indows

CP	1. Circular Polarization
	2. Control Point (networking)
	3. Connection Point
	4. Change Proposal (Inmarsat)
	5. Customer Premises
	6. Carrier Pulse
CP/M	Control Program for Microcomputers
CPA	1. Co-Polarization Attenuation
	2. Critical Path Analysis (ICT)
	3. Close Point of Approach (shipboard radar)
	4. Cost Per Action
	5. Center for Policy Alternatives (Internet)
	6. Computer Press Association
	7. Chip Protection Act
	8. Color Phase Alternation (television)
CPAF	Customer Premises Access Facility
CPAS	Cellular Priority Access Service
CPB	1. Corporation for Public Broadcasting
	2. Central Processing Board
CPC	1. Customer Port Controller
	2. Calling Party Control
	3. Calling Party Connected
CPCI	Compact Protocol Control Information
CPCN	Certificate of Public Convenience and Necessity (FCC)
CPCS	Common Part Convergence Sublayer (ATM)
CPD	Call Processing Data
CPDA	Compression Priority Demand Assignment
CPDLC	Controller-Pilot Data-Link Control (aviation)
CPE	1. Customer Premises Equipment (networking)
	2. Customer Provided Equipment (networking)
CPFSK	Continuous Phase FSK (modulation)
CPGA	Ceramic Pin-Grid Array (microchips)
cpi	characters per inch
CPI	1. Common Part Indicator (ATM)
	2. Common Programming Interface (computer)
	3. Common Physical Interface (ISDN)
	4. Computer to PBX Interface
	5. Cost Per Inquiry
CPL	1. Commercial Private Line
	2. Call Processing Language (programming)
	3. coupler
CPLD	Complex Programmable Logic Device (digital electronics)
cpm	cycles per minute

CPM	1. **C**ritical **P**ath **M**ethod (project control)
	2. **C**ustomer **P**remise **M**anagement
	3. **C**able **P**lant **M**anagement
	4. **C**ommunications **P**rocessor **M**odule
	5. **C**ontinuous **P**hase **M**odulation
CPN	1. **C**alling **P**arty **N**umber
	2. **C**ustomer **P**remises **N**etwork
	3. **C**omputer **PBX N**etwork
CPNI	**C**ustomer **P**roprietary **N**etwork **I**nformation
CPODA	**C**ompression **P**riority **D**emand **A**ssignment (protocol)
CPP	1. **C**alling **P**arty **P**ays (telephone call charge)
	2. **C**able **P**atch **P**anel (DEC)
CPR	**C**ontinuing **P**roperty **R**ecord (Telcordia Technologies)
cps	1. **c**haracters **p**er **s**econd (data transfer)
	2. **c**ycles **p**er **s**econd
CPS	1. **C**ellular **P**riority **S**ervice
	2. **C**all **P**rogress **S**ignal
	3. **C**assette **P**reparation **S**ystem (TV broadcasting)
	4. **C**o-**P**lanar **S**tripline
	5. **C**isco **P**owered **N**etworks
CPSI	**C**ustomer **P**remises **S**atellite **I**nterface
CPSK	**C**oherent **PSK** (modulation)
CPSR	**C**omputer **P**rofessionals for **S**ocial **R**esponsibility
CPSS	**C**ontrol **P**acket **S**witching **S**ystem
CPT	1. **C**all **P**rogress **T**one
	2. "**C**arriage **P**aid **T**o" (network language)
CPU	**C**entral **P**rocessing **U**nit (computer)
CPU-bound	Com**pu**tation-**bound**
CPUG	**C**all **P**ick**u**p **G**roup
CPW	**C**o-**P**lanar **W**aveguide (microwave)
CQ	General call to all stations (Morse code transmissions)
CQFP	**C**eramic **Q**uad **F**lat **P**ack (microchips)
CR	1. "**C**arriage **R**eturn" (control character of printers)
	2. **C**all **R**eference
	3. **C**all **R**egister (telephone switch)
	4. **C**all **R**equest
	5. **C**hannel **R**eliability
	6. **C**onnection **R**equest
	7. **C**ontrast **R**atio (screens)
	8. **C**ustomer **R**ecord
CR-LDP	**C**onstraint-based **R**outed-**L**abel **D**istribution **P**rotocol (MPLS)
CRAFT	**C**ooperative **R**esearch **A**ction **F**or **T**echnology
CRAG	**C**ellular **R**adio **A**dvisory **G**roup

CRATT	Cryptographic **Ra**dio **T**eletype equipment
CRB	Community **R**adio **B**roadcasters
CRC	1. **C**yclic **R**edundancy **C**heck (data error detection)
	2. **C**ommunications **R**esearch **C**enter (Canada)
	3. **C**ommunications **R**egulation **C**ommission (Bulgaria)
CRCS	**C**ontinuous **R**igid **C**able **S**upport
CRD	1. **C**luster **R**econciliation **D**escriptor
	2. **C**ontention **R**esolution **D**evice
CREAM	**C**osmic **R**adiation **E**ffects and **A**ctivation **M**onitors (microwave)
CREDFACS	**C**onduit, **R**isers, **E**quipment space, **D**ucts, and **Fac**ilities
CREG	**C**oncentrated **R**ange **E**xtension with **G**ain
CREN	**C**orporation for **R**esearch and **E**ducational **N**etworking (ICT)
CRF	1. **C**onnection-**R**elated **F**unction (ATM)
	2. **C**ell **R**elay **F**unction (ATM)
CRIS	**C**ustomer **R**ecord **I**nformation **S**ystem
CRISP	**C**AP **R**eal-time **I**TIP-Based **S**cheduling **P**rofile
CRITO	**C**enter for **R**esearch on **I**nformation **T**echnology and **O**rganizations (U.S.A.)
CRITCOM	**Cr**itical **I**telligence **Com**munications
CRM	1. **C**all **R**ecord **M**anagement
	2. **C**ustomer **R**elationship **M**anagement (ICT)
	3. **C**ell **R**ate **M**argin (ATM)
CRMAP	**C**yclic **R**eservation **M**ultiple **A**ccess **P**rotocol
CRO	1. **C**omplete with **R**elated **O**rder
	2. **C**athode-**R**ay **O**scilloscope (test instrument)
CROM	**C**ontrol and **R**ead-**O**nly **M**emory
CRP	1. **C**apacity **R**eserve **P**ool
	2. **C**abling **R**eference **P**anel
	3. **C**ommand **R**epeat
CRS	1. **C**oast **R**adio **S**tation
	2. **C**ell **R**elay **S**ervice (ATM)
CRST	**C**heyenne **R**iver **S**ioux **T**ribe (telephone authority)
CRT	**C**athode-**R**ay **T**ube (displays)
CRTC	**C**anadian **R**adio **T**elevision and **T**elecommunications **C**ommission
CRTSSDA	**C**ascaded **R**eactively **T**erminated **S**ingle-**S**tage **D**istributed **A**mplifier
CRV	**C**all **R**eference **V**alue (number)
Cryosar	**Cry**ogenic **S**witching by **A**valanche and **R**ecombination
CS	1. **C**luster **S**tartup
	2. **C**ell **S**election

C

	3. **C**hip **S**elect (electronics)
	4. **C**ontrol **S**trobe (logic signal)
	5. **C**onvergence **S**ublayer (B-ISDN and ATM)
	6. **C**ommunication **S**atellite (broadcasting)
	7. **C**apability **S**ets (INs)
	8. **C**oding **S**cheme
CS-ACELP	**C**onjugate **S**tructure-**A**lgebraic **C**ode **E**xcited **L**inear **P**rediction
CS-CDPD	**C**ircuit-**S**witched **C**ellular **D**igital **P**acket **D**ata
CSA	1. **C**all-path **S**ervices **A**rchitecture (IBM)
	2. **C**anadian **S**tandards **A**ssociation (Canada)
	3. **C**anadian **S**pace **A**gency (Canada)
	4. **C**omprehensive **S**ystem **A**ccounting
	5. **C**arrier **S**erving **A**rea (DSL)
CSAR	**C**ombat **S**earch **A**nd **R**escue (Motorola)
CSB	**C**able **S**ervices **B**ureau (FCC)
CSBT	**C**hinese users association for **S**atellite communications, **B**roadcasting and **T**elevision
CSC	1. **C**ommon **S**ignaling **C**hannel
	2. **C**ospas-**S**arsat **C**ouncil
	3. **C**ircuit **S**witching **C**enter
	4. **C**ustomer **S**ervice **C**enter (MCI)
	5. **C**ustomer **S**ervice **C**oordinator
	6. **C**ustomer **S**ervice **C**onsultant
	7. **C**ustomer **S**upport **C**enter
CSCD	**C**ircuit-**S**witched **C**ellular **D**ata
CSD	1. **C**luster **S**pecific **D**escriptor
	2. **C**ellular **S**ecurity **D**evice
	3. **C**ircuit-**S**witched **D**ata
CSDC	**C**ircuit-**S**witched **D**igital **C**apability (AT&T)
CSCF	**C**all **S**ession **C**ontrol **F**unction
CSDN	**C**ircuit-**S**witched **D**ata **N**etwork
CSE	1. **C**ommunications **S**ecurity **E**stablishment (Canada)
	2. **C**AMEL **S**ervice **E**nvironment (GSM)
CSEL	**C**able **Sel**ect (computer)
CSF	1. **C**lique **S**toring **F**acility
	2. **C**ell **S**ite **F**unction
CSFI	**C**ommunication **S**ubsystem **F**or **I**nterconnection (IBM)
CSI	1. **C**alled **S**ubscriber **I**dentification
	2. **C**ellular **S**pecialties **I**nc. (company)
	3. **C**omputo **S**ervice **I**nc. (company)
	4. **C**omputer **S**ecurity **I**nstitute
CSIRO	**C**ommonwealth **S**cientific and **I**ndustrial **R**esearch **O**rganization

CSID	Calling Station ID
CRIS	Customer Records Information System
CSC	Customer Service Center
CSL	1. Communications Services Limited
	2. Current-Sinking Logic (digital electronics)
	3. Current-Sourcing Logic (digital electronics)
CSLIP	Compressed Serial Link Internet Protocol
CSM	1. Communication Systems Management
	2. Communications Security Material
	3. Cell-Site Modem
	4. Clock Supply Module
	5. Combined Symbol Matching
CSMA	Carrier-Sense Multiple Access
CSMA-CA	Carrier-Sense Multiple Access with Collision Avoidance
CSMA-CD	Carrier-Sense Multiple Access with Collision Detection
CSMDR	Centralized Station Message Detail Reporting
CSMI	Call Screening, Monitoring and Intercept
CSN	Circuit-Switched Network
CSNet	Computer and Science Network (U.S.A.)
CSO	1. Central Services Organization
	2. Computing Services Office (Internet directory)
	3. Composite Second Order
CSOC	Convolutional Self-Orthogonal Code (error correction)
CSP	1. Certified Service Provider
	2. Competitive Service Provider
	3. Commerce Service Provider
	4. Content Service Provider
	5. Communications Service Provider
	6. Chip-Scale Package
CSPDN	Circuit-Switched Public Data Network
CSPF	Constrained Shortest Path First (routing protocol)
CSPP	Computer Systems Policy Project
CSQP	Customer/Supplier Quality Process
CSR	1. Cell Switch Router (ATM)
	2. Customer Service Record
	3. Customer Service Representative
	4. Customer Station Rearrangement
CSS	1. Cascading Style Sheets (Web design)
	2. Cellular Subscriber Station
	3. Coordinator, Surface Search
	4. Customer Support System (GSM)
	5. Content Scrambling System (broadcasting)
	6. Customer Service Segment

CST	1. **C**omputer-**S**upported **T**elephony (Siemens)
	2. **C**ompatible **S**ideband **T**ransmission
CSTA	**C**omputer-**S**upported **T**elephony **A**pplication
CSTP	**C**ustomer **S**pecific **T**erm **P**lan
CSU	1. **C**hannel **S**ervice **U**nit (networking)
	2. **C**hannel **S**haring **U**nit
	3. **C**ustomer **S**ervice **U**nit
	4. **C**ircuit **S**witching **U**nit
CSUA	**C**anadian **S**atellite **U**sers **A**ssociation
CSV	1. **C**ircuit-**S**witched **V**oice (call for voice)
	2. **C**omma **S**eparated **V**alue format (data storage)
CT	1. **C**onformance **T**est
	2. **C**ordless **T**elephone
	3. **C**omputer **T**elephony
	4. **C**omputed **T**omography (medicine)
	5. **C**all **T**ransfer
	6. **C**enter **T**ap (circuit diagrams)
CT Scan	**C**omputer **T**omographic **Scan**ning (medicine)
CT0	**Zero** Generation **C**ordless **T**elephony
CT1	**C**ordless **T**elephone Generation **1**
CT2	**C**ordless **T**elephone Generation **2**
CT2-CAI	**CT2** **C**ommon **A**ir **I**nterface
CT3IP	**C**hannelized **T3** **I**nterface **P**rocessor
CTA	1. **C**ordless **T**erminal **A**dapter (DECT)
	2. **C**ompetitive **T**elecommunications **A**ssociation (Canada)
CTAS	**C**arrier **T**est **A**ccess **S**witch
CTB	**C**omposite **T**riple **B**eat
CTC	**C**ounter/**T**imer **C**ircuit
CTCA	**C**anadian **T**elecommunications **C**onsultants **A**ssociation
CTCSS	**C**ontinuous **T**one **C**ontrolled **S**quelch **S**ystem (cellular networks)
CTD	1. **C**all **T**ransfer **D**elay (ATM)
	2. **C**lock **T**ime **D**eference
	3. **C**urrent **T**ransfer **D**evice
	4. **C**ontinuity **T**one **D**etector
CTE	1. **C**ircuit **T**erminating **E**quipment
	2. **C**hannel **T**ranslating **E**quipment
	3. **C**oefficient of **T**hermal **E**xpansion (microwave)
CTERM	**C**ommunications **Term**inal (protocol)
CTFM	**C**ontinuous-**T**ransmission **F**requency-**M**odulation
CTFT	**C**olor **T**hin-**F**ilm **T**ransistor (semiconductors)

CTI	1. Computer Telephony Integration (ICT)
	2. Computers in Testing Initiative (ICT)
	3. Critical Technologies Institute
	4. Call Technologies, Inc. (company)
CTIA	1. Cellular Telecommunications Industry Association (U.S.A.)
	2. CT Innovation Alliance
CTIP	Computer Telephony Interface Products
CTL	1. Complex Text Layout
	2. Complementary-Transistor Logic (digital electronics)
CTM	1. Cordless Terminal Mobility (DECT)
	2. Circuit Transfer Mode
	3. Conformal Transformation Method
CTN	1. Corporate Telecommunication Network
	2. Consumers' Telecommunication Network
	3. Call Tracking Number
CTO	1. Commonwealth Telecommunications Organization (U.K.)
	2. Chief Technology Officer (ICT)
CTOS	Convergent Technology Operating System
CTP	1. Connection Terminating Point
	2. Cordless Telephony Profile (Bluetooth)
CTR	Common Technical Regulations of Europe (ETSI)
Ctrl	Control key (keyboard)
CTS	1. Clear-To-Send (RS 232 signal)
	2. Communication Transport System
	3. Conformance Testing Service
	4. Carpal Tunnel Syndrome of wrist and hand (keyboard)
CTSS	Compatible Time-Sharing System
CTT	Central Traffic Terminal
CTTC	Coax To The Curb
CTTE	Common TDM Terminating Equipment
CTTU	Centralized Trunk Test Unit
CTU	1. Central Terminal Unit
	2. Command & Telemetry Unit
CTX	1. Clear-To-Transmit (RS 232 signal)
	2. centrex
CTY	Console Teletype
cu	cubic (SI Units)
CU	1. Carrier Unit
	2. Channel Unit
	3. Crosstalk Unit

C

CUA	Common User Access (computer)
CUD	could (Morse code transmissions)
CUG	Closed-User Group (GSM supplementary service)
CUI	Character User Interface (computer)
CUJT	Complementary Uni-Junction Transistor (semiconductors)
CUL	See You Later (Morse code transmissions)
CUSEEME	"See You, See Me" (Internet videoconferencing system)
CV	1. Code Violation (data encoding) 2. Checksum Value 3. Computer Vision 4. Continuous Vulcanization (cables) 5. Constant Voltage
CVBS	Composite Video Blanking and Synchronization (TV signals)
CVC	Computer Virus Catalog
CVD	Chemical Vapor Deposition process (semiconductors)
CVDL	Continuously Variable Delay Line
CVF	Compressed Volume File (Microsoft)
CVP	1. Cooperative Voice Processing (British) 2. Certified Vertical Partner
CVPO	Chemical Vapor Pulse Oxidation process (fiber optics)
CVR	Crystal Video Receiver
CVS	Computer Vision Syndrome
CVSD	Continuously Variable Slope Delta (modulation)
CW	1. Continuous Wave 2. Carrier Wave 3. Composite Wave 4. clockwise 5. Call Waiting Supplementary Service (GSM)
CVT	Constant Voltage Transformer
CWA	Communications Workers of America
CWC	Cable & Wireless Communications (broadcasting)
CWDM	1. Coarse Wavelength-Division Multiplexing (fiber optics) 2. Circular Waveguide Dual Mode (microwave)
CWIS	Campus-Wide Information System (universities)
CWML	Compact Wireless Markup Language (programming)
CWS	Communications Wiring System
CWSA	Constant Width Slot Antenna (microwave)
CWVA	Cable & Wireless Virtual Academy
CX	Change Context (networking)

CXR	**carrier**
CXML	**C**ommerce **XML** (programming language)
cybernaut	**cyber**netics astro**naut**
CYTA	**Cy**prus **T**elecommunication **A**uthority (GSM operator)
CZCS	**C**onstant **Z**one **C**olor **S**canner (remote sensing)

D d

d	1. Symbol for prefix **d**eci-, denoting one-tenth or 10^{-1}
	2. Symbol for **d**istance
D	1. Symbol for electrostatic flux **d**ensity
	2. Symbol for **D**rain (FETs)
	3. Symbol for **D**irectivity (antennas)
D-AMPS	**D**igital **AMPS**
D-Bit	**D**elivery confirmation **Bit** (X.25)
D-Channel	**D**ata **Channel** (ISDN)
D-CLEC	**D**igital **C**ompetitive **L**ocal **E**xchange **C**arrier
D-FET	**D**epletion-mode **FET** (semiconductors)
D-LINK	**D**iagonal **Link** (SS7)
D-VHS	**D**igital version of **VHS** recording or playback
D/A	**D**igital-**to**-**A**nalog (electronics)
D/C	**D**own **C**onverter (satcom)
D/I	**D**rop **and** **I**nsert
D/L	**D**own **L**ink (satcom)
D/S	**D**igital-**to**-**S**ynchro (electronics)
D2-MAC	**M**ultiplexed **A**nalog **C**omponent type **D2** (TV format)
D4	**D**igital **4**th Generation Channel Bank (transmission)
DA	1. **D**emand-**A**ssigned (access)
	2. **D**oesn't **A**nswer (phone)

	3. **D**istribution **A**rea (DSL)
	4. **D**irectory **A**ssistance
	5. **D**estination **A**ddress (LANs)
	6. **D**ifferential **A**ttenuation (Apple Macintosh)
	7. **D**esk **A**ccessory (software)
	8. **D**iscontinued **A**vailability (circuit)
DA/TDMA	**D**emand-**A**ssigned mode **TDMA** (access)
DAA	**D**ata **A**ccess **A**rrangement
DAB	1. **D**igital **A**udio **B**roadcasting (satcom)
	2. **D**ynamically **A**llocatable **B**andwidth
	3. **D**ata **A**cquisition **B**oard
DAC	1. **D**igital-to-**A**nalog **C**onverter (electronics)
	2. **D**ual **A**ttachment **C**oncentrator
DACC	1. **D**igital **A**ccess **C**ross-**C**onnect
	2. **D**irectory **A**ssistance **C**all **C**ompletion
DACCS	**D**igital **A**ccess **C**ross-**C**onnect **S**ystem (Intelsat)
DACD	**D**igital **A**utomatic **C**all **D**istributor
DACS	**D**igital **A**ccess and **C**ross-connect **S**ystem (DSL)
DAF	1. **D**estination **A**ddress **F**ield
	2. **D**ata **A**dministration **F**orum
	3. **D**ecrement **A**ll **F**rame (Motorola)
DAG	**D**istress **A**lert **G**enerator
DAI	**D**igital **A**udio **I**nterface
DAL	**D**edicated **A**ccess **L**ine
DAM	**D**ECT **A**uthentication **M**odule
DAMA	**D**emand-**A**ssignment **M**ultiple-**A**ccess
DAMPS	**D**igital **A**dvanced **M**obile **P**hone **S**ervice
DAN	**D**ECT **A**ccess **N**ode
DAND	**D**espun **A**ctive **N**utation **D**amper (Intelsat)
DAO	**D**isk **A**t **O**nce (data storage)
DAP	1. **D**irectory **A**ccess **P**rotocol (network management)
	2. **D**ata **A**ccess **P**rotocol (DECNet)
	3. **D**ata **A**ccess **P**oint (MCI)
	4. **D**ata **A**cquisition **P**rocessor
	5. **D**edicated **A**rray **P**rocessor (microchips)
	6. **D**emand-**A**ssigned **P**rocessor
DAQ	1. **D**elivered **A**udio **Q**uality
	2. **D**ata **A**c**q**uisition
DAR	1. **D**igital **A**udio **R**adio
	2. **D**ynamically **A**daptive **R**outing
DARPA	**D**efense **A**dvanced **R**esearch **P**roject **A**gency
DARPANet	**D**efense **ARPANet**
DARS	**D**igital **A**udio **R**adio **S**ystem (broadcast operator)

DAS	1. **D**ial **A**ccess **S**witching (Cisco)
	2. **D**ata **A**cquisition **S**ystem
	3. **D**ual **A**ttachment **S**tation
DASCPC	**D**emand **A**ssignment **SCPC** (access)
DASD	**D**irect-**A**ccess **S**torage **D**evice (semiconductors)
DASS	1. **D**emand **A**ssignment **S**ignaling & **S**witching unit
	2. **D**igital **A**ccess **S**ignaling **S**ystem
	3. **D**irect-**A**ccess **S**econdary **S**torage
DASS2	**D**igital **A**ccess **S**ignaling **S**ystem **No. 2**
.dat	Name extension for **dat**a files (computer)
DAT	1. **D**igital **A**udio **T**ape
	2. **D**ynamic **A**ddress **T**ranslation
Data CLEC	**Data C**ompetitive **L**ocal **E**xchange **C**arrier
DATE	**D**uly **A**uthorized **T**elecommunications **E**ntity (Intelsat)
DATTS	**D**ata **A**cquisition **T**elecommand and **T**racking **S**tation (Intelsat)
DATU	**D**irect-**A**ccess **T**est **U**nit
DAU	**D**umb **A**ss **U**ser
DAV	1. **D**istributed **A**uthoring and **V**ersioning
	2. **D**ata **Av**ailable (logic signal)
	3. **D**igital **A**udio-**V**isual
DAVIC	**D**igital **A**udio-**Vi**sual **C**ouncil (Italy)
DAVID	**D**igital **A**udio/**V**ideo **I**nteractive **D**ecoder (ITU)
DAWS	**D**igital **A**dvanced **W**ireless **S**ervice
dB	**D**ecibel (one-tenth of a **B**el)
dB/K	**D**eci**b**els relative to reciprocal temperature (in **K**elvin)
dBa	**D**eci**b**els **a**djusted
dBA	**D**ecibel **A**udio (sound level)
dBa(FIA)	**dBa** measured with **FIA**-line weighting (fiber optics)
dBa(HAI)	**dBa** measured with **HAI**-receiver weighting (fiber optics)
dBa0	**dBa** measured at **zero** transmission level point
dBc	**D**eci**b**els relative to unmodulated **c**arrier power level
dBd	**D**eci**b**els relative to a **D**ipole antenna gain
.dbf	Name extension for **d**ata**b**ase **f**iles (computer)
dBf	**D**eci**b**els above one **f**emtowatt
dBHz	**D**eci**b**els relative to one **H**ert**z**
dBi	**D**eci**b**els relative to an **i**sotropic antenna gain
dBic	**D**eci**b**els relative to an **i**sotropically radiated **c**arrier
dBk	**D**eci**b**els relative to one **k**ilowatt
dBm	**D**eci**b**els relative to one **m**illiwatt (0 dBm = 1 milliwatt)
DBM	**D**ouble-**B**alanced **M**ixer (microwave)
dBm0	**D**eci**b**els relative to one **m**illiwatt at a point of **zero**
dBm0p	**dBm0** measured with **p**sophometric weighting

D

dBmc	**dBm** measured with **C**-message weighting
dBmc0	**dBmc** measured at **0** Transmission Level Point
dBm(f$_1$ − f$_2$)	Flat noise power in **dBm**
dBmp	Decibels relative to one **m**illiwatt using **p**sophometric weighting
dBmV	Decibels relative to one **m**illi**v**olt across a specific impedance
dBnW	Decibels relative to one **n**ano**w**att (0 dBnW = 1 nW)
dB0	Decibels relative to **0**TLP
dBp	Decibels relative to one **p**icowatt
dBpsoph	Noise power in **dB**m measured with **psoph**ometric weighting
dBpW	Decibels referred to one **p**ico**w**att (0 dBpW = 1 pW)
dBq	Decibels relative to digital speech codec **Q**uality
dBr	Decibels referred to a given **r**eference
dBrn	Decibels relative to **r**eference **n**oise level
dBrnC	Decibels relative to **r**eference **n**oise using **C**-message weighting
dBrnC0	Decibels relative to **r**eference **n**oise using a type **C**-messaging channel adjusted for equivalence to a **0** dBm equivalent point
dBuV	Decibels relative to one **m**icro**V**olt
dBuW	Decibels relative to one **m**icro**W**att (μW)
dBV	Decibels relative to one **V**olt
dBW	Decibels relative to one **W**att
dBx	Decibels above the reference **coupling**
DB	1. **D**ata **B**urst
	2. **D**ummy **B**urst (GSM)
	3. **d**ata**b**ase
	4. **D**ata **B**us
DBA	1. **D**ynamic **B**andwidth **A**llocation (protocol feature)
	2. **D**irect **B**roadcast **A**ddress
	3. **D**atabase **A**dministrator (ICT)
DBC	**D**ense **B**inary **C**ode
DBD	1. **D**ata**b**ase **D**river
	2. **D**igital **B**earing **D**iscriminator
DBF	**D**igital **B**eam **F**orming (microwave)
DBFN	**D**igital **B**eam **F**orming **N**etwork
DBit	**D**elivery confirmation **B**it (X.25)
DBMS	**D**ata**b**ase **M**anagement **S**ystems (ICT)
DBP	**D**eutsche **B**undes**p**ost (German Communications)
DBPSK	**D**ifferential **B**inary **PSK** (modulation)
DBR	**D**eterministic **B**it **R**ate

DBS	1. **D**irect **B**roadcast by **S**atellite (TV service)
	2. **D**igital **B**roadcast **S**atellite
DBT	1. **D**eutsche **B**undespost **T**elecom (Germany)
	2. **D**efect-**B**ased **T**esting
DBTVS	**D**irect **B**roadcast **Tele**vision **S**atellite
DBU	**D**ial **B**acku**p**
DC	1. **D**irect **C**urrent (power)
	2. **D**isconnect **C**onform
	3. **Dis**connect (live internet chat)
DC/AC	**D**irect **C**urrent **to A**lternating **C**urrent (power)
DC/DC	**D**irect_**C**urrent-**to-D**irect_**C**urrent converter (electronics)
DCA	1. **D**ocument **C**ontent **A**rchitecture (IBM)
	2. **D**ynamic **C**hannel **A**llocation (DECT)
	3. **D**efense **C**ommunication **A**gency (U.S.A.)
	4. **D**irectory **C**lient **A**gent
	5. **D**ownload **C**onfiguration **A**rea (software downloading)
DCAS	**D**irect **C**arrier **A**dministration **S**ystem
DCASP	**D**igitally **C**ontrolled **A**nalog **S**ignal **P**rocessing
DCC	1. **D**ata **C**ommunications **C**hannel (MUX)
	2. **D**ata **C**ountry **C**ode (ATM)
	3. **D**igital **C**ompact **C**assette
	4. **D**igital **C**ross-**C**onnect (MUX)
	5. **D**igital **C**olor **C**ode
	6. **D**igital **C**ommand **C**enter
	7. **D**igital **C**ommunications **C**enter
	8. **D**igital **C**ommunications **C**onference
DCCH	**D**edicated **C**ontrol **Ch**annel (GSM)
DCD	1. **D**ata-**C**arrier-**D**etect (RS 232 signal)
	2. **D**ynamically **C**onfigurable **D**evice
	3. **D**uty **C**ycle **D**istortion
	4. **D**ocument **C**ontent **D**escription
DCDI	**D**irectional **C**orrelation **De**interlacing
DCE	1. **D**ata **C**ircuit-**T**erminating **E**quipment (networking)
	2. **D**ata **C**ommunications **E**quipment (networking)
	3. **D**istributed **C**omputing **E**nvironment (service)
DCF	**D**ispersion **C**ompensating **F**iber
DCFL	**D**irect-**C**oupled **FET** **L**ogic (digital electronics)
DCG	**D**ispersion **C**ompensation **G**rating (optical signals)
DCH	1. **D**-**C**hannel **H**andler
	2. **D**edicated **Ch**annel
DCI	1. **D**igital **C**ontrol **I**nterface
	2. **D**ata **C**ommunications Company of **I**ran
	3. **D**isplay **C**ontrol **I**nterface

DCLEC	A **CLEC** specialized in delivering only **D**SL services
DCM	1. **D**igital **C**ircuit **M**ultiplication
	2. **D**ecoder and **C**ontrol **M**odule
	3. **D**ispersion **C**ompensation **M**odule
	4. **D**ynamically **C**ontrollable **M**agnetic
DCME	**D**igital **C**ircuit **M**ultiplication **E**quipment
DCMI	**D**ublin **C**ore **M**etadata **I**nitiative
DCMS	**D**igital **C**ircuit **M**ultiplication **S**ystem
DCN	1. **D**ata **C**ommunication **N**etwork
	2. **Disco**nnect frame
DCNR	**D**igital **C**ontent **N**etwork **R**eceiver
DCOF	**D**oubly **C**ladded **O**ptical **F**iber
DCOM	**D**istributed **C**omponent **O**bject **M**odel (ICT)
DCP	1. **D**igital **C**ommunications **P**rotocol
	2. **D**evice **C**ontrol **P**rotocol
	3. **D**ata **C**ollection **P**latform (remote sensing)
DCPR	**D**etailed **C**ontinuing **P**roperty **R**ecords
DCPSK	**D**ifferentially **C**oherent **PSK** (modulation)
DCR	1. **D**irect-**C**onversion **R**eceiver
	2. **D**own-**C**onverting **R**eceiver
DCS	1. **D**igital **C**ellular **S**ystem (GSM)
	2. **D**igital **C**ommunication **S**ystem
	3. **D**igital **C**ross-connect **S**ystem (MUX)
	4. **D**igital **C**lock **S**ource
	5. **D**igital **C**oded **S**quelch
	6. **D**ata **C**ollection **S**ystem (remote sensing)
	7. **D**efense **C**ommunication **S**ystem
	8. **D**istributed **C**ommunications **S**ystem
DCS-1800	**D**igital **C**ellular **S**ystem at **1800 MHz** (GSM-1800)
DCT	1. **D**igital **C**arrier **T**ermination
	2. **D**iscrete **C**osine **T**ransform (mathematics)
DCTE	**D**ata **C**ircuit-**T**erminating **E**quipment
DCTI	**D**esktop **C**omputer **T**elephony **I**ntegration
DCTL	**D**irect-**C**oupled **T**ransistor **L**ogic (digital electronics)
DCTN	**D**efense **C**ommercial **T**elecommunications **N**etwork
DCTU	**D**igital **C**arrier **T**runk **U**nit
DCU	**D**istribution **C**ontrol **U**nit
DCV	**D**igital **C**ompressed **V**ideo
DCWV	**DC** **W**orking **V**oltage (electronics)
DCXO	**D**igitally-**C**ompensated **Crystal** **O**scillator
DD	1. **D**otted **D**ecimal notation
	2. **D**ouble **D**ensity (floppy diskette)
DDA	1. **D**omain **D**efined **A**ttribute (X.400)
	2. **D**igital **D**ifferential **A**nalyzer

DDB	1. **D**igital **D**ata **B**ank
	2. **D**evice **D**ependent **B**itmap
DDBMS	**D**istributed **D**atabase **M**anagement **S**ystem
DDC	1. **D**irect **D**epartment **C**alling (service)
	2. **D**irect **D**igital **C**ontrol (computer)
	3. **D**isplay **D**ata **C**hannel (VESA standard)
DDCMP	**D**igital **D**ata **C**ommunication **M**essage **P**rotocol
DDCP	**D**irect **D**igital **C**olor **P**roof (test sheet)
DDCPSK	**D**ouble **D**ifferential **C**oherent **PSK** (modulation)
DDE	**D**ynamic **D**ata **E**xchange (Windows feature)
DDEML	**D**ynamic **D**ata **E**xchange **M**anagement **L**ibrary
DDF	**D**igital **D**istribution **F**rame
DDI	1. **D**irect **D**igital **I**nterface
	2. **D**irect **D**ialing **I**nward
DDJ	**D**ata **D**ependent **J**itter
DDL	**D**ata **D**efinition/**D**escription **L**anguage (programming)
DDM	**D**istributed **D**ata **M**anagement architecture (networking)
DDN	1. **D**efense **D**ata **N**etwork (U.S.A.)
	2. **D**igital **D**ata **N**etwork
DDN-NIC	**D**efense **D**ata **N**etwork–**N**etwork **I**nformation **C**enter
DDNS	**D**ynamic **D**omain **N**aming **S**ystem (Internet)
DDoS	**D**istributed **D**enial-**of**-**S**ervice
DDP	1. **D**atagram **D**elivery **P**rotocol
	2. **D**istributed **D**ata **P**rocessing
	3. **D**ata **D**istribution **P**lan (IMO)
DDR	1. **D**isengagement **D**enial **R**atio
	2. **D**ouble **D**ata **R**ate (memory)
DDR-SDRAM	**D**ouble **D**ata **R**ate **S**ynchronous **D**ynamic **RAM** (memory)
DDS	1. **D**igital **D**ata **S**ervice (DSL)
	2. **D**igital **D**ata **S**torage (standard)
	3. **D**ataphone **D**igital **S**ervice (AT&T)
	4. **D**istributed **D**ata **S**ystem
	5. **D**irect **D**igital **S**ynchronous (interface)
	6. **D**irect **D**igital **S**ynthesizer
	7. **D**irect **D**ialing **S**ervice
DDSD	**D**elay **D**ial **S**tart **D**ial (protocol)
DDT	**D**igital **D**irty **T**rick (security feature)
De-emph	**De-emph**asis (frequency modulation)
DE	1. **D**iscard **E**ligibility bit (frame relay)
	2. **D**esignated **E**ntity
	3. **D**ifferentially **E**ncoded
DEB	**D**ifferentially **E**ncoded **B**aseband

DEBS	**D**ynamic **E**lectronic **B**ias **S**ystem
DEC	1. **D**igital **E**quipment **C**orporation
	2. **dec**oder
DECA	**D**igital **E**lectronics **C**ontrol **A**ssembly
DECNet	**D**igital **E**quipment **C**orporation's proprietary **Net**work architecture
DECPSK	**D**ifferentially **E**nhanced **C**oherent **PSK** (modulation)
DECT	1. **D**igital **E**nhanced **C**ordless **T**elecommunications (ETSI)
	2. **D**igital **E**uropean **C**ordless **T**elecommunications
DEF	**D**esign **E**xchange **F**ormat
Deg	**deg**ree
DEK	**D**ata **E**ncryption **K**ey
DEL	1. **Del**ete a file (MS DOS Command)
	2. **Del**ete key (PC keyboard)
DELNI	1. **D**igital **E**thernet **L**ocal **N**etwork **I**nterconnect
	2. **DEC** **L**ocal **N**etwork **I**nterface
DELTIC	**Del**ay-Line-**Ti**me-**C**ompression
DEM	1. **dem**odulator
	2. **D**igital **E**levation **M**odel (remote sensing)
DEMKO	**D**enmark **E**lektriske **M**aterie **Ko**ntrol
demo	**demo**nstration part (software package)
DEMOD	**demod**ulator
Demod/Remod	**Demod**ulation **and Remod**ulation
DEMPR	**DEC** **M**ulti**p**ort **R**epeater
DEMS	**D**igital **E**lectronic **M**anagement **S**ystem
DEMUX	**demu**ltiple**x**er
DEMUXing	**demu**ltiple**xing**
DEN	**D**irectory **E**nabled **N**etworking (network management)
DENet	**D**anish **E**thernet **Net**work
DEP	**D**egraded **E**rror **P**erformance
DEPS	**D**epartmental **E**ntry **P**rocessing **S**ystem
DEPSK	**D**ifferential **E**ncoded **PSK** (modulation)
dequeue	**d**ouble-ended **queue**
DES	1. **D**ata **E**ncryption **S**tandard (IBM algorithm)
	2. **D**estination **E**nd **S**tation (ATM)
	3. **D**ouble-**E**nded **S**ynchronization (networking)
DESC	**D**efense **E**lectronic **S**upply **C**enter
DET	**det**ector
Detem	Optical **det**ector/**em**itter combiner (transducer)
DEU	**D**ata **E**ncryption **U**nit
DF	1. **D**irection **F**inder (equipment)
	2. **D**istribution **F**rame
	3. **D**istress **F**requency

D

DFA	**D**oped **F**iber **A**mplifier (fiber-optic)
DFB	**D**istributed **F**eed**b**ack (laser)
DFC	**D**ata **F**low **C**ontrol
DFD	1. **D**ata **F**low **D**iagram (ICT)
	2. **D**igital **F**requency **D**iscriminator
DFE	**D**ecision **F**eedback **E**qualizer (adaptive filter)
DFES	**D**epartment **F**or **E**ducation and **S**kills (U.K. Government)
DFG	**D**iode **F**unction **G**enerator
DFI	**D**igital **F**acility **I**nterface
DFM	1. **D**igital **F**requency **M**odulation
	2. **D**esign **F**or **M**anufacturability
	3. **D**ispersive **F**ade **M**argin
DFS	1. **D**ynamic **F**requency **S**election
	2. **D**istributed **F**ile **S**ystem (Windows 2000 feature)
DFSK	**D**ouble **FSK** (modulation)
DFT	1. **D**irect **F**acility **T**ermination
	2. **D**iscrete **F**ourier **T**ransform (algorithm)
	3. **D**esign **F**or **T**est
DG	**D**irector **G**eneral
DGIS	**D**irect **G**raphics **I**nterface **S**pecification (computer)
DGNSS	**D**ifferential **G**lobal **N**avigation **S**atellite **S**ystems
DGPS	**D**ifferential **GPS**
DGPT	**D**epartment of **G**eneral **P**osts and **T**elecommunications, Vietnam
DGT	1. **D**irecsión **G**énérale des **T**élécommunicaciones (France)
DHACP	**D**ynamic **H**ost **A**utomatic **C**onfiguration **P**rotocol
DHBT	**D**ouble-**H**eterojunction **B**ipolar **T**ransistor
DHCP	**D**ynamic **H**ost **C**onfiguration **P**rotocol (TCP/IP)
DHN	**D**igital **H**ome **N**etwork
DHSD	**D**uplex **H**igh-**S**peed **D**ata
DHTML	**D**ynamic **HTML** (programming)
DI	1. **D**edicated **I**ntegrator
	2. **D**ielectric **I**solation
DIA	**D**ocument **I**nterchange **A**rchitecture (data transmission)
DIA/DCA	**DIA** **D**ocument **C**ontent **A**rchitecture
diac	**di**ode **AC** (semiconductor switch)
DIAG	**diag**nostic port
DIALAN	**D**MS **I**ntegrated **A**ccess **L**ocal **A**rea **N**etwork (Northern Telecom)

DIB	1. **D**ual **I**ndependent **B**us (Intel's Pentium II processor)
	2. **D**irectory **I**nformation **B**ase (X.500)
	3. **D**evice **I**ndependent **B**itmap (file format)
DiBEG	**Di**gital **B**roadcasting **E**xpert's **G**roup
DIC	**D**ubai **I**nternet **C**ity
DICODE-NRZ	**Di-Code-N**on-**R**eturn to **Z**ero level (data encoding)
DID	1. **D**estination **ID**
	2. **D**irect **I**nward **D**ialing (phone line)
	3. **D**etailed **I**nterface **D**ossier
DIEL	Advisory committee on telecommunications for **Di**sabled and **El**derly people
DIF	**D**ata **I**nterchange **F**ormat (computer)
DIFAR	**D**irectional **F**inding **A**nd **R**anging (submarine warfare)
DIFCOD	**Dif**ferential **Cod**er
DIFDECOD	**Dif**ferential **Decod**er
DiffServ	**Dif**ferentiated **Serv**ices (QoS)
DiffZL	**Dif**ferential **Z**iv–**L**empel (data encoding)
digicash	**digi**tal **cash**
DigiTAG	**Digi**tal **T**elevision **A**ction **G**roup (broadcasting)
DII	**D**efense **I**nformation **I**nfrastructure
DIL	1. **D**irect **I**n-**L**ine (office trunk)
	2. **D**ual **I**n-**L**ine (microchips)
DILL	Another form of saying "DLL"
DIM	1. **D**igital **I**nterface **M**odule
	2. **D**ocument **I**mage **M**anagement
DIMM	**D**ual **I**n-line **M**emory **M**odule (Macintosh PCs)
DIMS	**D**ocument **I**mage **M**anagement **S**ystem
DIN	**D**eutsches **I**nstitut für **N**ormung (German Standards)
DINA	1. **D**istributed **I**ntelligent **N**etwork **A**rchitecture
	2. **D**ynamic **I**ntelligent **N**etwork **A**rchitecture
	3. **D**irect **N**oise **A**mplification (jammer)
	4. **D**anish **I**nformatics **N**etwork in the **A**gricultural Science
DIOCES	**D**istributed **I**nteroperable and **O**perable **C**omputing **E**nvironments and **S**ystems
diode	A **two**-electr**ode** electron tube or semiconductor
DIP	1. **D**ual **I**nline **P**ackage (microchips)
	2. **D**ocument **I**mage **P**rocessing
	3. **D**edicated **I**nside **P**lant (local exchanges)
DIPS	**D**igital **I**mage **P**rocessing **S**ystem
dir	**dir**ectory (MS DOS Command)
DIS	**D**raft **I**nternational **S**tandard (ISO)

DISA	1. **D**irect **I**nward **S**ystem **A**ccess (PBX feature)
	2. **D**efense **I**nformation **S**ystem **A**gency
Disc	**disc**onnect
DISH	**D**ata **I**nterchange for **Sh**ipping
DISKCOPY	**Copy** from one floppy **Disk** to another (MS DOS Command)
DISN	**D**efense **I**nformation **S**ystems **N**etwork
DISNet	**D**efense **I**ntegrated **S**ecure **Net**work
DISOSS	**Di**stributed **O**ffice **S**upport **S**ystem (IBM)
DIT	**D**irectory **I**nformation **T**ree (X.500)
DITS	**D**igital **I**nformation **T**ransfer **S**ystem
DIU	1. **D**igital **I**nterface **U**nit
	2. **D**igital **I**ndoor **U**nit (VSAT terminals)
DIW	**D** type **I**nside **W**ire
DIX	**D**igital/**I**ntel/**X**erox
DL	1. **D**istance **L**earning (ICT)
	2. **D**ata **L**ink
	3. **D**igital **L**ibrary
	4. **D**ead **L**oad
DLC	1. **D**igital **L**oop **C**arrier (transmission equipment)
	2. **D**irect **L**ine **C**onsole
	3. **D**ynamic **L**oad **C**ontrol
	4. **D**ata **L**ink **C**ontrol (protocol)
	5. **D**ata **L**ink **C**onnection (frame relay)
	6. **D**ead-**L**ink **C**heck
	7. **D**ual **L**ayer **C**apacitor (electronics)
DLCI	1. **D**ata **L**ink **C**onnection **I**dentifier (frame relay)
	2. **D**igital **L**oop **C**arrier **I**nterface
DLE	**D**ata **L**ink **E**scape (character control code)
DLEC	**D**ata **L**ink **E**xchange **C**arrier
.dll	Name extension for **D**ynamic-**L**ink **L**ibrary files (computer)
DLL	1. **D**own **L**ine **L**oad
	2. **D**ynamic-**L**ink **L**ibrary (Windows feature)
	3. **D**ata **L**ink **L**ayer driver
	4. **D**igital **L**eased **L**ines
DLM	**D**ouble-**L**evel **M**etal (microchips)
DLP	**D**igital **L**ight **P**rocessing
DLPBC	**D**ual **L**oop **P**ort **B**ypass **C**ircuitry
DLPI	**D**ata **L**ink **P**rovider **I**nterface (ATM)
DLR	1. **D**estination **L**ocal **R**eference (GSM)
	2. **D**esign **L**ayout **R**eport
	3. **D**eutches **Z**entrum für **L**uft-und **R**aumfahrt (Germany)

D

DLS	1. **D**ata **L**ink **S**witching (IBM)
	2. **D**istance **L**earning **S**ection (e-learning)
	3. **D**igital **L**oud**s**peaker
DLSE	**D**ial **L**ine **S**ervice **E**valuation
DLSw	**D**ata **L**ink **S**witching (standard)
DLSW	**D**ata **L**ink **S**witching **W**orkgroup
DLT	1. **D**igital **L**inear **T**ape
	2. **D**esign **L**ife **T**ime (satellite)
DLTU	**D**ata **L**ink **T**runk **U**nit (AT&T)
DLVA	**D**etector **L**og **V**ideo **A**mplifier
DM	1. **D**elta **M**odulation
	2. **D**igital **M**odulation
	3. **D**elay **M**odulation
	4. **D**ifferential **M**odulation
	5. **D**etection **M**odule
	6. **D**isconnect **M**ode (X.25)
	7. **D**istress **M**essage
DMA	1. **D**irect **M**emory **A**ccess (computer)
	2. **D**eferred **M**aintenance **A**larm
	3. **D**ifferential **M**ode **A**ttenuation (AC power system)
	4. **D**ocument **M**anagement **A**lliance
	5. **D**ECT **M**ultimedia **A**ccess Profile
DMB	1. **D**igital **M**ultipoint **B**ridge
	2. **D**igital **M**ultimedia **B**roadcasting
DMBU	**D**ial **M**odem **B**ackup **U**nit
DMC	1. **D**ual voice grade **M**odem **C**ard
	2. **D**iscrete **M**emory **C**hannel
DMCA	**D**igital **M**illennium **C**opyright **A**ct
DMD	1. **D**ifferential **M**ode **D**elay (fiber-optic)
	2. **D**igital **M**icromirror **D**evice (display)
DME	1. **D**istributed **M**anagement **E**nvironment
	2. **D**istance **M**easuring **E**quipment
DMF	**D**ata **M**anagement **F**acility
DMG	**D**istress **M**essage **G**enerator (IMO)
DMH	**D**ata **M**essage **H**andler (cellular networks)
DMI	1. **D**esktop **M**anagement **I**nterface (networking)
	2. **D**igital **M**ultiplexed **I**nterface (AT&T)
DMI-BOS	**D**igital **M**ultiplexed **I**nterface-**B**it-**O**riented **S**ignaling
DML	1. **D**ata **M**anipulation **L**anguage (programming)
	2. **D**eclarative **M**arkup **L**anguage (programming)
	3. **D**evelopment **M**arkup **L**anguage (programming)
DMM	**D**igital **M**ultim**e**ter (test equipment)
DMO	1. **D**igital **M**odification **O**rder
	2. **D**ata **M**ode **O**peration

D

	3. **D**irect **M**ode **O**peration (cellular networks)
DMOS	**D**ouble-diffused **MOS** (semiconductors)
DMPDU	**D**erived **MAC PDU** (service)
DMR	1. **D**igital **M**icrowave **R**adio
	2. **D**ual-**M**ode **R**esonator (microwave)
DMS	1. **D**igital **M**ultiplex **S**ystem
	2. **D**igital **M**ultiplexed **S**witch (Northern Telecom)
	3. **D**igital **M**edia **S**olutions
	4. **D**ocument **M**anagement **S**ystem (ICT)
DMSP	**D**efense **M**eteorological **S**atellite **P**rogram
DMSU	**D**igital **M**ain **S**witching **U**nit
DMT	1. **D**iscrete **M**ulti**t**one (modulation)
	2. **D**igital **M**ultiplexed **T**runk
DMTF	**D**esktop **M**anagement **T**ask **F**orce (ICT)
DMUX	**dem**ultiple**x**er
DMW	**D**igital **M**icro**w**ave
DMX	**D**igital **M**usic **Ex**press
DMZ	**D**e**m**ilitarized **Z**one (networks)
DN	1. **D**irectory **N**umber (ISDN telephone)
	2. **D**igital **N**etwork
DNA	1. **D**igital **N**etwork **A**rchitecture
	2. **D**istributed **N**etwork **A**dministration
	3. **DEC N**etwork **A**rchitecture
	4. **D**ynamic **N**ode **A**ccess
DNAL	**D**edicated **N**etwork **A**ccess **L**ine
DNAR	**D**irectory **N**umber **A**nalysis **R**eporting
DNC	1. **D**istributed **N**etworking **C**omputing
	2. **D**ynamic **N**etwork **C**ontroller
	3. **D**irect **N**umerical **C**ontrol
	4. **D**istributed **N**etwork **C**omputing
DNCF	**D**irectory **N**umber **C**all **F**orwarding
DND	**D**o **N**ot **D**isturb (PBX)
DNI	**D**igital **N**on-**I**nterpolation
DNIC	**D**ata **N**etwork **I**dentification **C**ode (X.121)
DNID	**D**ata **N**etwork **Id**entification code (Inmarsat)
DNIS	**D**ialed **N**umber **I**dentification **S**ervice
DNL	**D**ynamic **N**on-**L**inearity (microwave)
DNP	**D**ial-up **N**etworking **P**rofile (Bluetooth)
DNPA	**D**ata **N**umbering **P**lan **A**rea (X.25)
DNPIC	**D**irectory **N**umber **P**rimary **I**nter**L**ATA **C**arrier
DNPS	**D**ivisional **N**etwork **P**roduct **S**upport
DNR	1. **D**ynamic **N**etwork **R**econfiguration (IBM)
	2. **D**ynamic **N**oise **R**eduction (satcom)

	3. **D**ialed **N**umber **R**ecorder
DNS	1. **D**omain **N**aming **S**ystem (Internet)
	2. **D**omain **N**ame **S**ervice (Internet)
	3. **D**istributed **N**etwork **S**ervice
DNSIX	**D**efense **N**etwork **S**ecurity **I**nformation **E**xchange
DNSSE	**D**omain **N**ame **S**ystem **S**ecurity **E**xtensions
DNTX	**D**o **N**ot **Transmit** (Morse code)
DOA	**D**ead **O**n **A**rrival (of an equipment)
DoB	**D**enial **of** **B**usiness
.doc	Name extension for Word Processor **doc**ument files (computer)
DoC	**D**epartment **of** **C**ommunications (Canadian Government)
DOC	**D**ynamic **O**verload **C**ontrol (switch)
DocObject	**Doc**ument **Object**
DoCoMo	**Do** **Co**mmunications over the **Mo**bile Network (operators)
DOCSIS	**D**ata-**O**ver-**C**able **S**ervice **I**nterface **S**pecifications
DoD	**D**epartment **of** **D**efense (U.S. army)
DoD-STD	**D**epartment **of** **D**efense **St**an**d**ard (U.S.A.)
DOD	1. **D**irect **O**utward **D**ialing (PBX feature)
	2. **D**epth **O**f **D**ischarge
DoDD	**D**epartment **of** **D**efense **D**irective (U.S.A.)
DoDI	**D**epartment **of** **D**efense **I**nstructions (U.S.A.)
DoDISS	**DoD** **I**ndex of **S**pecifications and **S**tandards (U.S.A.)
DOE	**D**irect **O**rder **E**ntry
DOM	**dom**estic
DOMSAT	**Dom**estic **Sat**ellite communication system
DOP	**D**edicated **O**utside **P**lant (local exchanges)
Doran	**Do**ppler **ran**ging system
DoS	**D**enial-**o**f-**S**ervice (network attack)
DOS	**D**isk **O**perating **S**ystem (computer)
DOSA	**D**istributed **O**pen **S**ignaling **A**rchitecture (IP telephony)
DOT memory	**D**omain-**T**ip **memory**
DOV	**D**ata **O**ver **V**oice
DOVAP	**Do**ppler **V**elocity **A**nd **P**osition system
DP	1. **D**ata **P**ort
	2. **D**ata **P**arity
	3. **D**ata **P**rocessing
	4. **D**eflate **P**rotocol
	5. **D**ial **P**ulse
	6. **D**istribution **P**artner (Inmarsat)
DPA	1. **D**emand **P**rotocol **A**rchitecture
	2. **D**igital **P**ort **A**dapter (Northern Telecom)

3. **D**igital **P**erformance **A**rchive
4. **D**earborn **P**rotocol **A**dapter (PC board)
5. **D**ifferential **P**ower **A**nalysis
6. **D**ocument **P**rinting **A**pplication

DPBC   **D**ouble-**P**ole **B**ack **C**onnected
DPBX   **D**igital **P**rivate **B**ranch **E**xchange
DPC
1. **D**estination **P**oint **C**ode (routing)
2. **D**ata **P**rocessing **C**enter (remote sensing)
3. **D**igital **P**ort **C**luster

DPCCH   **D**edicated **P**hysical **C**ontrol **Ch**annel
DPCH   **D**edicated **P**hysical **Ch**annel
DPCM   **D**ifferential **PCM** (modulation)
DPCS   **D**igital **P**ersonal **C**ommunications **S**ervices
DPDCH   **D**edicated **P**hysical **D**ata **Ch**annel
DPDT   **D**ouble-**P**ole **D**ouble-**T**hrow (switches)
DPE   **D**istributed **P**rocessing **E**nvironment
DPFC   **D**ouble-**P**ole **F**ront **C**onnected
DPHEMT   **D**ouble-heterojunction **P**seudomorphic **H**igh-**E**lectron **M**obility **T**ransistor
dpi   **d**ots **p**er **i**nch (scanners)
DPI
1. **D**edicated **P**air-**I**n (telephone wire pair)
2. **D**irect **P**ickup **I**nterference
3. **D**ots **P**er **I**nch (image resolution)
4. **D**ata **P**rocessing Company of **I**ran

DPLB   **D**igital **P**rivate **L**ine **B**illing
DPLL   **D**igital **P**hase-**L**ocked **L**oop
DPLX
1. **Du**ple**x**
2. **Du**ple**x**er

DPM
1. **D**igital **P**anel **M**eter (test equipment)
2. **D**igital **P**hase **M**odulation
3. **D**ual-**P**recedence **M**essage

DPMA
1. **D**ata **P**rocessing **M**anagement **A**ssociation
2. **D**ynamic **P**ower **M**anagement **A**rchitecture (ICT)

DPMI   **D**OS **P**rotected **M**ode **I**nterface
DPMS
1. **D**igital **P**erformance **M**onitoring **S**ystem
2. **D**isplay **P**ower **M**anager **S**ignaling (VESA standard)

DPNSS   **D**igital **P**rivate **N**etwork **S**ignaling **S**ystem
DPO
1. **D**edicated **P**air **O**ut (telephone wire pair)
2. **D**irect **P**ublic **O**ffering
3. **D**igital **P**ulse **O**rigination

DPP
1. **D**istributed **P**rocessing **P**eripherals
2. **D**istributed **P**rogramming **P**latform
3. **D**istributed **P**ipe **P**rotocol
4. **D**igitally **P**rogrammable **P**otentiometer

	5. Direct Print Protocol
	6. Demand Priority Protocol
DPPB	Digital Primary Patch Bay (fiber optics)
DPRAM	Dual Port RAM (memory)
DPRS	Domestic Public Radio Service
DPS	1. Differential Phase Shift
	2. Digital Passage Service
	3. Digital Protection Switch
DPSK	Differential PSK (modulation)
DPST	Double-Pole Single-Throw (switches)
DPT	Digital Pulse Termination
DPU	1. Digital Processing Unit
	2. Dynamic Path Update (IBM)
	3. Director of Public Utilities
DPX	Data Path loop Extension
DQDB	Distributed Queue Dual Bus (architecture)
DQPSK	Differential Quadrature PSK (modulation)
DQTV	Distribution Quality Television
DR	1. Dual Rate
	2. Design Review
	3. Digital Radio
DRA	Direct Radiating Array (antennas)
DRAM	Dynamic RAM (memory)
DRAW	Direct Read After Write (optical disks)
DRB	Digital Radio Broadcasting
DRC	Design Rule Checking
DRCS	Digital Radio Concentrator System
DRDRAM	Direct Rambus Dynamic RAM (memory)
DRDW	Direct Read During Write (optical disks)
DRFM	Digital Radio Frequency Memory
DRiP	Duplicate Ring Protocol (Cisco)
DRM	Digital Rights Management
DRO	1. Dielectric Resonator Oscillator (microwave)
	2. Dielectric Resonator Oscillator (electronics)
DRP	1. Digital Receiver Processor (Inmarsat)
	2. Distribution and Replication Protocol
DRRS	Digital Radio Relay System
DRS	1. Data Relay Satellite
	2. Digital Reference Signal
	3. Digital Radio System
	4. Digital Reconfiguration Service
	5. Direct Receiving Station (satcom)
	6. Disengagement Request Signal

DRT	**DS**3 **R**edundancy and **T**ermination
DRTL	**D**ial **R**epeating **T**ie **L**ine
DRTS	**D**ata **R**elay and **T**est **S**atellite
DRTT	**D**ial **R**epeating **T**ie **T**runk
DRU	1. **DACS R**emote **U**nit
	2. **D**igital **R**emote **U**nit (NEC)
.drv	Name extension for **dr**iver files (computer)
.drw	Name extension for Micrografx **DRAW** format files (graphics)
DRX	**D**iscontinuous **R**eception (GSM mobile station)
DS	1. **D**ispersion **S**hifted (optical fiber)
	2. **D**own**s**tream
	3. **D**igital **S**ervice
	4. **D**igital **S**ignal
	5. **D**ata **S**trobe (logic signal)
DS-0	**D**igital **S**ignal Level **0**
DS-1	**D**igital **S**ignal Level **1** (synonym for T1)
DS-2	**D**igital **S**ignal Level **2** (synonym for T4)
DS-3	**D**igital **S**ignal Level **3** (synonym for T3)
DS-4	**D**igital **S**ignal Level **4** (synonym for T4)
DS-5	**D**igital **S**ignal Level **5** (synonym for T54)
DS-CDMA	**D**irect **S**equence **CDMA** (access)
DSA	1. **D**AMA **S**ignaling **A**nalyzer
	2. **D**istributed **S**ystem **A**rchitecture (networking)
	3. **D**ata **S**ervice **A**dapter
	4. **D**irectory **S**ervice **A**gent (software)
	5. **D**irect **S**elling **A**ssociation
	6. **D**igital **S**ignature **A**lgorithm
	7. **D**ial **S**ervice **A**ssistance
DSAP	**D**estination **S**ervice **A**ccess **P**oint
DSAT	1. **D**igital **S**upervisory **A**udio **T**ones
	2. **D**igital **Sat**ellite Television (broadcasting)
DSB	1. **D**ouble-**S**ide**b**and (modulation)
	2. **D**igital **S**ound **B**roadcasting
DSB-EC	**D**ouble-**S**ide**b**and **E**mitted **C**arrier (modulation)
DSB-RC	**D**ouble-**S**ide**b**and **R**educed **C**arrier (modulation)
DSB-SC	**D**ouble-**S**ide**b**and **S**uppressed **C**arrier (modulation)
DSB-TC	**D**ouble-**S**ide**b**and **T**ransmitted **C**arrier (modulation)
DSC	1. **D**ata **S**atellite **C**hannel
	2. **D**ifferential **S**canning **C**alorimetry
	3. **D**igital **S**ervice **C**hannel
	4. **D**igital **S**elective **C**alling (shipping)
	5. **D**igital **S**ubscriber **C**ontroller
	6. **D**igital **S**till **C**amera

D

	7. Digital-to-Synchro Converter
	8. Distributed Statistical Computing
	9. descrambler
DSCP	Differential Service Code Point
DSCS	Defense Satellite Communication System
DSDC	Direct Service Dialing Capability
DSE	1. Distributed Single-Layer Embedded (test method)
	2. Distributed Systems Engineering
	3. Data Switching Exchange
	4. Data Switching Equipment
	5. Deep Sky Exploration society
	6. Digital Subtitle Encoder
	7. Dynamic System Estimation Library
DSF	Dispersion Shifted Fiber
DSH	Double Super Heterodyne
DSI	1. Digital Speech Interpolation (voice compression)
	2. Digital Switching Information
	3. Detailed Spectrum Investigation
DSL	1. Digital Subscriber Line (access)
	2. Digital Subscriber Loop (AT&T)
	3. Design System Language (programming)
DSLAM	DSL Access Multiplexer (network device)
DSM	1. Delta–Sigma Modulation
	2. Direct Sequence Modulation
	3. Direct-Spread Modulation
	4. Design Structure Matrix
	5. Digital Switching Matrix
	6. Dependency Structure Matrix
DSMCC	Digital Storage Media Command and Control (protocol)
DSN	1. Distributed Systems Network
	2. Defense Switch Network
	3. Double Shelf Network
	4. Deep Space Network
DSNG	Digital Satellite News Gathering
DSO	1. Digital Storage Oscilloscope
	2. Digital Sampling Oscilloscope
DSOM	Distributed System Object Model (IBM)
DSP	1. Digital Signal Processor (chip)
	2. Display Systems Protocol
	3. Domain-Specific Part (ATM)
DSR	1. Data-Set-Ready (RS 232 signal)
	2. Data Sampling Rate
	3. Data Signaling Rate

4. **D**igital **S**atellite **R**adio (broadcasting)

5. **D**isengagement **S**uccess **R**atio (fiber optics)

DSRC   **D**edicated **S**hort-**R**ange **C**ommunications

DSRI   **D**estination **S**tation **R**outing **I**ndicator

DSRR   **D**igital **S**hort-**R**ange **R**adio

DSS

1. **D**igital **S**witched **S**ervice

2. **D**igital **S**ignature **S**tandard

3. **D**igital **S**ignaling **S**ystem

4. **D**igital **S**atellite **S**ystem (STB hardware)

5. **D**irect **S**tation **S**elector (PBX hardware)

6. **D**irect **S**atellite **S**ystem

7. **D**ecision **S**upport **S**ystem (set of programs)

DSS1   **D**igital **S**ubscriber **S**ignaling **S**ystem **1**

DSS2   **D**igital **S**ubscriber **S**ignaling **S**ystem **2**

DSSCS   **D**efense **S**pecial **S**ervice **C**ommunications **S**ystem

DSSLL   **D**ifferentiated **S**ervices over **S**pecific **L**ink **L**ayer

DSSS   **D**irect **S**equence **S**pread **S**pectrum (transmission)

DSSSL   **D**ocument **S**tyle **S**emantics and **S**pecification **L**anguage (ISO)

DST

1. **D**irect **S**ounding **T**ransmission

2. **D**ispersion **S**upported **T**ransmission

DSTB   **D**igital **S**et-**T**op-**B**ox

DSTE

1. **D**igital **S**ubscriber **T**erminating **E**quipment

2. **D**ata **S**ubscriber **T**erminating **E**quipment

DSTN   **D**ouble-layer **S**uper-**T**wisted **N**ematic (display)

DSTO   **D**efense **S**cience and **T**echnology **O**rganization (Australia)

DSTP   **D**ouble-**S**hielded **T**wisted **P**air

DSTR   **distr**ibution

DSU   **D**ata **S**ervice **U**nit (networking)

DSUB   **D**ata **S**ervice (SCPC) **Sub**-band Signaling Channel

DSVD   **D**igital **S**imultaneous **V**oice and **D**ata (modems)

DSX   **D**igital **S**ignal **Cross-Connect** system

DT

1. **D**eutsche **T**elekom

2. **D**irect **T**ermination

DTAB   **D**emountable **T**ape-**A**utomated **B**onding (IC)

DTAD   **D**igital **T**elephone **A**nswering **D**evice

DTAN   **D**espun **T**orque **A**ctuated **N**utation (Intelsat)

DTAP   **D**irect **T**ransfer **A**pplication **P**art (GSM)

DTC

1. **D**igital **T**runk **C**ontroller

2. **D**igital **T**ransmit **C**ommand

3. **D**igital **T**raffic **C**hannel (cellular networks)

DTCM   **D**ifferential **T**rellis **C**oded **M**odulation

DTCP	Dual Tape-Carrier Package (microchips)
DTD	Document Type Definition (XML)
DTE	Data Terminal Equipment (networking)
DTF	1. Dynamic Tracking Filter
	2. Distance-To-Fault (microwave)
DTH	Direct To Home (broadcasting)
DTI	Digital Trunk Interface
DTL	1. Diode–Transistor Logic (digital electronics)
	2. Designated Transit List (ATM)
	3. Dial Tone Line
	4. Distance Teaching and Learning (e-learning)
DTL/TTL	DTL input to TTL output and vice versa (digital electronics)
DTLU	Digital Trunk and Line Unit
DTM	1. Dynamic synchronous Transfer Mode
	2. Digital Terrain Model (remote sensing)
DTMF	Dual-Tone Multifrequency (telephone sets)
DTMX	Digital Trunk Manual (telephony) Exchange
DTN	1. Data Transfer Network
	2. Data Transmission Network
DTO	1. Dial Tone Office
	2. Digital Test Oscilloscope
	3. Digitally Tuned Oscillator
	4. Direct Termination Overflow
DTP	1. Desktop Publishing (ICT)
	2. Distributed Transaction Processing (computer networks)
DTR	1. Data-Terminal-Ready (RS 232 signal)
	2. Data Transfer Rate
	3. Discrete Tone Relation
DTRS	1. Digital Trunked Radio System
	2. Digital Tape Recording System
	3. Digital Telemetry Recording System
	4. Data Transfer Request Signal
DTS	1. Digital Termination System (microwave)
	2. Digital Television Standard
	3. Digital Theater System (home audio systems)
	4. Distributed Time Service
	5. Decoding Time Stamp
DTSP	Digital Television Station Project, Inc.
DTSR	1. Dial Tone Speed Recording
	2. Digital Tape System Recording
	3. Digital Temporary Storage Recording
DTT	1. Digital Trunk Testing

2. **D**igital **T**ape **T**ransfer
3. **D**igital **T**ie **T**runk
4. **D**igital **T**elecommunication **T**runk
5. **D**igital **T**errestrial **T**elevision (broadcasting)

DTTB **D**igital **T**errestrial **T**elevision **B**roadcasting
DTTL **D**ata **T**ransition **T**racking **L**oop
DTU 1. **D**igital **T**est **U**nit
2. **D**igital **T**runk **U**nit
3. **D**igital **T**ransmission **U**nit
DTV **D**igital **T**ele**v**ision (broadcasting)
DTVC **D**igital **T**ele**v**ision by **C**able
DTVE **D**igital **T**rans**v**ersal **E**qualizer
DTX **D**iscontinuous **Transmission** (cellular networks)
DU 1. **D**ata **U**nit
2. **D**ispersion-**U**nshifted (optical fiber)
DUA **D**irectory **U**ser **A**gent (software)
DUAL **D**istributed **U**pdate **Al**gorithm
DUART **D**ual **U**niversal **A**synchronous **R**eceiver–**T**ransmitter
DUE **D**umb **U**ser **E**rror
DUN **D**ial-**U**p **N**etworking (PC communications)
DUP **D**ata **U**ser **P**art (SS7)
DUT **D**evice **U**nder **T**est (equipment)
DUV **D**ata **U**nder **V**oice (AT&T)
DV **D**igital **V**ideo
DVB **D**igital **V**ideo **B**roadcasting (organization)
DVB-C **DVB C**able transmission standard
DVB-H **DVB-H**andheld
DVB-IRD **DVB I**ntegrated **R**eceiver **D**ecoder
DVB-RCS **DVB-R**eturn **C**hannel by **S**atellite system
DVB-S **DVB S**atellite transmission standard
DVB-SI **DVB S**ervice **I**nformation
DVB-T **DVB T**errestrial (standard)
DVBG **D**igital **V**ideo **B**roadcasting **G**roup
DVC 1. **D**igital **V**ideo **C**ompression
2. **D**igital **V**ideo **C**assette
DVCD **D**ouble **V**ideo **C**ompact **D**isk
DVD 1. **D**igital **V**ideo **D**isk (ICT)
2. **D**igital **V**ersatile **D**isk (ICT)
DVD-E **D**igital **V**ideo **D**isk-**E**rasable
DVD-R **D**igital **V**ideo **D**isk-**R**ecordable
DVD-RAM **D**igital **V**ideo **D**isk-**R**andom **A**ccess **M**emory (rewritable)
DVD-ROM **D**igital **V**ideo **D**isk-**R**ead **O**nly **M**emory
DVD-R/W **D**igital **V**ideo **D**isk **R**e-**W**ritable

DVD2	Digital Video Disk 2nd Generation
DVI	1. Digital Video Interface (standard)
	2. Digital Video Interactive
DVM	1. Digital Voltmeter (test equipment)
	2. Data and Voice Multiplexer
	3. Data and Voice Multiplexing
DVMRP	Distance Vector Multicast Routing Protocol (IETF)
DVOM	Digital Volt–Ohm-Milliammeter (electronics)
DVP	Digital Video Processor
DVR	Digital Video Recorder
DVRN	Dense Virtual Routed Network
DVS	Digital Video Service
DVSI	Digital Voice Systems, Inc. (company)
DVST	Direct View Storage Tube
DVTS	Desktop Videoconferencing Telecommunications System
DWA	Data Warehousing Alliance (Microsoft)
DWC	Digital Wireless Communications
DWDM	Dense Wavelength-Division Multiplexing (fiber optics)
DWG	Digital Working Group
DWI	Data Warehousing Institute
DWMT	Discrete Wavelet Multitone
DWS	Dialable Wideband Service
DWT	Discrete Wavelet Transform (mathematics)
DX	1. Data Exchange
	2. Direct Current Signaling
	3. duplex
DXB	Drawing Interchange Binary (file extension)
DXC	Digital Cross-Connect (MUX)
DXer	A long-Distance listener of short wave radio stations (Ham Radio)
.dxf	Name extension for AutoCAD 2-D format files (graphics)
DXF	Drawing interchange Format (CAD files)
DXI	Data Exchange Interface
DXing	The hobby of receiving and listening to a long Distant short wave radio station (Ham Radio)
dyn	dyne (unit of force)

E e

e	1. Symbol for basic unit of **e**lectron charge
	2. Symbol for instantaneous value of an alternating voltage
	3. Symbol for **e**lectronic version
e-bomb	**e**-mail **bomb**
e-book	**e**lectronic **book**
e-business	**e**lectronic **business**
e-cash	**e**lectronic **cash** (e-commerce)
e-check	**e**lectronic **check** (e-commerce)
e-commerce	**e**lectronic **commerce**
e-credit	**e**lectronic **credit**
e-Fax	**e**lectronic **Fac**simile
e-form	**e**lectronic **form**
e-government	**e**lectronic **government**
e-law	**e**lectronic **law**
e-learning	**e**lectronic **learning** (e-learning)
e-mail	**e**lectronic **mail**
e-money	**e**lectronic **money**
e-rate	**e**lectronic **rate**
e-signature	**e**lectronic (digital) **signature**
e-text	**e**lectronic **text**
e-stamp	**e**lectronic **stamp**
e-zine	**e**lectronic Maga**zine**

E	1. Symbol for **E**mitter (transistor circuit diagrams)
	2. Symbol for **E**lectric field strength (antennas)
	3. Symbol for prefix **e**xa-, denoting 10^{18} or 2^{60}
E band	Optical Frequency **band** ranging from 1360 to 1460 nm
E Channel	Echo **Channel**
E Link	Extended **Link**
E Port	Expansion **Port**
E&I	**E**ngineering **and I**nstallation
E&M	"**E**ar **and M**outh" (signaling)
E-1	The **E**uropean equivalent of American T-**1** transmission line operating at 2.048 Mbps
E-2	The **E**uropean equivalent of American T-**2** transmission line operating at 8.448 Mbps
E-3	The **E**uropean equivalent of American T-**3** transmission line operating at 34.368 Mbps
E-BCCH	Extended-**B**roadcast **C**ontrol **Ch**annel
E-Carrier	European **Carrier** (ITU-T)
E-FET	**E**nhancement-mode **FET** (semiconductors)
E-IDE	**E**nhanced **IDE** (interface)
E-LMI	**E**nhanced **L**ocal **M**anagement **I**nterface (frame relay)
E-Nose	**E**lectronic **Nose**
E-O	**E**lectrical-**O**ptical
E-OTD	**E**nhanced **O**bserved **T**ime **D**ifference
E-TDMA	**E**xtended **TDMA** (access)
E-time	**E**xecution **time**
E/A	**E**vents **and A**larms subsystem (Intelsat)
E/S	**E**arth **S**tation (satcom)
E2PROM	**E**lectronically **E**rasable **P**rogrammable **ROM** (memory)
EA	1. **E**xtended **A**ddressing (frame relay)
	2. **E**qual **A**ccess
	3. **E**arly **A**ssignment
	4. **E**ffective **A**vailability
EAA	**E**lectronic **A**fzar **A**zma Company (Iran)
EAC	1. **E**uropean **A**stronaut **C**enter
	2. **E**ngineering **A**ssistance **C**ontract (Intelsat)
EACA	**E**uropean **A**ssociation of **C**ommunications **A**gencies
EACEM	**E**uropean **A**ssociation of **C**onsumer **E**lectronics **M**anufacturers
EADAS	**E**ngineering and **A**dministrative **D**ata **A**cquisition **S**ystem
EADP	**E**uropean **A**ssociation of **D**irectory and database **P**ublishers
EAE	**E**xtended **A**rithmetic **E**lement
EAEO	**E**nd **A**ccess **E**nd **O**ffice

EAFC	Enhanced Automatic Frequency Control (electronics)
EAGE	Electrical Aerospace Ground Equipment (Intelsat)
EAGLE	Extended Area GPS Location Enhancement
EAI	1. External Alarm Indicator
	2. Enterprise Application Integration
EAN	European Article Numbering (bar coding)
EANTC	European Advanced Networking Test Center (Germany)
EAP	Extensible Authentication Protocol (OSI model)
EAR	Echo Attenuation Ratio (transmission)
EARC	Extraordinary Administrative Radio Conference
EARN	European Academic and Research Network
EAROM	Electrically Alterable ROM (memory)
EARP	Ethernet Address Resolution Protocol
EARSeL	European Association of Remote Sensing Laboratories
EAS	1. Emergency Alert System (cable TV broadcast)
	2. Events and Alarms System (Intelsat)
	3. Extended Area Service (network feature)
	4. Electronic Acquisition System
	5. Electronic Articles Surveillance
	6. Engineering and Applied Sciences
	7. Enterprise Application Server
	8. European Astronomical Society
EASI	ETSI ATM Services Interoperability
EAX	Electronic Automatic Exchange
EB	1. Expansion Buffer
	2. Errored Block
	3. Exa-Byte, which is 2^{60} or 1024 times petabytes
Eb/No	Energy per bit-to-Noise power spectral density (ratio)
EBA	1. European Broadcasting Area
	2. Eventual Byzantine Agreement
EBAM	Electron-Beam-Accessed Memory
EBCDIC	Extended Binary-Coded Decimal Interchange Code (computer)
EBD	Effective Bill Date
EBDI	Electronic Business Data Interchange
EBE	Embedded Base Equipment (Bell)
EBGA	Enhanced Ball-Grid Array (microchips)
EBHC	Equated Busy Hour Call (Intelsat)
EBICON	Electron Bombardment Induced Conductivity
EBIOS	Enhanced BIOS (computer)
EBITDA	Earnings Before Interest, Taxes, Depreciation and Amortization

EBL	1. Electronic **B**earing **L**ine (navigation)
	2. Electron **B**eam **L**ithography (semiconductors)
EBO	Embedded **B**ase **O**rganization (Bell)
EBONE	European network back**bone** (service)
EBPP	Electronic **B**ill **P**resentation and **P**ayment
EBR	Electronic **B**eam **R**ecorder
EBS	1. Emergency **B**roadcast **S**ystem (cable TV)
	2. Electron-**B**ombarded **S**emiconductor (device)
EBSD	Electron **B**ack-**S**catter **D**iffraction (microwave)
EBU	European **B**roadcasting **U**nion
EBX	Electronic **B**ranch **E**xchange
EC	1. **E**xchange **C**arrier
	2. **E**lectronic **C**arrier
	3. **E**nergy **C**ommunications (PBX feature)
	4. **E**lectro-optic **C**oefficient (fiber optics)
	5. **E**lement **C**ontroller
	6. **E**quipment **C**lock
	7. **E**nd **C**hain
	8. **E**cho **C**anceller
ECA	**E**xchange **C**arrier **A**ssociation
ECAC	**E**uropean **C**ivil **A**viation **C**onference
ECB	1. **E**rror **C**orrection **B**it
	2. **E**nhanced **C**ellular **B**ase Station
ECC	1. **E**rror **C**hecking and **C**orrection code
	2. **E**mbedded **C**ontrol **C**hannel (MUX)
	3. **E**lectronic **C**ommon **C**ontrol
	4. **E**lectronic **C**ommerce **C**anada Inc.
	5. **E**lectronically **C**ontrolled **C**oupling
	6. **E**mergency **C**ommunications **C**enter
	7. **E**xchange **C**arrier **C**ode (Telcordia Technologies)
	8. **E**lliptic **C**urve **C**ryptography (wireless)
	9. **E**gypt **C**yber **C**enter
ECC RAM	**E**rror **C**orrecting **C**ode **RAM** (memory)
ECCA	**E**uropean **C**able **C**ommunications **A**ssociation (Belgium)
ECCKT	**E**xchange **C**ompany **C**ircuit
ECCM	1. **E**lectronic **C**ounter-**C**ounter**m**easure (interference)
	2. **E**lectromagnetic **C**ounter-**C**ounter**m**easure (interference)
ECCO	**E**quatorial **C**onstellation **C**ommunications **O**rganization
ECDIS	**E**lectronic **C**hart **D**isplay Information **S**ystem
ECF	**E**mbedded **C**ode **F**ormatting
ECG	**E**lectro**c**ardio**g**ram (medicine)

ECGA	Enhanced Color Graphics Adapter (640×400 pixel monitors)
ECH	1. Enhanced Call Handling
	2. Echo-Cancelled Hybrid (DSL)
ECHO	European Commission Host Organization
ECI	1. End Chain Indicator
	2. External Call Interface
ECIS	1. European Committee for Interoperable Systems
	2. European Computer Industry research Center
	3. European Conference on Information Systems
ECITC	European Committee for Information Technology and Certification
ECL	Emitter-Coupled Logic (digital electronics)
ECM	1. Echo Canceller Module
	2. Error Correction Mode (G3 fax machine)
	3. Electronic Countermeasures (radar)
	4. Electromagnetic Countermeasures (interference)
	5. Entitlement Control Message (broadcasting)
ECMA	European Computer Manufacturers Association (Switzerland)
ECMEA	Enhanced Cellular Messaging Encryption Algorithm
ECMP	Equal Cost Multiple Routing (traffic)
ECMR	Equal Cost Multipath Routing
ECMS	Egyptian Company for Mobile Services (operator)
ECN	1. Electronic Communications Network
	2. Emergency Call Network
	3. Explicit Congestion Notification (IP standard)
ECO	Electron-Coupled Oscillator (electronics)
ECOC	European Conference on Optical Communication
ECOMEG	E-Commerce Experts Group
ECP	1. Enhanced Call Processing (voice mail)
	2. Extended Capabilities Port (computer)
	3. Encryption Control Protocol
	4. Electronic Commerce Platform
	5. Executive Cellular Processor
ECPA	Electronic Communications Privacy Act
ECPE	Embedded Customer-Premises Equipment (telephone company)
ECS	1. European Communication Satellite
	2. Electronic Commerce Service
	3. Electronic Chart System (navigation)
ECSA	Exchange Carriers Standards Association
ECSD	Enhanced Circuit-Switched Data
ECSSB	Exalted Carrier SSB (modulation)

E

ECT	1. Enterprise Caching Technology
	2. Explicit Call Transfer
ECTA	European Competitive Telecommunications Association
ECTEL	European Telecommunications and Professional Electronics Industry
ECTF	1. Enterprise Computer Telephony Forum (U.S.A.)
	2. European Community Telework Forum
ECTRA	European Committee for Telecommunications Regulatory Affairs
ECTUA	European Council of Telecommunications Users Association
ED	1. Electronic Directory
	2. Ending Delimiter (LANs)
ED RAM	Enhanced Dynamic RAM (memory)
EDA	1. Electronic Design Automation (ICT)
	2. Electronic Directory Assistance
	3. Erbium-Doped Amplifier (fiber optics)
EDAC	Error Detection and Correction (transmission)
EDACS	Enhanced Digital Access Communications System
EDAM	Enhanced Digital Announcement Machine
EDC	1. Error Detection and Correction
	2. Error Detection Code (transmission)
EDCH	Enhanced D-Channel Handler
EDD	Enhanced Disk Drive
EDDA	European Digital Dealers Association
EDF	Erbium-Doped Fiber
EDFA	Erbium-Doped Fiber Amplifier (fiber optics)
EDGAR	Electronic Data Gathering Archiving and Retrieval
EDGE	Enhanced Data-Rates for GSM Evolution
EDH	Electronic Document Handling
EDI	Electronic Data Interchange (ICT)
EDIA	Electronic Data Interchange Association
EDIDAG	Electronic Data Interchange of Dangerous Goods information
EDIF	Electronic Data Interchange Format (ICT)
EDIFACT	EDI for Administration, Commerce And Transport (ICT)
EDIMAR	Electronic Data Interchange of Maritime documents (ICT)
EDIS	Emergency Digital Information System/Service
EDIU	Expansion Digital Interface Unit
EDL	Environment Data Link
EDLIN	Editor of Lines (outdated MS-DOD text editor)

EDM	1. Electronic Document Management
	2. Electronic Discharge Machining
EDMS	Engineering Document Management System (ICT)
EDO	1. Extended Data Output (memory)
	2. Enhanced Data Output (memory)
	3. Equipment Design Objectives
EDO DRAM	Extended Data Output DRAM (memory)
EDO RAM	Extended Data Output RAM (memory)
EDP	1. Electronic Data Processing (computer)
	2. Exchange Delivery Point
EDPS	Electronic Data Processing System
EDRS	European Data Relay Satellite
EDS	Extended Data Service
EDSAC	Electronic Delay Storage Automatic Computer
EDSL	Extended DSL (access)
EDSS1	European Digital Signaling System No. 1
EDTR	Effective Data Transfer Rate (transmission)
EDTV	1. Enhanced-Definition Television (broadcasting)
	2. Extended-Definition Television (broadcasting)
EDVAC	Electronic Discrete Variable Automatic Computer
EE	End-to-End signaling
EEC	1. Error Checked and Corrected
	2. Electromagnetic Emission Control
EEG	Electroencephalogram (medical equipment)
EEHLLAPI	Entry Emulator High-Level Language Applications Programming Interface
EEHO	Either End Hop Off
EEI	1. External Environment Interface
	2. Edison Electric Institute
EEL	Enhanced Extended Link
EEMA	European Electronic Messaging Association
EEMS	Enhanced Expanded Memory Specification (memories)
EEP	1. Electroencephalophone
	2. Electromagnetic Emission Policy
EEPG	Enhanced Electronic Program Guide
EEPROM	Electronically Erasable Programmable ROM (memory)
EER	Effective Earth Radius
EES	1. Earth Exploration Satellite
	2. Escrow Encryption Standard (U.S. Department of Justice)
EESN	Expanded Electronic Serial Number
EESS	Earth Exploration Satellite Service (ITU)

E

EESSI	European Electronic Signature Standardization Initiative (ICT)
EETDN	End-to-End Transit Delay Negotiation
EF	Entrance Facility
EF&I	Engineer, Furnish, and Install (purchase method)
EFCI	Explicit Forward Congestion Indicator (ATM)
EFD	Event Forwarding Discriminator (wireless)
EFF	Electronic Frontier Foundation (standards organization)
EFI&T	Engineer, Furnish, Install, and Test
EFIS	Electronic Flight Information System
EFL	1. Emitter-Follower Logic (digital electronics)
	2. Equivalent Focal Length
EFMA	Ethernet in the First Mile Alliance
EFOC	European Fiber Optics and Communications (conference)
EFR	Enhanced Full Rate (codec)
EFS	1. Error Free Seconds
	2. Enhanced Full-rate Service
	3. Electronic Filing System (Intelsat)
EFT	Electronic Funds Transfer (ICT)
EFTA	European Free Trade Association (Geneva)
EFTPOS	Electronic Funds Transfer at Point Of Sale (ICT)
EFTS	Electronic Funds Transfer System
EG	European Graphics Association
EGA	Enhanced Graphics Adapter (640×350 pixel monitors)
EGC	1. Enhanced Group Calling (Inmarsat)
	2. Equal Gain Combiner
EGI	Embedded Global Positioning Satellite Inertial Navigation System
EGM	Extraordinary General Meeting (Intelsat)
EGNOS	European Geostationary Navigation Overlay Service (satellite)
EGP	Exterior Gateway Protocol (routing)
EGPRS	Enhanced General Packet Radio Service
EGSM	Extended (frequency range) **GSM**
EGW	Edge Gateway
EHF	1. Extremely High Frequency (30–300 GHz range)
	2. Error Hold File
EHR	Enhanced Half-Rate codec
EHT	Extremely High Tension
EHTPS	Extremely High-Tension Power Supply
EHV	Extra-High Voltage (power)
EHz	Exa-Hertz (10^{18} Hz)

EI	Error Interval
EIA	Electronic Industries Alliance (standards organization)
EIAJ	Electronic Industries Association of Japan
EICS	Emergency Integrated Communication System
EICTA	European Information and Consumer Electronics Technology Industry Association
EID	Equipment Identifier (mobile radio)
EIDE	Enhanced Integrated Drive Electronics (interface)
EIDQ	European International Directory Enquiries
EIF	Electronic Industries Foundation
EIG	Electronic Information Group
EIGRP	Enhanced Interior Gateway Routing Protocol (Cisco)
EIIA	Embedded Industrial Internet Appliance
EIM	1. Ethernet Inverse Mapper
	2. Emirates Internet & Multimedia (U.A.E.)
EIMF	European Interactive Media Federation
EIN	European Information Network
EIP	Early Implementers Program (Novell, Inc.)
EIPA	1. Electronic Information and Communication for Pedagogical Academies (Austria)
	2. European Information Provider Association
EIR	1. Excess Information Rate (frame relay)
	2. Equipment Identity Register (GSM)
EIRENE	European Integrated Railway Radio Enhanced Network
EIRP	Effective Isotropically Radiated Power (microwave)
EIS	1. Executive Information System (ICT)
	2. Expanded Interconnection Service
	3. Epidemic Intelligence Service
EISA	Extended Industry-Standard Architecture (PC bus)
EIT	Event Information Table (broadcasting)
EITA	European Information Technology Association (U.K.)
EIU	Ethernet Interface Unit
EIUF	European ISDN Users' Forum
EJTAG	Enhanced JTAG
EKE	1. Encrypted Key Exchange (protocol)
	2. Electronic Key Exchange
EKMS	Electronic Key Management System
EKTS	Electronic Key Telephone Service (ISDN)
EL	1. elevation
	2. Electro Luminescent
ELAN	Emulated LAN (ATM)
ELE	Extremely Low-frequency Emission (monitors)
ELEC	Enterprise Local Exchange Carrier

E

ELED	Edge-emitting Light-Emitting Diode
ELF	Extremely Low Frequency (30–300 Hz range)
ELFEXT	Equal Lever For End Cross-Talk
ELG	European Launching Group (broadcasting)
ELINT	Electronic Intelligence
ELIU	Electrical Line Interface Unit
ELOT	Greek Organization for Standardization (Greece)
ELS NetWare	Entry Level System NetWare (networking)
ELSEC	Electronic Security
ELSR	Edge Label Switch Router
ELSSE	Electronic Sky Screen Equipment
ELSU	Ethernet LAN Service Unit
ELT	Emergency Locator Transmitter (COMSAR)
EM	1. "End of Medium" (character control code)
	2. Engineering Model
	3. Element Manager
	4. Expanded Memory
EMA	1. Electronic Messaging Association (U.S.A.)
	2. Electronic Missile Acquisition
EMACS	Editor MACros (text editor)
EMAG	ETSI MIS Advisory Group
Email	Electronic mail
EMBARC	Electronic Mail Broadcast to A Roaming Computer (Motorola)
EMC	Electromagnetic Compatibility (interference)
EMCON	Emission Control (propagation)
EMD	1. Equilibrium Mode Distribution (propagation)
	2. Electromechanical Dialing (telephone exchange)
EME	1. Electromagnetic Energy
	2. Electromagnetic Environment
	3. Externally Mounted Equipment
	4. Earth-Moon-Earth (radio communications)
EMEA	Europe, Middle East and Africa (region)
.emf	Name extension for Enhanced Metafile format (graphics)
EMF	1. Electromotive Force (parameter)
	2. Electromagnetic Force (parameter)
	3. Element Management Function
	4. Enhanced Metafile (printer file)
	5. European Multimedia Forum
EMG	Electromyogram (medicine)
EMI	1. Electromagnetic Interference (noise)
	2. Exchange Message Interface
EML	Element Management Layer (networking)

EMM	1. **E**ntitlement **M**anagement **M**essage (broadcasting)
	2. **E**xpanded **M**emory **M**anager (device driver)
EMMA	**E**nhanced **M**ulti**m**edia **A**rchitecture (broadcasting)
EMMSEC	**E**uropean **M**ultimedia **M**icroprocessor **S**ystems and **E**lectronic **C**ommerce (ICT)
EMOS	**E**quipment **M**anagement **O**perating **S**ystem
EMP	**E**lectro**m**agnetic **P**ulse
EMPD	**E**quilibrium **M**odal-**P**ower **D**istribution (propagation)
Emph	**Emph**asis (FM transmission)
EMPS	**E**lectronic **M**obile **P**ayment **S**ervices (project)
EMR	1. **E**xchange **M**essage **R**ecord (Telcordia Technologies)
	2. **E**lectro**m**agnetic **R**adiation (hazard)
EMRP	**E**ffective **M**onopole-**R**adiated **P**ower (Intelsat)
EMS	1. **E**xpanded **M**emory **S**pecification (standard)
	2. **E**uropean **M**obile **S**atellite
	3. **E**lement **M**anagement **S**ystem
	4. **E**lectronic **M**essage **S**ervices
	5. **E**nterprise **M**essaging **S**erver
	6. **E**nhanced **M**essage **S**ervice (GSM)
	7. **E**lectro**m**agnetic **S**usceptibility
EMT	**E**lectrical **M**etal **T**ubing
EMTUG	**E**uropean **M**anufacturing **T**echnology **U**ser **G**roup
EMU	**E**lectro**m**agnetic **U**nit (parameter)
EMUG	**E**uropean **MAP** **U**sers **G**roup
EMUT	**E**nhanced **M**anpack **U**ltrahigh-frequency **T**erminal
EMV	**E**lectro**m**agnetic **V**ulnerability
EMW	1. **E**lectro**m**agnetic **W**ave
	2. **E**lectro**m**agnetic **W**arfare
EN	**E**quivalent **N**ode
ENA	1. **E**nterprise **N**etwork **A**ddressing
	2. **E**xtended **N**etwork **A**ddressing
ENC	1. **E**lectronic **N**avigation **C**hart
	2. **enc**oder
ENCB	**E**stonian **N**ational **C**ommunication **B**oard
Endec	**En**coder–**dec**oder
ENET	1. **E**ther**net**
	2. **E**nhanced **Net**work
ENFIA	**E**xchange **N**etwork **F**acility for **I**nterstate **A**ccess
ENG	**E**lectronic **N**ews **G**athering
ENI	**E**mbedded **N**etwork **I**nterface
ENIAC	**E**lectronic **N**umerical **I**ntegrator **A**nd **C**omputer
ENID	**E**GC (closed-) **N**etwork **Id**entification code (Inmarsat)
ENN	**E**mergency **N**ews **N**etwork
ENOB	**E**ffective **N**umber **O**f **B**its (ADC)

E

ENOS	Enterprise Network Operating System (Sun Microsystems)
ENQ	"**enq**uiry" (character control code)
ENR	Excess Noise Ratio
ENS	1. Emergency Number Service
	2. Enterprise Network Services (software)
ENSO	ETSI National Standardization Organization
ENT	Equivalent Noise Temperature
ENTELEC	Energy Telecommunications and electrical Association
ENUM	Electronic Number
EO	1. End Office (networking)
	2. Erasable Optical drive
EOA	End-Of-Address (signal)
EOB	End-Of-Block (signal)
EOC	1. Embedded Overhead Communication Channel (transmission)
	2. Electro-Optic Coefficient
EOD	End-Of-Data (signal)
EOE	Electronic Order Exchange
EOF	End-Of-File (control character)
EOL	1. End-Of-Line (control character)
	2. End-Of-Orbit life (satcom)
EOP	1. End-Of-Procedure frame
	2. End-Of-Program
EOS	1. Earth Observation System (remote sensing)
	2. Earth Observation Satellite
	3. End-Of-Selection (control character)
	4. Electro-Optical System
EOT	1. End-Of-Transmission (control signal)
	2. End-Of-Tape (control signal)
	3. End-Of-Text (flag or marker)
EOTC	1. European Organization for Testing and Certification
	2. Electro-Optic Technology Center
EOW	Engineering Order-Wire (MUX)
EP-DVB	European Parliament Digital Video Broadcasting
EPABX	Electronic PABX
EPAC	Enhanced Perceptual Audio Coder
EPC	Electric Power Conditioner
EPD	Early Packet Discard (ATM)
EPF	Electronic Payment Forum (www.epf.net)
EPFD	1. Electronic Position-Fixing Device (GPS)
	2. Equivalent Power Flux Density (microwave)
EPG	Electronic Program Guide (broadcasting)
EPI	Electronic Position Indicator

EPIC	1. Electronic Privacy Information Center
	2. Explicitly Parallel Instruction Computing (ICT)
	3. European Project on Information infrastructure Co-ordination
EPIISG	European Project on Information Infrastructure Starter Group
EPIRB	Emergency Position-Indicating Radio Beacon (COMSAR)
EPLANS	Engineering, Planning and Analysis System
EPLD	Electronically Programmable Logic Device (digital electronics)
EPLSR	Enhanced Position Location Reporting System
epndB	effective perceived noise decibel
EPN	1. Expansion Port Network
	2. Esteeman Paging Network (Iran)
EPOC	Electronic Piece Of Cheese (operating system)
EPOS	Electronic Point Of Sale (ICT)
EPP	1. Enhanced Parallel Port (computer)
	2. European Polar Platform
EPPA	European Public Paging Association
EPPI	Expanded Plan-Position Indicator
EPR	Electron Paramagnetic Resonance (electronics)
EPRI	Electric Power Research Institute
EPRML	Extended Partial Response/Maximum Likelihood
EPROM	Erasable Programmable ROM (memory)
.eps	Name extension for Encapsulated PostScript files (graphics)
EPS	1. Encapsulated Post-Script (image format)
	2. Endless Phase Shifter
EPSCS	Enhanced Private-Switched Communications Service
EPSF	Encapsulated Post-Script File
EPSN	Enhanced Private-Switched Network
EPU	Expanded Processing Unit
EQ	Equalizer, Equalization
EQL	equalizer
EQTV	Extended Quality Television
ER	1. Error Ratio
	2. Equipment Room
	3. Explicit Rate (ATM)
ERA	European Research Area
ERB	Emergency Radio Beacon
ERC	1. European Radiocommunications Committee
	2. Electromagnetic Radiation Control
ERD	Entity Relationship Diagram (ICT)

E

ERL	Echo Return Loss (transmission)
ERM	1. Enterprise Resource Management
	2. Entity Relationship Model (ICT)
ERMA	1. Electronic Recording Method Accounting
	2. Electronic Recording Machine Accounting
ERMES	Enhanced Radio Messaging System (ETSI)
ERN	Early Routing Node
ERO	European Radiocommunications Office (CEPT)
ERP	1. Effective Radiated Power (antenna parameter)
	2. Enterprise Resource Planning (ICT)
ERS	3. European Remote-sensing Satellite
ERSS	European Remote Sensing Satellite
ERTMS	The European Railway Traffic Management System
ERTS	Earth Resources Technology Satellite (remote sensing)
Es/No	Energy **per s**ymbol **to N**oise power spectral density (ratio)
ES	1. Earth Station
	2. Electronic Switch
	3. Expert Systems
	4. Errored Seconds (MUX)
	5. End System (ATM)
	6. Exchange Service (BellSouth)
	7. Enhanced Services
ES-IS	End System to Intermediate System (protocol)
ESA	1. European Space Agency (France)
	2. Emergency Stand-Alone
	3. Electronic Spectrum Analyzer
	4. Electronically Steered Array (microwave antennas)
ESAS	Electronic Service Activation System (Inmarsat)
ESBAR	Epitaxial Schottky Barrier (diode)
Esc	1. Escape key (PC keyboard)
	2. Engineering Service Circuits
ESC	Egyptian Space Communications (Egypt)
ESCA	1. European Speech Communication Association
	2. Electron Spectroscopy for Chemical Analysis
ESCD	Extended System Configuration Data (computer BIOS)
ESCON	Enterprise System Connectivity (service)
ESD	1. Electrostatic Discharge
	2. Electrostatic Sensitive Device
	3. Electronic Software Distribution
	4. Error Signal Degrade (MUX)
ESDA	Electronic System Design Automation (ICT)
ESDI	Enhanced Small Device Interface (hard disk)
ESDRAM	Enhanced Synchronized Dynamic RAM (memory)

ESF	1. Extended Super-Frame format (T-1 line)
	2. European Science Foundation
ESFMU	Extended Super-Frame Monitoring Unit
ESG	Event Schedule Guide (broadcasting)
ESH	End System Hello packet
ESI	1. Enhanced Serial Interface (computer)
	2. End System Identifier (ATM)
	3. Equivalent Step Index (profile)
	4. Environmental Sensing Instrument
ESL	Equivalent Series **inductance** (electronics)
ESLNT	Equivalent Satellite Link Noise Temperature
ESM	1. Extended Subscriber Module
	2. Electronic warfare Support Measure
	3. Electronic Support Measure
	4. Element and Subnetwork Manager
ESMA	Extended Subscriber Module-100 **A**
ESMR	Enhanced Specialized Mobile Radio (cellular networks)
ESMTP	Extended Simple Mail-Transfer Protocol
ESMU	Extended Subscriber Module-100 URBAN
ESN	1. Electronic Security Number (GSM)
	2. Emergency Services (telephone) Number
	3. Electronic-Switched Network
ESnet	Energy Science **net**work
ESNUG	E-mail Synopsis User Group
ESOs	European Standardization Organizations
ESOC	European Space Operations Center
ESP	1. Enhanced Serial Port (computer)
	2. Enhanced Service Provider
	3. Encapsulating Security Payload
	4. Ethernet Service Provider
	5. Earth Surface Potential
ESPA	1. European Selective Paging Association
	2. Educational Software Publishers Association (e-learning)
ESPAN	Enhanced Switch Port Analyzer
ESPRIT	European Strategic Program for Research in Information Technology
ESQ	End System Query packet
ESR	1. Equivalent Series Resistance (electronics)
	2. Electronic Spin Resonance (microwave)
	3. Errored Seconds Ratio

E

ESS	1. Electronic Switching System (Lucent Technologies)
	2. Event Synchronization System
EST	Electroshock Therapy
ESTEC	European Space Research and Technology Center (Netherlands)
ESTO	European Science and Technology Observatory
ESU	Electrostatic Unit (parameter)
ESZ	Emergency Service Zone
ET	Exchange Termination
ETA	Electronic Technicians Association
ETACS	Extended TACS (British cellular)
ETB	1. End of Transmission Block (control character)
	2. Electronic Term Book
	3. Electronic Test Bed
	4. Engineers Tool Box
ETC	1. Enhanced Throughput Cellular (protocol)
	2. Earth Terminal Complex
	3. Electronic Toll Collecting
ETCP	Electrolytic Tough-pitch Copper (wires)
ETDM	Electrical Time-Division Multiplexing
ETDMA	Extended TDMA (access)
ETE	Equivalent Telephone Erlangs
ETF	European Telecsonferencing Federation
ETFTP	Enhanced Trivial FTP (networking)
EtherLEC	Ethernet Local Exchange Carrier
ETI	Electronic Telephone Interface
ETIS	European Telecommunication Informatics Services
ETITO	Electro-Technology Industry Training Organization
ETL	Embedded Transmission Line (microwave)
ETM	1. Electronic Ticketing Machine
	2. Enhanced Thermic Mapper (remote sensing)
ETN	Electronic Tandem Network
ETNOA	European Telecommunications Network Operations Association
ETNS	Emergency Telephone Number System
ETO	European Telecommunications Office
ETPI	Eastern Telecommunications of Philippines Inc.
ETR	1. ETSI Technical Report
	2. Effective Transmission Rate
	3. Estimated Time of Repairing
ETRA	Egyptian Telecommunications Regulatory Authority
ETRF	Electrically Tuned RF (filter)
ETRI	Electronics & Telecommunications Research Institute (Korea)

ETS	1. **E**uropean **T**elecommunication **S**tandard (GSM)
	2. **E**uropean **T**echnology **S**ervices
	3. **E**ngineering **T**est **S**atellite (Japan)
	4. **E**ffective **T**ransmission **S**peed
	5. **ETS**I **T**echnical **S**pecification
	6. **E**lectronic **T**andem **S**witching
	7. **E**thernet **T**erminal **S**erver
	8. **E**lectronic **T**echnology **S**ystems
ETSA	**E**uropean **T**elecommunication **S**ervices **A**ssociation
ETSI	**E**uropean **T**elecommunications **S**tandardization Institute (France)
ETT	**E**lectro **T**hermal **T**hruster
ETTM	**E**lectronic **T**oll and **T**raffic **M**anagement
ETV	1. **E**ducational **TV** (e-learning)
	2. **E**uropean **T**raining **V**illage (e-learning)
ETX	**E**nd-of-**Transmission** (control character)
EU	**E**uropean **U**nion
EUC	**E**xtended **U**NIX **C**ode
EUCL	**E**nd **U**ser **C**ommon **L**ine charge
EUI-48	**E**xtended **U**nique **I**dentifier-**48** (bits)
EUI-64	**E**xtended **U**nique **I**dentifier-**64** (bits)
EULA	**E**nd-**U**ser **L**icense **A**greement (softwares)
Eumetsat	**Eu**ropean **Met**eorological **Sat**ellite (organization)
EUNet	**E**uropean **U**NIX **Net**work
EURESCOM	**Eu**ropean institute for **Re**search and **S**trategic studies in Tele**com**munications
Euro ISDN	**Euro**pean **ISDN**
EUROBIT	**Euro**pean Association of **B**usiness Machine Manufacturers and **I**nformation **T**echnology
EUROCAE	**Eur**opean **O**rganization for **C**ivil **A**viation Electronics
EUT	**E**quipment **U**nder **T**est
EUTOTELDEV	**Euro**pean **Tel**ecommunication **Dev**elopment
Eutelsat	**Eu**ropean **Tel**ecommunications **Sat**ellite (organization)
EUTP	**E**nhanced **U**nshielded **T**wisted **P**air
EUV	**E**xtreme **U**ltra-**V**iolet
eV	**e**lectron**v**olt (unit of energy)
EV	**E**uropean **V**ideotelephony
EVPRT	**Ev**ent **Pr**inter
EV-DO	**Ev**olution **D**ata **O**nly services
EVA	**E**lectronic **V**alue **A**dded
EVC	**E**nhanced **V**ideo **C**onnection (standard)
EVDV	**Ev**olution **D**ata **V**oice
EVM	**E**rror **V**ector **M**agnitude
EVR	**E**lectronic **V**ideo **R**ecording

E

EVRC	Enhanced Variable Rate voice Coder
EW	Electronic Warfare (interference)
EWOS	European Workshop for Open Systems
EWP	Electronic White Pages
EWSD	Elektronisches Wahl System Digital (Siemens Switching system)
EWSM	Electronic Warfare Support Measure
EWTIS	European Waters Traffic Information System
EXCA	Exchangeable Card Architecture
EXCSA	Exchange Carrier Standards Association
.exe	Filename extension for executables programes (computer)
EXM	Exit Message
EXOR	Exclusive OR (digital gate)
EXOS	1. Extension Outside
	2. Exospheric Satellite
EXP	expansion
ext	external
EXT	extension
EXTN	extension
EXZ	Excessive Zeros (data encoding)
EYP	Electronic Yellow Pages
EZ	East Zone beam

F f

f	1. Symbol for prefix **f**emto-, denoting 10^{-15}
	2. Symbol for **F**requency
F	1. Symbol for **F**arad (unit of capacitance)
	2. Symbol for **F**ilament of tubes (electronics)
	3. Symbol for **F**ocal length
	4. Symbol for **F**ixed service (ITU-T)
F key	Function **key**
F Port	Fabric **Port**
F-BCCH	Fast Broadcast Control Channel
F-ES	Fixed End System
F/D	Focal Length-**to-D**iameter (ratio)
F/O	Fiber Optics
F/V	Frequency-**to-V**oltage (converter)
F/V-V/F	Frequency-**to-V**oltage **or V**oltage-**to-F**requency (converter)
F33	Fleet **33** (Inmarsat terminal)
F55	Fleet **55** (Inmarsat terminal)
F77	Fleet **77** (Inmarsat terminal)
fA	**f**emtoampere
FA	Fault Alarm
FAA	1. Federal Aviation Administration
	2. Frequency Assignment Authority
FAC	1. Forced Authorization Code
	2. Final Assembly Code (GSM mobile station)

FACC	Ford Aerospace & Communications Corporation (company)
FACCH	Fast Associated Control Channel (GSM)
FACOM	Fully Automatic Computer
FACS	Facilities Access Control System
FACT	1. Fabrication And Continuity Test Program (Intelsat)
	2. Fairchild Advanced CMOS Technology (electronics)
FACTE	Fujitsu Access & Transport Equipment
FAD	Function Access Domain
FADS	Force Administration Data System
FAL	Frame Alignment Loss (MUX)
FALU	Floating-point Arithmetic Logic Unit (computer)
FAM	Fast Access Memory
FAMA	Fixed-Assigned Multiple Access
FAN	1. Flexible Access Network
	2. Far Area Network
FANE	Flexible Access Network Element
FANP	Flow Attribute Notification Protocol (routing)
FANS	Future Air Navigation System (aviation)
fanzine	An online magazine for fans of a particular group of persons
FAP	1. Fuse Alarm Panel
	2. Formats And Protocols
FAQ	Frequently Asked Questions (ICT)
FAR	1. Federal Aviation Regulation
	2. False Alarm Rate
FARNet	Federation of American Research Network
FAS	1. Frame Alignment Signal (MUX)
	2. Frame Acquisition and Synchronization
	3. Facility-Associated Signaling
FASIC	Function and Algorithm-Specific Integrated Circuit
FASSET	Functional Advanced Satellite System for Evaluation and Test
FAST	Fairchild Advanced Schottky TTL (electronics)
FAT	1. File Allocation Table (computer)
	2. Factory Acceptance Tests
FATAM	File Application, Transfer, Access, and Management
FAU	Fixed Access Unit (wireless)
FAW	Frame Alignment Word (MUX)
FAX	short term for Facsimile Equipment/Machine
FB	1. Framing Bit
	2. Fine Business (Morse code transmissions)
	3. Frequency Correction Burst (GSM)
FBBC	Full Baud Bipolar Coding

FBC	Facility-**B**ased **C**arrier
FBG	Fiber **B**ragg **G**rating
FBO	Facilities-**B**ased **O**perator (VPNs)
FBSS	Full **B**and **S**pread **S**pectrum
FBT	Fused **B**iconic **T**aper
FBU	1. Functional **B**usiness **U**nit
	2. Failed **B**efore **U**tilization (Bellsouth)
FBus	Frame Transport **Bus**
FBW	1. Forward **B**and **W**orking (Intelsat)
	2. Fractional **B**and**w**idth
FBWA	Fixed **B**roadband **W**ireless **A**ccess
FBX	**F**ujitsu **B**roadband **Cross-connect** Node
FC	1. **F**eedback **C**ontrol
	2. **F**rame **C**ontrol (networking)
	3. **F**iber-optic **C**onnector (NTT)
	4. **F**iber **C**hannel
	5. **F**orecast **C**enter Station (Intelsat)
FC and PC	**F**ace **C**ontact **and** **P**oint **C**ontact
FC Switch	**F**iber **C**hannel **Switch**
FC-0	**F**iber **C**hannel level **0**
FC-1	**F**iber **C**hannel level **1**
FC-2	**F**iber **C**hannel level **2**
FC-3	**F**iber **C**hannel level **3**
FC-4	**F**iber **C**hannel level **4**
FC-AL	**F**iber **C**hannel-**A**rbitrated **L**oop (networking)
FC-EL	**F**iber **C**hannel-**E**nhanced **L**oop (networking)
FC-PGA	**F**lip **C**hip-**P**in **G**rid **A**rray (microchips)
FC-PH	**F**iber **C**hannel **Ph**ysical Standard
FCA	1. **F**iber **C**hannel **A**ssociation (U.S. standards organization)
	2. **F**lip-**C**hip **A**ssembly
	3. **F**ree **Ca**rrier
FCAPS	**F**ault, **C**onfiguration, **A**ccounting, **P**erformance, and **S**ecurity
FCB	**F**ile **C**ontrol **B**lock (MS-DOS)
FCBGA	**F**lip-**C**hip **B**all **G**rid **A**rray (microchips)
FCC	1. **F**ederal **C**ommunications **C**ommission (standards)
	2. **F**lexible **C**ontrol **C**able
FCCH	**F**requency **C**orrection **Ch**annel (GSM)
FCCRP	**F**ederal **C**ommunications **C**ommission **R**egistration **P**rogram
FCD	**F**acility **C**ompletion **D**ate
FCI	**F**orward **C**all **I**ndicator

F

FCIA	Fiber Channel Industry Association
FCIP	Fiber Channel over **IP** (Cisco)
FCLC	Fiber Channel Loop Community
FCN	function
FCOS	Feature Class Of Service
FCOT	Fiber Control Office Terminal
FCS	1. Frame Checking Sequence (transmission)
	2. Fraud Control System (MCI)
	3. Facility Control System
	4. Federation of Communication Services (U.K.)
	5. Fiber Channel Specification
FCSI	Fiber Channel System Initiative (IBM)
FCT	1. Fixed Cellular Terminal
	2. Fast CMOS Technology
FCW	1. Federal Computer Week (magazine)
	2. Fiber Cutoff Wavelength (fiber optics)
FD	Functional Description (Intelsat)
FDB	1. Fahrenheit Dry Bulb
	2. Functional Description Block (Intelsat)
FDC	1. Flight Dynamics and Commanding (Intelsat)
	2. Floppy Disk Controller (computer)
FDCC	Forward Control Channel
FDCCH	Forward Digital Control Channel (cellular networks)
FDD	1. Floppy Diskette Drive (computer)
	2. Frequency-Division Duplexing (transmission)
FDDI	Fiber Distributed Data Interface (LANs)
FDE	Frequency Domain Equalizer
FDF	Fiber Distribution Frame
FDFD	Finite Difference Frequency Domain
FDHM	Full Duration at Half Maximum
FDHP	Full Duplex Handshaking Protocol
FDI	Feeder Distribution Interface (switching)
FDL	Facilities Data Link (T-1)
FDM	Frequency-Division Multiplexing
FDMA	Frequency-Division Multiple Access
FDP	1. Fiber-optic Distribution Panel
	2. Fractional Degradation of Performance
FDPL	Foreign Data Processor Link (PBX)
FDR	Frequency Domain Reflectometry
	(microwave)
FDS	Frequency-Division Switching
FDTD	Finite Difference Time Domain
FDX	Full-Duplex (transmission)

FE	1. Functional Entity
	2. Format Effector (control character)
FEA	1. Field-Emitter Array (displays)
	2. Finite-Element Array (microchips)
FEBE	Far-End Block Error (MUX)
FEC	1. Forward Error Correction (transmission)
	2. Forwarding Equivalence Class
	3. Far-End Cross talk
	4. Future ECM Computer
	5. Flexible Etched Circuits
FECC	Federal Emergency Communications Coordinator
FECCOD	FEC Encoder
FECDEC	FEC Decoder
FECM	Future Entitlement Control Message
FECN	Forward Explicit Congestion Notification (frame relay)
FED	1. Field Emission Display (TV and monitor)
	2. Fire Emitting Diode
	3. Frequency Error Detector
FEFO	First Ended, First Out
FEI	Federation of the Electronics Industry (United Kingdom)
FEL	Free Electron Laser
FEM	Finite-Element Method
FEMA	Federal Emergency Management Agency
FEN	Feeder Echo Noise (antenna)
FENS	Fujitsu-Enhanced Network Management System
FEP	1. Front-End Processor (network computer)
	2. Fluorinated Ethylene Propylene-insulated wire
FER	Frame Error Rate
FERF	Far-End Remote Failure alarm (MUX)
FERL	Further Education Resource For Learning (ICT)
FERM	Forward Explicit Rate Management (ATM)
FES	1. Fixed Earth Station
	2. Frequency Exchange Signaling
	3. Field-Emission Spectroscopy (electronics)
	4. Fixed End System
FET	Field-Effect Transistor (semiconductors)
FETA	FET Amplifier
FEX	Foreign Exchange Trunk
FEXT	Far-End Cross-Talk (transmission)
FF	1. "Form Feed" (printer function)
	2. Flip-Flop
	3. Fast Forward
FFDI	Fast Fiber Data Interface

F

FFL	Front Focal Length (fiber optics)
FFM	Flat Fade Margin
FFO	Fixed-Frequency Oscillator
FFOL	Fiber Flow On LAN
FFP	Federation of Functional Processor
FFS	Flash File Standard
FFSK	Fast FSK (modulation)
FFT	1. Fast Fourier Transform (algorithm)
	2. Fast File Transfer
FFTDCA	Final-Form Text Document Content Architecture
FG	Functional Group (ATM)
FGA	Feature Group A
FGB	Feature Group B
FGC	Feature Group C
FGC-EA	Feature Group C and Equal Access
FGCPW	Finite Groundplane Coplanar Waveguide (microwave)
FGD	Feature Group D
FGD-EA	Feature Group D and Equal Access
FGDC	Federal Geographic Data Committee
FGM	Fixed Gain Mode
FGS	Felicity Grounding System
FH	Frequency-Hopping (transmission)
FH-CDMA	Frequency-Hopping—CDMA (access)
FHMA	Frequency-Hopping Multiple Access
FHSS	Frequency-Hopping Spread Spectrum (transmission)
FIA	Fiber-optic Industry Association
FIB	Forward Indicator Bit (SS7 message)
FIC	Flight Information Center (aviation)
FICON	Fiber Connectivity (IBM)
FID	Field Identifier (ISDN)
FIF	Fractal Image Format
FIFO	"First In, First Out" (buffer memories)
FIGS	Figure Shift
FILO	"First In, Last Out" (memory and data communication)
FIN	Field Inspection Notice
FIPS	Federal Information Processing Standards
FIR	1. Finite Impulse Response (filter)
	2. Fast Infrared (port)
FIRM	Functionally Integrated Resource Manager
FIRMR	Federal Information Resource Management Regulation
FIRST	Forum of Incident Response and Security Teams (organization)
FIS	Forms Interchange Standard
FISINT	Foreign Instrumentation Signal Intelligence

F

FISK	**F**ax a d**isk** (sending information on phone line)
FISO	**F**ast **I**n, **S**low **O**ut (memories)
FISU	**F**ill-**I**n **S**ignal **U**nit (SS7 message)
FIT	**F**ailures **I**n **T**ime
FITE	**F**orward **I**nterworking **T**elephony **E**vent
FITL	**F**iber-**I**n-**T**he-**L**oop
FIU	**F**acsimile **I**nterface **U**nit
FIX	**F**ederal **I**nternet E**x**change
FK	**F**oreign **K**ey
FL	1. **F**iber **L**ine
	2. **F**ault **L**ocating
FLAG	**F**iber-optic **L**ink **A**round the **G**lobe (Cable Consortium)
FLB	**F**acility **L**oop **B**ack
FLC	1. **F**iber **L**oop **C**arrier
	2. **F**erroelectric **L**iquid **C**rystal
FLCD	**F**erroelectric **L**iquid-**C**rystal **D**isplay
FLEC	**F**orward-**L**ooking **E**conomic **C**ost
FleetBB	**Fleet B**road-**B**and (Inmarsat)
.fli	Name extension for animation files in the **FLI** file format
FLINK	Combination of a **Fl**ash and a **W**i**nk** signal
FLIR	**F**orward-**L**ooking **I**nfra**r**ed unit
FLL	**F**iber-in-the-**L**oop-**L**ocal
FLM	**F**ujitsu **L**ightwave **M**ultiplexer
FLOP	**Fl**oating-point **Op**eration
FLOPS	**Fl**oating-point **O**perations **P**er **S**econd (computer speed)
FLR	**F**orward-**L**ooking **R**adar
FM	1. **F**requency **M**odulation
	2. **F**ull **M**odulation
	3. **F**ault **M**anagement
FM-band	**F**requency **M**odulation **band** ranging from 88 to 108 MHz
FM/AM	**A**mplitude **M**odulation of a carrier **by** subcarriers that are **F**requency-**M**odulated by information
FM/PM	**P**hase **M**odulation of a carrier **by** subcarriers that are **F**requency-**M**odulated by information
FMA	**F**ixed-**M**ount **A**ntenna
FMAS	**F**acility **M**aintenance and **A**dministration **S**ystem
FMB	**F**rame **M**arker **B**us
FMBS	**F**rame **M**ode **B**earer **S**ervice
FMC	1. **F**ixed **M**obile **C**onvergence (service)
	2. **F**orward **M**otion **C**ompensator (remote sensing)
FMCW	**F**requency-**M**odulated **C**ontinuous **W**ave (radar)

FMFB	Frequency Modulation Feedback
FMI	Fixed Mobile Integration
FMIC	Flexible MVIP Interface Circuit
FMIF	FM Improvement Factor
FMLB	First Make/Last Break (connectors)
FMS	1. Fleet Management Service
	2. Flight Management System
	3. Flexible Manufacturing System (ICT)
FMT	Fade Migration Technique
FMV	Fair Market Value
FN	Frame Number (GSM)
FNA	Functional Network Architecture
FNC	Federal Networking Council (NASA)
FNPA	Foreign Numbering Plan Area
FNPRM	Further Notice of Proposed Rule-Making (FCC)
FNR	1. Fixed Network Reconfiguration
	2. Faculty Network Resources
FNS	1. Fiber Network Systems
	2. Fiber Network Service
FO	Fiber Optics
FOA	1. Fiber-Optic Amplifier
	2. Fiber-Optic Association, Inc.
	3. First Office Application
FOB	Free Onboard (contracts)
FOBP	Fractional Out-of-Band Power
FOC	1. Firm Order Confirmation
	2. Fiber-Optic Combiner
	3. Fiber-Optic Connector
	4. Full Operational Compatibility
FOCC	Fiber-Optic Cable Component
FOCUS	Federation On Computing in the United States
FOD	Fax-On-Demand
FODA	FIFO-Ordered Demand Assignment
FODU	Fiber-Optic Distribution Unit
FOG	Fiber-Optic Around the Golf (cable consortium)
FOI Act	Freedom Of Information Act
FoIP	Fax over IP
FOIRL	Fiber-Optic Inter-Repeater Link (IEEE)
FOL	Fiber-Optic Link
FOM	Fiber-Optic Modem
FOMA	Freedom Of Multimedia Access
FORTRAN	Formula Translation (programming language)
FOSDIC	Film Optical Sensing Device for Input to Computer
FOSSIL	Fido/Opus/Seadog Standard Interface Layer

FOT	1. **F**iber-**O**ptic **T**ransmitter
	2. **F**requency **o**f **O**ptimum **T**ransmission
FOTM	**F**iber-**O**ptic **T**est **M**ethod
FOTP	**F**iber-**O**ptic **T**est **P**rocedure
FOTS	1. **F**iber-**O**ptic **T**ransmission **S**ystem
	2. **F**iber-**O**ptic **T**elemedicine **S**ystem
FOV	**F**ield **O**f **V**iew (remote sensing)
FOWM	**F**iber-**O**ptic **W**ell **M**onitoring
FOX	**F**iber-**O**ptic e**x**tender
FP	1. **F**eature **P**ackage (software)
	2. **F**ile **P**rocessor
	3. **F**loating **P**oint
FPBGA	**F**ine-**P**itch **B**all **G**rid **A**rray (microchips)
FPCB	**F**ield-**P**rogrammable **C**ircuit **B**oard
FPD	1. **F**ull **P**age **D**isplay
	2. **F**lat **P**anel **D**isplay
FPDL	**F**oreign **P**rocessor **D**ata **L**ink (Rockwell)
FPG	**F**eature **P**lanning **G**uide
FPGA	1. **F**ield-**P**rogrammable **G**ate **A**rray (digital electronics)
	2. **F**ine-**P**itch **G**rid **A**rray (microchips)
FPH	**F**ree **Ph**one (INs)
FPI	**F**ormal **P**ublic **I**dentifier
FPIC	**F**ield-**P**rogrammable **I**nterconnect **C**hip
FPID	**F**ield-**P**rogrammable **I**nterconnect **D**evice
FPIS	**F**orward **P**ropagation **I**onospheric **S**catter (fiber optics)
FPLA	**F**ield-**P**rogrammable **L**ogic **A**rray (digital electronics)
FPLMTS	**F**uture **P**ublic **L**and **M**obile **T**elecommunication **S**ystem (ITU)
FPM DRAM	**F**ast **P**age **M**ode **D**ynamic **RAM** (memory)
FPN	**F**loating-**P**oint **N**umber
FPODA	**F**ixed **P**riority-**O**riented **D**emand **A**ssignment (access technique)
FPP	1. **F**ixed **P**ath **P**rotocol
	2. **F**iber-optic **P**atch **P**anel
fps	**f**rames **p**er **s**econd (video images)
FPS	1. **F**ast **P**acket **S**witching
	2. **F**ault **P**rotection **S**ubsystem (power)
	3. **F**ocus **P**rojection and **S**canning
	4. **F**irst **P**erson **S**hooter (video games)
	5. **F**rames **P**er **S**econd (video or 3D games)
FPSLIC	**F**ield-**P**rogrammable **S**ystem-**L**evel **IC**
FPT	**F**orced **P**erfect **T**erminator (IBM)
FPU	**F**loating-**P**oint **U**nit (circuit)
FQDN	**F**ully **Q**ualified **D**omain **N**ame (Internet)

FQFP	Fine-pitch **Q**uad **F**lat **P**ack
FR	1. **F**rame **R**elay (access protocol)
	2. **F**ull **R**ate codec
	3. **F**lat **R**ate Service
FRA	**F**ixed **R**adio **A**ccess (WLLs)
FRAD	1. **F**rame **R**elay **A**ccess **D**evice
	2. **F**rame **R**elay **A**ssembler/**D**isassembler
FRAM	**F**erroelectric **RAM** (memory)
FRBS	**F**rame **R**elaying **B**earer **S**ervice
FRC	**F**unctional **R**edundancy **C**hecking
FRCID	**F**rame **R**elay **C**able **I**nterface **D**evice
FRD	**F**ire **R**etard**d**ant (cables)
FRED	**F**ast **R**ecovery **E**pitaxial **D**iode
FREE	**F**orum for **R**esponsible and **E**thical **E**-mail
FreeBSD	**Free** **B**erkeley **S**oftware/Standard **D**istribution (ICT)
Freq	1. **Freq**uency
	2. File **req**uest
FRF	1. **F**rame **R**elay **F**orum
	2. **F**requency **R**esponse **F**unction
FRL	**F**acility **R**estriction **L**evel (AT&T)
FRLP	**F**orward and **R**eturn **L**ink **P**air (Inmarsat)
FRM	1. **F**ocus–**R**otation **M**ount
	2. **F**ixed-**R**eference **M**odulation
FRMR	**Fra**me **R**eject
FRND	**F**rame **R**elay **N**etwork **D**evice
FRP	**F**ast **R**eservation **P**rotocol
FRS	1. **F**lat **R**ate **S**ervice
	2. **F**rame **R**elay **S**ervice
	3. **F**amily **R**adio **S**ervice
	4. **F**ield **R**outing **S**ystem
FRSE	**F**rame **R**elay **S**witching **E**quipment
FRSP	**F**rame **R**elay **S**ervice **P**rovider
FRST	**F**oundation for **R**esearch **S**cience & **T**echnology (New Zealand)
FRTE	**F**rame **R**elay **T**erminal **E**quipment
FRTT	**F**ixed **R**ound-**T**rip **T**ime
FRU	**F**ield **R**eplacement **U**nit
fs	**f**emto**s**econd (10^{-15} s)
FS	1. **F**ixed **S**ervice (Intelsat)
	2. **F**ix **S**tuff (MUX)
	3. **F**ile **S**eparator
	4. **F**ull **S**cale (test equipments)
	5. **F**ield **S**trength
	6. **F**rame **S**tatus (networking)

FSAA	Full Screen Anti-Aliasing (ICT)
FSAN	Full Services Access Network (ATM-based optical network)
FSAPT	Faculty of Scientific-Applied Post and Telecommunications (Iran)
FSB	Front Side Bus (computer)
FSBS	Frame Switching Bearer Service
FSC	1. Fiber-Switched Capable (GMPLS)
	2. Frequency Spectrum Congestion (transmission)
FSCS	Frequency-Selective Conducting Surface
FSD	Frequency Spectrum Designation (propagation)
FSDPSK	Filtered Symmetric Differential **PSK** (modulation)
FSF	1. Free Software Foundation
	2. Frequency Scaling Factor
FSK	Frequency-Shift Keying (modulation)
FSL	1. Fiber Subscriber Loop
	2. Free Space Loss
	3. Flexible Service Logic
	4. Facsimile Signal Level
FSN	1. Full Service Network
	2. Forward Sequence Number (GSM)
FSO	1. Foreign Switching Office
	2. Free Space Optics
FSOQ	Frequency Shift Offset Quadrature (modulation)
FSP	1. File Service Protocol
	2. Fiber-optic Splice Panel
FSR	1. Full-Scale Range
	2. Full Sheet Resonance
FSS	1. Fixed Satellite Service frequency band (ITU)
	2. Frequency-Selective Surface
FST	Frequency-Shift Telegraphy
FSTC	Financial Services Technology Consortium
FSTS	Federal Secure Telephone System
ft	foot
ft²	**square** foot
ft/min	foot **per min**ute
ft³/min	**cubic** foot **per min**ute
ft/s	foot **per s**econd
FT	France Telecom (company)
FT-1	Functional **T-1**
FT-3	Functional **T-3**
FTA	Federal Telecommunications Act
FTAM	File Transfer, Access & Management (protocol)
FT Bus	Frame Transport **Bus**

F

FTC	Fast Time Constant (algorithm)
FTE	1. Facsimile Terminal Equipment
	2. Full-Time Equivalent (call center)
FTIP	Fiber Transport Inside Plant
FTL	Flash Translation Layer (PCMCIA standard)
FTMSC	France Telecom Mobile Satellite Communications
FTNS	Fixed Telecommunications Network Service
FTP	1. File Transfer Protocol (networking)
	2. Foil Twisted Pair (cables)
FTS	1. File Transfer Support
	2. Finite Time Stability (ATM)
FTSC	1. Federal Telecommunication Standards Committee
	2. FidoNet Technical Standards Committee
FTT	Fault-Tracing Time
FTTB	1. Fiber-To-The-Building (fiber optics)
	2. Fiber-To-The-Business (fiber optics)
FTTC	Fiber-To-The-Curb (fiber optics)
FTTCab	Fiber-To-The-Cabinet (fiber optics)
FTTH	Fiber-To-The-Home (fiber optics)
FTTL	Fiber-To-The-Loop (fiber optics)
FTTN	Fiber-To-The-Neighborhood (fiber optics)
FTTO	Fiber-To-The-Office (fiber optics)
FTTP	Fiber-To-The-Premise (fiber optics)
FTTS	Fiber-To-The-Subscriber (fiber optics)
FU	Frame Unit
FUNI	Frame-based User–Network Interface (ATM)
FUT	Fiber Under Test
fV	femtovolt
FVC	1. Forward Voice Channel
	2. Frequency-to-Voltage Converter (electronics)
FVO	Field Verification Office
FVR	Flexible Vocabulary Recognition
fW	femtowatt
FWA	1. Fixed Wireless Access
	2. Forward-Wave crossed-field Amplifier (electronics)
FWHM	Full Width at Half Maximum (pulse)
FWLL	Fixed Wireless Local Loop
FWM	Four-Wave Mixing (fiber optics)
FX	1. Foreign Exchange (service)
	2. Fixed service
FXC	Fiber switch Cross-Connect
FXO	Foreign Exchange Office
FXS	Foreign Exchange Station
FZA	Fresnel Zone Antenna

G g

g	1. Symbol for acceleration due to Earth's **g**ravity (9.81 m/s^2)
	2. Symbol for **g**ram (unit of weight)
G	1. Symbol for prefix **g**iga-, denoting 10^9 or 2^{30}
	2. Symbol for **G**auss (unit of magnetic induction)
	3. Symbol for **G**rid of tubes (electronics)
	4. Symbol for **G**ravitational force
	5. Symbol for **G**ate (FETs)
	6. Symbol for electrical conductance
	7. Symbol for **G**ain (antennas and amplifiers)
G-M counter	**G**eiger–**M**üeller **counter** (electronics)
G-MSC	**G**ateway **M**obile **S**witching **C**enter (GSM)
G/A	**G**round-**to-A**ir communication
G/G	**G**round-**to-G**round communication
G/T	**G**ain-**to**-noise **T**emperature (ratio)
G2	Symbol for Intelligence, meaning secret information (U.S. Army)
G2B	**G**overnment-**to-B**usiness (e-commerce)
G2C	**G**overnment-**to-C**onsumer (e-commerce)
G2G	**G**overnment-**to-G**overnment (e-commerce)
G3 Fax	**G**roup **3 Fax** system
G4 Fax	**G**roup **4 Fax** system

GA	1. **G**o **A**head (telex)
	2. **G**eneral **A**ssembly (ETSI)
	3. **G**enerally **A**vailable (contracts)
GaALAs	**Ga**llium–**Al**uminum–**As**senide (semiconductors)
GAAP	**G**enerally **A**ccepted **A**ccounting **P**rinciples
GaAs	**Ga**llium–**As**enide (semiconductors)
GaAsFET	**Ga**llium–**As**enide **FET** (semiconductors)
GaAsP	**Ga**llium–**As**enide **P**hosphide (semiconductors)
GAB	1. **G**roup **A**ccess **B**ridging (service)
	2. **G**roup **A**udio **B**ridging
	3. **G**roup **A**synchronous **B**rowsing
GAC	1. **G**lobal **A**rea **C**overage
	2. **G**overnment **A**dvisory **C**ommittee (Internet)
GAGAN	**G**PS **A**nd **GEO A**ugmentated **N**avigation (India)
GAL	**G**eneric **A**rray **L**ogic (digital electronics)
GaInAs	**Ga**llium–**In**dium–**As**senide (semiconductors)
GAM	**G**PS-**A**ided **M**unitions
GaN	**Ga**llium **N**itride (semiconductors)
GAN	**G**lobal **A**rea **N**etwork (Inmarsat)
GaP	**Ga**llium **P**hosphide (semiconductors)
GAP	1. **G**eneric **A**ccess **P**rofile (DECT)
	2. **G**round-to-**A**ir **P**aging
	3. **G**eneric **A**ccess **P**rofile (Bluetooth)
GARP	1. **G**eneric **A**ttribute **R**egistration **P**rotocol (VLANs)
	2. **G**lobal **A**tmospheric **R**esearch **P**rogram
	3. **G**rowth **A**t a **R**easonable **P**rice
GARS	**G**eological **A**pplications for **R**emote **S**ensing
GAT	1. **G**eneric **A**ddressing and **T**ransport Protocol
	2. **G**o **A**head **T**one (fiber optics)
GATM	**G**lobal **A**ir **T**raffic **M**anagement (aviation)
GAW	**G**lobal **A**tmosphere **W**atch
GAZEL	**G**lobal **A**rizona **E**-**L**earning (E-learning)
Gb	1. **G**iga**b**it, which is 1,000,000,000 bits (transmission)
	2. **G**il**b**ert (unit of magnetomotive force)
Gb/s	**G**iga**b**it **per s**econd
GbE	**G**iga**b**it **E**thernet
Gbps	**G**iga**b**its **p**er **s**econd (transmission)
GB	**G**iga**b**yte, which is 2^{30} or 1024 times Megabyte (computer)
GBAS	**G**round-**B**ased **A**ugmentation **S**ystem
GBCS	**G**lobal **B**usiness **C**ommunications **S**ystems (AT&T)
GBH	**G**roup **B**usy **H**our (MUX)
GBIC	**G**iga**b**it **I**nterface **C**onverter (Ethernet)
GBM	**G**ulf **B**usiness **M**achines (Abu Dhabi)

GBN	**G**PRS **B**ackbone **N**etwork
GBR	**G**round-**B**ased **R**adar
GC	**G**lobal **C**overage
GCA	1. **G**ame **C**ontrol **A**dapter (computer)
	2. **G**round-**C**ontrolled **A**pproach (radar system)
	3. **G**ain-**C**ontrolled **A**mplifier
GCAC	**G**eneric **C**onnection **A**dmission **C**ontrol (ATM)
GCD	**G**reat **C**ircle **D**istance
GCE	**G**round **C**ommunications **E**quipment
GCF	**G**lobal **C**ertifications **F**orum (GSM)
GCI	1. **G**raphic **C**haracter **I**nternal (Siemens)
	2. **G**round-**C**ontrolled **I**nterception (radar)
GCMD	**G**lobal **C**hange **M**aster **D**irectory (NASA)
GCOM	**G**lobal **C**limate **O**bservation **M**ission (remote sensing)
GCPW	**G**rounded **C**oplanar **W**aveguide (microwave)
GCRA	**G**eneric **C**all **R**ate **A**lgorithm (ATM)
GCS	1. **G**round **C**ontrol **S**tation
	2. **G**lobal **C**ommunications **S**ervice
	3. **G**overnmental **C**ommunication **S**ystems
	4. **G**ate-**C**ontrolled **S**witch (electronics)
GCT	**G**reenwich **C**ivil **T**ime (UTC)
GD	**G**raceful **D**iscard (frame relay)
GDDM	**G**raphical **D**ata **D**isplay **M**anager (SNA)
GDE	**G**roup-**D**elay **E**qualizer
GDF	**G**roup **D**istribution **F**rame (MUX)
GDG	**G**lobal **D**evelopment **G**ateway (World Bank)
GDHS	**G**round **D**ata **H**andling **S**ystem
GDI	**G**raphics **D**evice **I**nterface (MS Windows)
GDMO	**G**uidelines for the **D**efinition of **M**anaged **O**bjects
GDN	**G**raphic **D**isplay **N**etwork
GDOP	**G**eometric **D**ilution **O**f **P**recision
GDPS	**G**lobal **D**ata **P**rocessing **S**ystem
GE	1. **G**igabit **E**thernet
	2. **G**ood **E**vening (Morse code transmissions)
GEA	**G**igabit **E**thernet **A**lliance
GEANet	**G**igabit **E**uropean **A**cademic **Net**work (fiber network)
GEDCOM	**Ge**nealogical **D**ata **Com**munication (data exchange format)
GEM	1. **G**raphics **E**nvironment **M**anager
	2. **G**ateway of **E**ducational **M**aterials (E-learning)
GEMS	**G**lobal **E**nvironmental **M**onitoring **S**ystem (United Nations)
GEN	**gen**erator
Gencam	**Gen**eric **C**omputer-**A**ided **M**anufacturing (ICT)

G

GENTEX	**Gen**eral **Te**lex
GENIAC	**Gen**ius **A**lmost-Automatic **C**omputer
GENIE	**G**eneral **E**lectric **N**etwork for **I**nformation **E**xchange
GEO	1. **Ge**ostationary **E**arth **O**rbiting (satellite)
	2. **Ge**osynchronous **E**arth **O**rbiting (satellite)
GEOS	**Ge**oworks **O**perating **S**ystem
GEOSS	**G**lobal **E**arth **O**bservation **S**ystem of **S**ystems (remote sensing)
GEQ	**G**ain **Eq**ualizer
GERAN	**G**SM–**E**dge **R**adio **A**ccess **N**etwork (GSM)
GES	**G**round **E**arth **S**tation (Inmarsat)
GETS	**G**overnment **E**mergency **T**elecommunications **S**ervice
GeV	**G**iga**e**lectron**v**olt (10^9 eV)
GF	**G**old **F**ranc (billing)
GFC	**G**eneric **F**low **C**ontrol (ATM)
GFCI	**G**round **F**ault **C**ircuit **I**nterrupter (circuit breaker)
GFE	**G**overnment **F**urnished **E**quipment
GFI	1. **G**roup **F**ormat **I**dentifier (packet switching)
	2. **G**round-**F**ault **I**nterrupter (circuit breaker)
GFLOPS	**One billion Fl**oating-point **O**perations **P**er **S**econd (computer)
GFM	**G**raphics **F**ile **M**anager
GFP	1. **G**eneric **F**raming **P**rocedure
	2. **G**lobal **F**unction **P**lane (INs)
	3. **G**round **F**ault **P**rotector
GFSK	**G**aussian **FSK** (modulation)
GFXO	**G**round start **FXO**
GFXS	**G**round start **FXS**
GGP	**G**ateway-to-**G**ateway **P**rotocol (DARPA)
GGRF	**G**SM **G**lobal **R**oaming **F**orum
GGSN	**G**ateway **G**PRS **S**upport **N**ode (cellular networks)
GH Effect	**G**ordon-**H**ous **Effect** (transmission)
GHOST	**G**SM **H**osted **SMS** **T**eleservices
GHz	**G**iga**h**ertz (10^9 Hz)
GI	**G**rade **I**ndex (fiber optics)
GIAC	**G**lobal **I**ncident **A**nalysis **C**enter
GIC	1. **G**eneral **I**nstrument **C**orporation
	2. **G**igabit **I**nterface **C**onverter
	3. **G**eomagnetically **I**nduced **C**urrent
GID	**G**ARP **I**nformation **D**eclaration
GIDEP	**G**overnment and **I**ndustries **D**ata **E**xchange **P**rogram
.gif	Name extension for **GIF** format files (graphics)
GIF	**G**raphical **I**nterchange **F**ormat (imaging)

Gig	1. **Gig**abits per second
	2. **Gig**ahertz per second
Gig-E	**Gig**abit-**E**thernet
GIGAMO	**Giga**byte-class **M**agneto-**O**ptical (storage technology)
GIGO	**G**arbage **I**n, **G**arbage **O**ut (computer)
GII	**G**lobal **I**nformation **I**nfrastructure
GIIC	**G**lobal **I**nformation **I**nfrastructure **C**ommission
GILC	**G**lobal **I**nternet **L**ibrary **C**ampaign
GILS	**G**overnment **I**nformation **L**ocator **S**ervice
GIM	**G**roup **I**dentification **M**ark (cellular networks)
GIMP	**G**NU **I**mage **M**anipulation **P**rogram (software)
GIP	1. **G**SM/DECT **I**nterworking **P**rofile
	2. **G**raded-**I**ndex **P**rofile (fiber optics)
	3. **G**ARP **I**nformation **P**rotocol
GIS	1. **G**eographical **I**nformation **S**ystem (ICT)
	2. **G**eeky **I**nternal **S**tuff (database)
	3. **G**eneral **I**nformation **S**ervices
	4. **G**eneral **I**nformation **S**essions
	5. **G**lobal **I**nformation **S**ociety
	6. **G**eospatial **I**nformation **S**ystem
GISIS	**G**lobal **I**ntegrated **S**hipping **I**nformation **S**ystem (IMO)
GITS	**G**overnment **I**nformation **T**echnology **S**ervice
GIX	**G**lobal **I**nternet **E**xchange
GJISI	**G**lobal **J**ustice **I**nformation **S**haring **I**nitiative
GJXDM	**G**lobal **J**ustice **X**ML **D**ata **M**odel (U.S.A.)
GKP	**G**lobal **K**nowledge **P**artners
GKS	**G**raphical **K**ernel **S**ystem
GL	**G**raphics **L**ibrary
GLIS	**G**lobal **L**and **I**nformation **S**ystem
GLONASS	**Gl**obal **O**rbiting **Na**vigation **S**atellite **S**ystem (Russia)
GLR-L	**G**lobal **L**ocation **R**egister—**L**ocal (mobile networks)
GLR-N	**G**lobal **L**ocation **R**egister—**N**etwork (mobile networks)
GM	**G**ood **M**orning (Morse code transmissions)
Gmax	**max**imum **G**ain
GMCR	**G**lobe-**M**ackay **C**able and **R**adio Corporation
GMD	**G**uaranteed **M**essage **D**elivery
GMDSS	**G**lobal **M**aritime **D**istress and **S**afety **S**ystem (COMSAR)
GMES	**G**lobal **M**onitoring for **E**nvironmental and **S**ecurity
GMII	**G**igabit **M**edia-**I**ndependent **I**nterface
GMP	**G**lobal **M**anaged **P**latform
GMPCS	**G**lobal **M**obile **P**ersonal **C**ommunication **S**ystem (ITU-T)
GMPLS	**G**eneralized **M**ulti**p**rotocol **L**abel **S**witching
GMR	**G**iant **M**agneto **R**esistance effect
GMRP	**G**ARP **M**ulticast **R**egistration **P**rotocol (networking)

G

GMS	1. **G**eostationary **M**eteorological **S**atellite
	2. **G**eneric **M**aintenance **S**ystem
GMSC	**G**ateway **M**obile **S**witching **C**enter
GMSK	**G**aussian-filtered **MSK** (GSM modulation)
GMT	**G**reenwich **M**ean **T**ime
GN	**G**ood **N**ight (Morse code transmissions)
GNCT	**G**eneralized **N**o **C**ircuit **T**reatment
GND	**G**rou**nd** (electronics)
GNE	**G**ateway **N**etwork **E**lement (SONET)
GNMC	**G**lobal **N**etwork **M**anagement **C**enter
GNN	**G**lobal **N**etwork **N**avigator (Web-based information service)
GNOME	**GN**U **O**bject **M**odel **E**nvironment
GNSS	**G**lobal **N**avigational **S**atellite **S**ystem (U.S.A.)
GNU	**G**NU's **N**ot **U**nix Operating System
GO-MVIP	**G**lobal **O**rganization for **MVIP**, Inc. (U.S.A.)
GOES	1. **G**eostationary **O**perational **E**nvironmental **S**atellite
	2. **G**eostationary **O**rbiting **E**arth **S**atellite
GOF	**G**lass **O**ptical **F**iber
GOGO	**G**overnment-**O**wned, **G**overnment-**O**perated
GΩ	**G**iga**ohm** (electronics)
GOLD	**G**lobal **O**n**l**ine **D**irectory
GOM	**G**eneric **O**bject **M**odel
GOME	**G**lobal **O**zone **M**onitoring **E**xperiment
GOMOS	**G**lobal **O**zone **Mo**nitoring **S**ystem (remote sensing)
GOMS	**G**eostationary **O**perational **M**eteorological **S**atellite
GOPS	**One billion O**perations **P**er **S**econd (computer)
GORIZONT	The Russian geostationary telecommunications satellite
GORS	**G**eneral **O**rganization of **R**emote **S**ensing (Syria)
GoS	**G**rade **o**f **S**ervice
GOS	**G**lobal **O**bserving **S**ystem
GOSAT	**G**reenhouse gas **O**bserving **Sat**ellite
GOSIP	**G**overnment **O**pen **S**ystem **I**nterconnection **P**rofile (OSI model)
GPA	**G**eneral **P**urpose **A**dapter (AT&T)
GPC	**G**ateway **P**rotocol **C**onverter
GPCL	**G**eneral **P**remises **C**abling **L**icense
GPF	**G**eneral **P**rotection **F**ault (MS Windows)
.jpg	Name extension for **JPEG** format files (graphics)
GPI	1. **G**TI **P**hysical **I**nterface
	2. **G**ammaFax **P**rogrammers **I**nterface
	3. **G**lide **P**ath **I**ndicator
	4. **G**round **P**osition **I**ndicator
GPIB	**G**eneral-**P**urpose **I**nterface **B**us (computer)

GPIRS	**G**lobal **P**ositioning **I**nertial **R**eference **S**ystem
GPL	1. **G**eneral **P**ublic **L**icense (Free Software Foundation)
	2. **G**raphical **P**rogramming **L**anguage
GPM	**G**lobal **P**recipitation **M**easurement (remote sensing)
GPoP	**G**lobal **P**oint **o**f **P**resence
GPR	**G**round-**P**enetrating **R**adar
GPRS	**G**eneral **P**acket **R**adio **S**ervice (cellular network)
GPS	1. **G**lobal **P**ositioning **S**ystem
	2. **G**lobal **P**osition **S**potting
GPTC	**G**eneral **P**ost and **T**elecom **C**ompany (Libya)
GPU	**G**raphics **P**rocessing **U**nit (computer)
GQFP	**G**uard-ring **Q**uad **F**lat **P**ack
GR	**G**eneric **R**equirements (Telcordia Technologies)
GRC	**G**round **R**eturn **C**ircuit
GRE	**G**eneric **R**outer **E**ncapsulation (Internet)
GREJ	**G**roup **Rej**ect
GREP	**G**eneralized **R**egular **E**xpression **P**arser (Unix utility)
GRIC	**G**lobal **R**each **I**nternet **C**onnection
GRID	**G**lobal **R**esource **I**nformation **D**atabase (remote sensing)
GRIN	1. **Gr**adient **In**dex (fiber optics)
	2. **Gr**aded **In**dex (fiber optics)
GRM	**G**eneral **R**elationship **M**odel
GRSU	1. **G**eneric **R**emote **S**witch **U**nit
	2. **G**eographic **R**emote **S**witch **U**nit
GRX	**G**PRS **R**oaming **Ex**change
GS	**G**roup **S**eparator
GS Trunk	**G**round **S**tart **Trunk**
GSA	**G**eneral **S**ervices **A**dministration (U.S.A.)
GSC	**G**lobal **S**tandards **C**ollaboration (ETSI)
GSDN	**G**lobal **S**oftware **D**efined **N**etwork (AT&T)
GSF	**G**eneric **S**ervices **F**ramework (software design)
GSI	1. **G**rand-**S**cale **I**ntegration (microchips)
	2. **G**lide **S**lope **I**ndicator
GSM	**G**lobal **S**ystem for **M**obile Communications (Europe)
GSM-900	**GSM** at **900** MHz
GSM-R	**GSM** Standard for **R**ailway Applications
GSMA	**GSM** **A**ssociation
GSMP	**G**eneral **S**witch **M**anagement **P**rotocol (ATM)
GSMPLMN	**GSM** **P**ublic **L**and **M**obile **N**etwork
GSN	1. **G**ateway **S**witching **N**ode
	2. **G**igabyte **S**ystem **N**etwork
	3. **G**lobal **S**ubscriber **N**umber
GSO	**G**eostationary **O**rbit (satellite)

GSO-FSS	Geostationary Satellite Orbit—Fixed Satellite Service
GSOC	German Space Operations Center
GSR	Gigabit Switch Router
GSS	Group Switch Selector
GSSA	Generic Security Service Application
GSSAP	Generic Security Service Application
GSTN	Global-Switched Telephone Network (ITU-T)
GT	1. Global Title
	2. Gain Transfer
GTA	1. Government Telecommunications Association (U.S.A.)
	2. Guam Telephone Authority
GTE	1. General Telephone and Electronics (company)
	2. Group Translating Equipment
GTI	Generic Traffic Interface
GTL	1. Global Tele-Systems Ltd (India)
	2. Gunning Transceiver Logic (CMOS circuits)
GTLD	Generic Top-Level Domain (Internet)
GTL+	Gunning Transceiver Logic **Plus** (CMOS circuits)
GTLP	Gunning Transceiver Logic Plus (electronics)
GTMOSI	General Teleprocessing Monitor for **OSI**
GTN	Global Transaction Network (AT&T)
GTO	1. Geostationary Transfer Orbit
	2. Gate Turn-Off
	3. General Telecommunications Organization (Oman)
GTOC	General Telephone Operating Company
GTP	1. General Telemetry Processor
	2. Generalized Trunk Protocol
	3. GNOME Translation Project
	4. Green Transport Plans
	5. Government Telecommunications Program
GTS	1. Global Telecommunications System
	2. Global TeleSystems (company)
	3. Government Telecommunications System
GTSS	Global Telecom Solutions Sector (Motorola)
GTT	Global Title Translation
GUI	Graphical User Interface (computer)
GUID	Global Unique Identification (component object model)
GUM	Grand Unified Multicast (IETF)
GVF	Global VSAT Forum (England)
GVI	Global Vegetation Index (remote sensing)
GVNS	Global Virtual Network Service

GVRP	**G**ARP **VL**AN **R**egistration **P**rotocol (VLANs)
GVTel	**G**arden **V**alley **Tel**ephone (company)
GW	1. **G**ate**w**ay
	2. **G**iga**w**att (10^9 W)
.gz	Name extension for files compressed with the UNIX **gz**ip
Gzip	**G**NU **zip** (compression software)

G

H h

h	1. Symbol for **h**our
	2. Symbol for **h**ecto-, meaning one hundred or 10^2
	3. Symbol for Planck's constant (electronics)
H	1. Symbol for magnetic field strength (electronics)
	2. Symbol for **H**enry (unit of inductance)
	3. Symbol for **h**orizontal polarization
H PAD	**H**ost **P**acket **A**ssembler/**D**isassembler
H-Channel	**H**igh-Speed **Channel** (ISDN)
H-MVIP	**H**igh-density **MVIP**
H-pol	**H**orizontal **pol**arization
H-sync	**H**orizontal **sync**hronization
Ha-Dec	**H**our **a**ngle–**Dec**lination angle
HA	**H**orn **A**lert of car (for mo bile ringing)
HAARP	**H**igh-frequency **A**ctive **A**uroral **R**esearch **P**rogram
HAAT	**H**eight **A**bove the **A**verage **T**errain (antenna)
HACBSS	**H**omestead **A**nd **C**ommunity **B**roadcasting **S**atellite **S**ervice
HAD	**H**alf **A**mplitude **D**uration
HAL	1. **H**ardware **A**bstraction **L**ayer (operating system)
	2. **H**ardwired **A**rray **L**ogic (digital electronics)
HALE	**H**igh-**A**ltitude **L**ong **E**ndurance
HALT	**H**ighly **A**ccelerated **L**ife **T**esting (satellite)
HAM Radio	**H**ome **Am**ateur **Radio**

HAN	Home Area Network
HAPS	High-Altitude Platform Systems (application)
HARM	High-speed Anti-Radiation Missile
HASP	Houston Automated Spooling Program (transmission protocol)
HAVI	Home Audio–Video Interoperability architecture
HBA	Host Bus Adapter (computer)
HBC	Half-band Bipolar Coding
HBFG	Host Behavior Functional Group (ATM)
HBS	Home Base Station (wireless)
HBT	Heterojunction Bipolar Transistor (electronics)
HC	High-speed CMOS (as in 74 **HC**245 IC numbers)
HCA	Host Channel Adapter (fiber optics)
HCI	1. Host Command Interface
	2. Human Computer Interface standards
	3. High Council of Informatics (Iran)
HCIT	High Council of Information Technology (Iran)
HCL	Hardware Compatibility List (Microsoft)
HCN	1. Hierarchical Computer Network
	2. Hybrid Communications Network
HCO	Hearing Carry Over (TRS)
HCPLD	High-Capacity Programmable Logic Device (microchips)
HCS	1. Hundred Call Seconds
	2. Hard Clad Silica (optical fiber)
	3. Hierarchical Cell Structure (cellular networks)
	4. High Council of Space (Iran)
HCSS	High-Capacity Storage System (data storing)
HCT	1. Hardware Compatibility Test (Microsoft)
	2. High-speed CMOS with TTL thresholds (as in 74 **HCT**245 IC numbers)
HD	1. Half Duplex (operation)
	2. High-Definition (TV set)
	3. Hard Disk
	4. Hard Drive
	5. High-Density
HD-MAC	High-Definition MAC (TV standard)
HDB	High-Density Bipolar code
HDB3	High-Density Bipolar **three zeros** Substitution (data encoding)
HDBMS	Hierarchical Data-Base Management System
HDC	Hard Disk Controller (computer)
HDCT	Hybrid Discrete Cosine Transform
HDD	Hard Disk Drive

HDDR	High-**D**ensity **D**igital **R**ecorder
HDDT	High-**D**ensity **D**igital **T**ape
HDF	Hierarchical **D**ata **F**ormat (computer)
HDFSS	High-**D**ensity **F**ixed **S**atellite **S**ervice
HDL	Hardware **D**escription **L**anguage
HDLC	1. High-level **D**ata-**L**ink **C**ontrol (protocol)
	2. Hardware **D**escription **L**anguage **C**
HDML	Handheld **D**evices **M**arkup **L**anguage (programming)
HDO	Hearing **D**estination **O**rder (FCC)
HDPCM	Hybrid **D**ifference **PCM** (modulation)
HDRSS	High-**D**ata-**R**ate **S**torage **S**ystem
HDSL	High-bit-rate **DSL** (access)
HDSL2	High-bit-rate **DSL** version **2** (access)
HDT	Host **D**igital **T**erminal (cable TV)
HDTMOS	High-**D**ensity **T**echnique **MOS** (semiconductor)
HDTP	Handheld **D**evice **T**ransport **P**rotocol (wireless)
HDTV	High-**D**efinition **Tele**vision (broadcasting)
HDX	Half-**D**uple**x** (operation)
HDWDM	Hyper **D**ense **W**avelength-**D**ivision **M**ultiplexing
HE	Head **E**nd (cable TV)
HEAO	High-**E**nergy **A**stronomy **O**bservatory
HEC	1. Header **E**rror **C**heck (ATM)
	2. Hong Kong **E**lectric **C**ompany
HEHO	Head-**E**nd **H**op **O**ff (traffic)
HEL	Hardware **E**mulation **L**ayer (operating systems)
HEM	Hughes **E**ngineering **M**odule
HEM wave	Hybrid **E**lectro**m**agnetic **wave**
HEMP	High-altitude **E**lectro**m**agnetic **P**ulse
HEMT	High **E**lectron-**M**obility **T**ransistor (semiconductors)
HeNe laser	Helium–**Ne**on **laser**
HEO	Highly **E**lliptical **O**rbit (satellites)
HEOS	Highly **E**ccentric **O**rbit **S**atellite
HEP	High-**E**nergy **P**hysics
HEPA	High-**E**fficiency **P**articulate **A**rrester (filter)
HEPIC	High-**E**nergy **P**hysics **I**nformation **C**enter
HEPNet	High-**E**nergy **P**hysics **Net**work
HER	High-**E**fficiency **R**ed (LEDs)
HERALD	**H**arbor **E**cho **R**anging **A**nd **L**istening **D**evice (sound detection)
HERF	High-**E**nergy **R**adio **F**requency
HERO	**H**azards of **E**lectromagnetic **R**adiation to **O**rdnance (military)
HERP	**H**azards of **E**lectromagnetic **R**adiation to **P**ersonnel (military)

H

HET	1. **H**all-**E**ffect **T**ransducer (semiconductors)
	2. **Het**erodyne
Hex	**Hex**adecimal
HEXFET	**Hex**agonal **FET** (semiconductors)
HF	1. **H**igh **F**requency (3–30 MHz range)
	2. **H**ands **F**ree (PBX and Wireless)
HFAARS	**H**igh-**F**requency **A**daptive **A**ntenna **R**eceiving **S**ystem
HFAI	**H**ands **F**ree **A**nswer on **I**ntercom
HFC	**H**ybrid-**F**iber **C**oax (network architecture)
HFCPN	**H**igh-**F**requency **C**onditioned **P**ower **N**etwork
HFDF	**H**igh-**F**requency **D**istribution **F**rame
HFET	**H**eterostructure **FET** (semiconductors)
HFS	**H**ierarchical **F**ile **S**ystem (computer)
HFSCD	**H**ierarchical **F**ile **S**ystem format for **C**ompact **D**isk (networking)
HFSS	**H**igh-**F**requency **S**tructure **S**imulator
HFU	**H**ands **F**ree **U**nit
HG	**H**ome **G**ateway (LANs)
HGA	1. **H**ercules **G**raphics **A**dapter (720×348 pixel monitors)
	2. **H**igh-**G**ain **A**ntenna
Hi-Cap	**H**igh-**Cap**acity system
Hi-Fi	**H**igh-**Fi**delity system
Hi-Tech	**H**igh-level **Tech**nology
HI-OVIS	**H**ighly **I**nteractive **O**ptical **V**isual **I**nformation **S**ystem
Hi-res	**H**igh-**res**olution system
Hi8 Video	**H**igh-quality extension of the **Video 8** format
HIC	**H**ybrid **I**ntegrated **C**ircuit (electronics)
HIC	**H**igh-tech **I**ndustries **C**enter (Iran)
HIF	**H**ost **I**nter**f**ace node
HiLAT	**H**igh-**Lat**itude Satellite (U.S.A)
HIO	**H**ighly **I**nclined **O**rbit
HIPAA	**H**ealth **I**nsurance **P**ortability and **A**ccountability **A**ct (U.S.A.)
HiperLAN	**H**igh-**per**formance **LAN**
HiPOT	**H**igh-**Pot**ential (testing)
HiPPI	**H**igh-**P**erformance **P**arallel **I**nterface (fiber optic)
HiRAN	**H**igh-precision SHO**RAN**
HIRS	**H**igh-resolution **I**nfra**r**ed **S**ounder (remote sensing)
HITS	1. **H**igh-**I**ntensity **T**ransient **S**ignal
	2. **H**awaiian **I**nter-islands **T**elecommunications **S**ystem
	3. **H**awaiian **I**nformation **T**ransfer **S**ystem
HIVR	**H**ost **I**nteractive **V**oice **R**esponse

HJBT	Hetero-Junction Bipolar Transistor (semiconductors)
HJFET	Hetero-Junction FET (semiconductors)
HKIX	Hong Kong Internet Exchange
HKSW	Hook Switch
HLC	High-Level Committee (ITU)
HLD	1. High-Level Design
	2. High-Level Dispatcher
HLES	Home Land Earth Station (Inmarsat)
HLF	High-Level Function
HLL	High-Level Language
HLLAPI	High-Level Language Applications Programming Interface (IBM)
HLP	High-Level Protocol
HLR	Home-Location Register (cellular networks)
HLS	1. Hue—Lightness–Saturation (color model)
	2. Hitless Switch
HLUT	Half Look-Up Table
HMA	High-Memory Area (IBM PCs)
HMD	Head-Mounted Display
HMG	Hyper Master Group
HMI	1. Human-Machine Interface
	2. Hub Management Interface (networking)
HMIC	Hybrid Microwave Integrated Circuit
HMM	Hidden Markov Method (algorithm)
HMMP	Hypermedia Management Protocol (network management)
HMS	1. High-performance Management System
	2. Hydrological and Meteorological Station
HNDS	Hybrid Network Design System
HNF	High-performance Network Forum
HNPA	Home Numbering Plan Area
HNS	Hughes Network Systems (company)
HO	Handover (GSM)
HO Tone	Handoff Tone (cellular networks)
HOBIC	Hotel Billing Information Center (AT&T)
HOBIS	Hotel Billing Information System (AT&T)
HOC	Highest Outgoing Channel
HOI	Higher Order Interface
HOL	High-Order Language (programming)
HOMC	Human-Oriented Media Computing
HomePNA	Home Phone-line Networking Alliance (standards organization)
HomeRF	Home Radio Frequency
HOMO	Home Office Mobile Office

H

HOP	Higher Order Path
HOPA	Higher Order Path Adaptation
HOPC	Higher Order Path Connection
HOPCM	High-Order PCM (modulation)
HOPS	Hardwire Order Processing System
HOPSN	Higher Order Path Sub-Network
HOPT	Higher Order Path Termination
HOR	Horizontal polarization
HOVC	High-Order Virtual Container (MUX)
hp	horsepower
HP	Hewlett-Packard (telecom company)
HPA	1. High-Power Amplifier (microwave)
	2. High-Performance Addressing (LCD displays)
	3. Hi-vision Promotion Association (broadcasting)
HPAD	Host Packet Assembler/Disassembler
HPBW	Half-Power Beam-Width (antennas)
HPC	Handheld Personal Computer
HPF	High-Pass Filter
HPFS	High-Performance File System (OS/2)
HPGL	Hewlett-Packard Graphics Language (programming)
HPIB	Hewlett-Packard Interface Bus
HPIIS	High-Performance International Internet Services
HPIL	Hewlett-Packard Interface Loop (as in HP41CX calculator)
HPM	High-Power Microwave
HPO	High-Performance Option
HPPI	High-Performance Parallel Interface
HPPM	High-Power Propulsion Module
HPR	High-Performance Routing (LANs)
HPT	1. High-order Path Termination
	2. Host Processing Time
HQ	Headquarter
HR	1. Half-Rate
	2. Here (Morse code transmissions)
HRC	1. Hypothetical Reference Circuit
	2. Hybrid Ring Control
	3. High-Rupturing Capacity
HRDP	Hypothetical Reference Digital Path
HRDS	Hypothetical Reference Digital Section
HRFWG	Home RF Working Group (US standards organization)
HRP	Hypothetical Reference Path
HRPT	High-Resolution Picture Transmission
HRSC	High-Resolution Stereo Color (remote sensing)

HRV	High-Resolution Visible (remote sensing)
HRVIR	High-Resolution Visible and Middle Infrared (remote sensing)
HRX	Hypothetical Reference Connection (ISDN)
HSA	High-Speed Access
HSB	1. Hue–Saturation–Brightness (color model)
	2. Hot Standby
HSC	1. Hierarchical Storage Controller
	2. High-Speed Channel
HSCH	High-Speed Channel
HSCI	High-Speed Communications Interface (Cisco)
HSCS	High-Speed Circuit-Switched
HSCSD	High-Speed Circuit-Switched Data (wireless)
HSD	1. High-Speed Data (Inmarsat)
	2. Home Satellite Dish
HSDA	High-Speed Data Access
HSDL	High-speed Subscriber Data Line (Telcordia Technologies)
HSDPA	High-Speed Downlink Packet Access
HSDRAM	High-speed Synchronized Dynamic RAM (memory)
HSDU	High-Speed Data Unit
HSF	Heat-Sink Fan (electronics)
HSI	Hue–Saturation–Intensity (color model)
HSL	Hue–Saturation–Luminance (color model)
HSLN	High-Speed Local Network
HSM	1. Hierarchical Storage Management (file storage)
	2. Hardware-Specific Module (networking)
HSN	1. Hopping Sequence Number (GSM)
	2. Hierarchically Synchronized Network (networking)
HSOP	High-power Small-Outline Package (microchips)
HSPD	High-Speed Packet Data
HSPSD	High-Speed Packet-Switched Data (wireless)
HSR	Harmonic Suppression Reactor
HSRP	Hot Standby Routing Protocol (routing)
HSSI	High-Speed Serial Interface (Cisco)
HSSP	High-Speed Switched Port
HST	High-Speed Technology
HSTL	1. High-Speed Transceiver Logic (digital electronics)
	2. High-Speed Transistor Logic (digital electronics)
HSTP	High-Speed Transport Protocol
HSTR	High-Speed Token-Ring
HSV	Hue–Saturation—Value (color model)
hSymbols/s	hecto (100) Symbols per seconds (TV)
HSYNC	Horizontal Sync (video signals)

H

HT	1. "**H**orizontal **T**abulation" (character control code)
	2. **H**igh **T**ension
HTC	**H**igh-**T**ech **C**omputer
HTCC	**H**igh-**T**emperature **C**o-fiber **C**eramic
HTCP	**H**yper **T**ext **C**atching **P**rotocol
HTG	**H**unt **G**roup
HTL	**H**igh-**T**hreshold **L**ogic (digital electronics)
.htm	The MS-DOS/Windows 3.x name extension for **HTM**L files
.html	Name extension for **HTML** files
HTML	**H**yper **T**ext **M**arkup **L**anguage (programming)
HTML+	**HTML Plus** (programming)
HTR	**H**ard-**T**o-**R**each
HTRB	**H**igh-**T**emperature **R**everse **B**ias
HTS	**H**igh- **T**emperature **S**uperconductor
HTTL	**H**igh-power **T**ransistor-to-**T**ransistor **L**ogic (digital electronics)
HTTP	**H**yper **T**ext **T**ransfer **P**rotocol (World Wide Web)
HTTP-NG	**H**yper **T**ext **T**ransfer **P**rotocol **N**ext **G**eneration
HTTPD	**H**yper **T**ext **T**ransfer **P**rotocol **D**aemon
HTTPS	**HTTP-S**ecure **C**onnection
HTU-C	**H**DSL **T**erminal **U**nit—**C**entral (HDSL modem or line card)
HTU-R	**H**DSL **T**erminal **U**nit—**R**emote (HDSL modem or PC card)
HUD	**H**ead-**U**p **D**isplay (combat aircraft)
HUT	1. **H**igh-**U**sage **T**runk
	2. **H**opkins **U**ltraviolet **T**elescope
HV	1. **H**igh **V**oltage
	2. **H**a**v**e (Morse code transmissions)
HVAC	1. **H**eating, **V**entilating, and **A**ir **C**onditioning system
	2. **H**igh-**V**oltage **AC**
HVDC	**H**igh-**V**oltage **D**irect **C**urrent (power)
HVP	**H**ub **V**oice **P**ort
HVPS	**H**igh-**V**oltage **P**ower **S**upply
HVQ	**H**ierarchical **V**ector **Q**uantization (video compression)
HWCS	**H**armony **W**ireless **C**ommunications **S**ystem (Motorola)
HyCoS	**H**ypermedia **Co**mmunication **S**erver
HYTELNet	**H**ypertext-browser for **TELNet** Accessible sites
Hz	**H**ert**z** (unit of frequency)

I i

I	1. Symbol for electrical **current** (electronics) 2. Symbol for **I**ntensity (electronics) 3. Symbol for "**On**" position of rocker switches (electronics)
I Package	Installation **Package**
I&C	Installation **and C**heckout
I&M	Installation **and M**aintenance
I&R	Installation **and R**epair (telephone company department)
I-4s	**Fourth** Generation of **I**nmarsat **S**atellite**s** (Inmarsat)
I-beam	Name of cursor when resembles the capital letter **I** (computer)
I-CASE	**I**ntegrated **C**omputer-**A**ided **S**oftware **E**ngineering
I-CF	**I**SDN **C**all **F**orwarding
I-CFDA	**I**SDN **C**all **F**orwarding **D**on't **A**nswer
I-CFDAIO	**I**SDN **C**all **F**orwarding **D**on't **A**nswer **I**ncoming **O**nly
I-CFIB	**I**SDN **C**all **F**orwarding **I**nterface **B**usy
I-CFIBIO	**I**SDN **C**all **F**orwarding **I**nterface **B**usy **I**ncoming **O**nly
I-CFIG	**I**SDN **C**all **F**orwarding **I**ntra-**G**roup
I-CFIO	**I**SDN **C**all **F**orwarding **I**ncoming **O**nly
I-CFPF	**I**SDN **C**all **F**orwarding over **P**rivate **F**acilities
I-CFV	**I**SDN **C**all **F**orwarding **V**ariable
I-CFVCG	**I**SDN **C**all **F**orwarding **V**ariable facilities for **C**ustomer **G**roup

I-CNIS	ISDN Calling Number Information Service
I-Commerce	Internet Commerce
I-EDI	Internet-based Electronic Data Interchange
I-ETS	Interim European Telecommunications Standard
I-frame	Intra-coded frame (MPEG animation)
I-HC	ISDN Hold Capability
i-LINK	Another name for FireWire Technology
i-Mode	Internet-Mode (GSM service)
I-MUX	Inverse Multiplexer
I-Order	Installation Order
I-picture	Intra-coded picture (MPEG animation)
I-PNNI	Integrated PNNI (ATM)
i-Pod	Internet Pod (Apple)
I-time	Instruction time
I-TV	Interactive Television
I-Way	Information Superhighway
I/F	Interface
I/O	1. Input/Output (electronics)
	2. In-route/Out-route
I/P	Input
I/R	Installation and Repair
I2	Internet 2
I^{2}C	Inter-Integrated Circuit
I^{2}ICE	Integrated Instrumentation and In-Circuit Emulation (electronics)
I^{2}L	Integrated-Injection Logic (digital electronics)
I^{2}O	Intelligent Input/Output (standard)
I^{2}S^{2}C^2	International Information Systems Security Certification Consortium
I^{2}T	Intelligent Interface Technology
i386SL	A version of Intel's 386 family of microprocessors
i486DX	An advanced version of Intel's 80386 microprocessor
i486SL	A low-power consumption version of Intel's 80 486DX
i486SX	A low-cost alternative to the Intel's 80 486DX
IA	1. International telegraph Alphabet
	2. Internet Appliance
	3. Instrumentation Amplifier (electronics)
	4. Implementation Agreement
	5. Intelligent Agent
	6. Intel Architecture
IA-32	Intel Architecture 32 bit
IA-64	Intel Architecture 64 bit
IA2	International telegraph Alphabet No. 2
IA5	International telegraph Alphabet No. 5

IAAB	Inter-American Association of Broadcasting
IAB	1. Internet Architecture Board (DARPA)
	2. Internet Activities Board
IAC	1. Information Access Company
	2. Information Analysis Center (U.S. defense department)
	3. Interactive Synchronous Communications
	4. Inter-Application Communications (protocol)
	5. Internet Access Coalition (organization)
	6. Institute for Advanced Commerce (IBM)
	7. Industry Advisory Council
IADs	Integrated Access Devices (Internet)
IADP	International Assistance & Development Program (Intelsat)
IAEA	International Atomic Energy Agency
IAEEE	Iranian Association of Electrical & Electronics Engineers
IAF	International Astronautical Federation
IAGA	International Association of Geomagnetism and Aeronomy
IAGC	Instantaneous Automatic Gain Control (Intelsat)
IAHC	International Ad Hoc Committee
IAI	Initial Address Information
IAL	Intel Architecture Labs (Intel)
IAM	1. Initial Access Message (SS7)
	2. Incoming Address Message (SS7)
	3. Intermediate Access Memory
IAMSAR	International Aeronautical and Maritime Search And Rescue (IMO)
IANA	Internet Assigned Numbers Authority (organization)
IAO	Intra-Office SONET signal
IAP	1. Internet Access Provider
	2. Intelligent Access Point
IAP1	Initial Acquisition Phase 1 code
IAP2	Initial Acquisition Phase 2 code
IAPP	Inter-Access Point Protocol
IARL	International Amateur Radio League
IARU	International Amateur Radio Union
IASA	Integrated AUTODIN System Architecture
IASG	Internetwork Address Sub-Group
IAT	1. Intermediate Attitude Trim
	2. International Atomic Time
IATA	International Air Transport Association
IATE	International Accounting and Traffic Analysis Equipment

IAU	1. Initial Acquisition Unit
	2. International Astronomical Unit
IB	Input Buffer
IBA	Independent Broadcasting Authority
IBC	1. International Broadcasting Convention
	2. Information Bearer Channel
	3. Integrated Broadband Communications
	4. Interface Buffer Controller
	5. Internal Bus Controller
	6. Initial Billing Company
IBCN	Integrated Broadband Communications Network
IBDN	Integrated Building Distribution Network (fiber optics)
IBG	Inter-Block Gap (storage disks or tapes)
IBI	1. Inter-Burst Interference
	2. Intergovernmental Bureau for Informatics
	3. International Broadcast Institute
IBIC	Interface Bus Interactive Control
IBL	Initial Binary Load
IBM	International Business Machines Corporation
IBM AT	IBM's personal computer Advanced Technology
IBM PC	IBM Personal Computers
IBM-ERS	IBM Emergency Response Service
IBN	Institut Belge de Normalisation (Belgium)
IBND	Interim Billed Number Database
IBNPR	In-Band Noise Power Ratio
IBO	Input Back-Off
IBOC	In-Band On-Channel (broadcasting)
IBPD	In-Band Power Difference
IBS	1. Intelsat Business Service
	2. Intelligent Battery System
	3. International Broadcasting Station
IBT	1. Internet-Based Training (e-learning)
	2. Isochronous Burst Transmission (data network)
IBW	Instantaneous Band-Width
IBX	Integrated Business Exchange (same as PBX)
IC	1. Integrated Circuit (electronics)
	2. International Channel
	3. Intercom
	4. Interexchange Carrier
	5. Intermediate Cross-connect
IC DRAM	Integrated Circuit DRAM (memory)
ICA	1. International Communications Association (U.S. standards)
	2. Integrated Communications Adapter (fiber optics)

iCal	Internet **Cal**endaring and scheduling (Apple Inc.)
Ical	An X-based **cal**endaring and scheduling program
ICAL	Internet Community At Large
iCalendar	Internet **Calendar**ing and scheduling (Apple Inc.)
ICALEP	International Conference on Accelerator and Large Experimental Physics
ICAM	Integrated Conditional Access Module
ICANN	Internet Corporation for Assigned Names and Numbers
ICAO	International Civil Aviation Organization
ICAP	Intellectual **Cap**ital
ICAPI	International Call **API**
ICASA	Independent Communications Authority of South Africa
ICB	Individual Case Basis (service)
ICC	International Calling Card (service)
ICCAPI	International Call Control **API**
ICCB	Internet Configuration Control Board
ICCC	Internet Channel Commerce Connectivity protocol
ICCF	1. Interexchange Carriers Compatibility Forum
	2. International Civic Communication Forum (Ukraine)
	3. International Correspondence Chess Federation
ICCID	Integrated Circuit Card **ID**
ICD	1. International Code Designator
	2. Interface Control Document
	3. Interconnect Diagram
ICDL	International Computer Driving License
ICE	1. In-Circuit Emulator (electronics)
	2. Information Content and Exchange (protocol)
	3. Information Communications and Entertainment
ICE Age	Information Communication and Entertainment **Age**
ICEA	Insulated Cable Engineers Associations, Inc. (U.S.A.)
ICFA	International Computer Facsimile Association
ICI	1. Inter-Channel Interference
	2. Inter-Carrier Interface
	3. Interface Control Information (ISO)
	4. Incoming Call Identification
	5. International Commission on Illumination
ICIA	Information and Communications Industry Association, Ltd
ICII	Iranian Communication Industries Incorporation
ICIT	International Center for Information Technology (MCI)
Icky PIC	St**icky P**lastic Insulated Conductor
ICM	1. Integrated Call Management
	2. Image Color Matching (scanning)

ICMP	Internet Control Message Protocol (IETF)
ICMPv6	Internet Control Message Protocol version 6
ICN	Idle-Channel Noise
ICNI	Integrated Communications, Navigation, and Identification
ICNIRP	International Commission of Non-Ionizing Radiation Protection
ICO	Intermediate Circular Orbit (satellite)
ICOMP	Intel Comparative Microprocessor Performance index (TEST)
ICONET	Satellite Communication Network of ICO Global Communications
ICONTEC	Instituto Colombiano de Normas Técnicas
ICP	1. Internet Connection Provider
	2. Internet Cache Protocol
	3. Integrated Communications Provider
	4. Integrated Communications Platform
	5. Independent Communications Provider
	6. Intelligent Call Processor/Processing (AT&T service)
ICQ	"I Seek You" (instant messaging client)
ICR	1. Initial Cell Rate (ATM)
	2. Internet Call Routing node (Telcordia Technologies)
	3. In-Circuit Reconfiguration
ICS	1. Interactive Call Setup
	2. Integrated Communications System (Northern Telecom)
	3. Internet Connection Sharing
ICSA	International Computer Security Association (U.S.A.)
ICSC	1. Interexchange Customer Service Center
	2. Infrared Communication Systems Study Committee
ICSTD	Information, Communication, and Space Technology Division (UN)
ICT	Information & Communications Technology
ICTA	1. International Computer-Telephony Association
	2. International Center for Technology Assessment
	3. International Commission on Technology and Accessibility
	4. Independent Cable & Telecommunications Association
ICTC	Information and Communications Technology Center
ICTSB	ICT Standards Board (ETSI)
ICTUS	ICT Unified Submissions
ICV	Integrity Check Value
ICW	Interrupted Continuous Wave (modulation)

ID	1. **Id**entification
	2. **Id**entifier
	3. **I**dentification **D**igit(s)
	4. **I**nput **D**evice
	5. **I**ntermediate **D**evice
	6. **I**ntegrated **D**ispatch
IDA	1. **I**ntegrated **D**ata **A**ccess
	2. **I**ntegrated **D**igital **A**ccess
	3. **I**nterchange of **D**ata between **A**dministrations
IDAB	**I**mproved **D**igital **A**udio **B**roadcast (technology)
IDAL	**I**nternational **D**edicated **A**ccess **L**ine
IDAPI	**I**ntegrated **D**atabase **A**pplication **P**rogramming Interface
IDC	1. **I**nternet **D**ata **C**enter
	2. **I**ntermediate-rate **D**igital **C**arrier
	3. **I**nternational **D**ata **C**orporation (company)
	4. **I**nsulation **D**isplacement **C**onnector (DSL)
IDCMA	**I**ndependent **D**ata **C**ommunications **M**anufacturers **A**ssociation
IDCS	**I**ntegrated **D**igital **C**ommunications **S**ystem
IDCSP	**I**nitial **D**efense **C**ommunications **S**atellite **P**rogram
IDD	**I**nternational **D**irect **D**ial (code)
IDDD	**I**nternational **D**irect **D**istance **D**ialing (feature)
IDDS	**I**nstallable **D**evice **D**river **S**erver
IDE	1. **I**ntegrated **D**evice **E**lectronics (computer interface)
	2. **I**ntelligent **D**erive **E**lectronics (disk)
	3.**I**ntegrated **D**evelopment **E**nvironment (Windows program)
IDEA	**I**nternational **D**ata **E**ncryption **A**lgorithm
IDECM	**I**ntegrated **D**efense **E**lectronic **C**ounter**m**easure (military)
IDEF	**I**ntrusion **D**etection **E**xchange **F**ormat
IDEN	**I**ntegrated **D**igital **E**nhanced **N**etwork (wireless)
IDev	**I**nput **Dev**ice
IDF	**I**ntermediate **D**istribution **F**rame
IDFT	**I**nverse **D**iscrete **F**ourier **T**ransform (algorithm)
IDI	**I**nitial **D**omain **I**dentifier (ATM)
IDL	**I**nterconnect **D**escription **L**anguage
IDL	1. **I**nteractive **D**istance **L**earning (e-learning)
	2. **I**nterface **D**esign **L**anguage (programming)
IDLC	**I**ntegrated **D**igital **L**oop **C**arrier
IDM	**I**sochronous **D**e**m**odulation
IDML	**I**nternational **D**evelopment **M**arkup **L**anguage (programming)

IDN	1. Integrated Digital Network
	2. Information Digit Node (Sprint)
	3. International Domain Name (Internet)
	4. Identification Number
IDNX	1. International Digital Network Exchange
	2. Integrated Digital Network Exchange
IDP	1. Integrated Data Processing
	2. Internetworking Datagram Protocol (Internet)
IDR	1. Intermediate Data Rate (Intelsat)
	2. Isochronous Distortion Ratio (transmission)
IDRP	Inter-Domain Routing Protocol (Novell NetWare)
IDS	1. Intrusion Detection System
	2. Internal Data Services
	3. **Current** of Drain-to-Source (electronics)
IDSCP	Initial Defense Communication Satellite Program
IDSL	ISDN DSL (access)
IDSS	Saturated **current** of Drain-to-Source (electronics)
IDSU	Intelligent Data Service Unit
IDT	1. Inter-machine Digital Trunk
	2. Inter-DXC Trunk
	3. Interactive Data Transaction
	4. Integrated Digital Terminal
IDTS	Integrated Data Test System (software program)
IDTV	Improved-Definition Television (broadcasting)
IDU	1. In-Door Unit (VSAT)
	2. Interface Data Unit
	3. Indefeasible Right of Use (optical fiber)
IE	1. Information Engineering
	2. Internet Explorer (software)
IEC	1. Inter-Exchange Carrier (see IXC)
	2. Internet Exchange Center
	3. International Electrotechnical Commission (Geneva)
	4. International Engineering Consortium (U.S.A.)
IECEJ	Institute of Electronic and Communication Engineers of Japan
IEE	Institute of Electrical Engineers (U.K.)
IEEE	Institute of Electrical and Electronic Engineers, Inc. (standards)
IEI	Iran Electronic Industries
IEN	Internet Experimental Note
IEPG	1. Internet Engineering and Planning Group (ISPs)
	2. Intelligent Electronic Program Guide
IES	1. Information Exchange Service
	2. Illuminating Engineering Society

	3. Inter-Enterprise Systems
IESNA	Illuminating Engineering Society of North America
IESG	Internet Engineering Steering Group (Internet society)
IESS	Intelsat Earth Station Standard
IETF	Internet Engineering Task Force (IAB subcommittee)
IF	Intermediate Frequency
IFAC	International Federation of Automation Control
IFax	Internet-interfaced **Fac**simile device
IFB	Indirect Feedback
IFC	Information From Controller
IFCC	Internet Fraud Compliant Center
IFCM	Independent Flow Control Message
IFD	Image File Directory
IFE	In-Flight Entertainment system
IFEA	Internet Free Expression Alliance (organization)
.iff	Name extension for files having **IFF** format (computer)
IFF	1. Interchange File Format
	2. Identification, Friend, or Foe (aircraft detection equipment)
IFFM	Incoming First Fail to Match (telephone company)
IFFT	Inverse Fast Fourier Transform (algorithm)
IFG	Inter-Frame Gap (frame relay)
IFIP	International Federation for Information Processing
IFL	1. Inter-Facility Link (fiber optics)
	2. International Frequency Library
IFM	1. Intermediate Frequency Module
	2. Instantaneous Frequency Measurement
IFMP	Ipsilon Flow Management Protocol (IETF)
IFOC	Integrated Fiber Optic Circuit
IFOV	Instantaneous Field Of View (remote sensing)
IFP	Internet Fax Protocol
IFR	Instrument Flight Rules (aviation)
IFRB	International Frequency Registration Board (ITU-R)
IFS	1. International Freephone Service (ITU-T)
	2. Installable File System Manager
	3. Ionospheric Forward Scatter (propagation)
IFSK	Incoherent **FSK** (modulation)
IFSS	Intermediate Frequency Sub-System
IFTS	International Toll Free Service
IFWP	International Forum on the White Paper
IG	1. Isolated Ground
	2. International Gateway
	3. Insertion Gain (transmission)
IGA	Intermediate-Gain Antenna

I

IGBT	Insulated-Gate Bipolar Transistor (semiconductors)
IGC	Intelligent Graphics Controller
IGES	Initial Graphics Exchange Specification (file format)
IGFET	Insulated-Gate FET (semiconductors)
IGMP	Internet Group Management Protocol (networking)
IGO	Inter-Governmental Organization
IGP	1. Interior Gateway Protocol (Cisco)
	2. Integrated Graphics Processor
IGRP	Interior Gateway Routing Protocol (Cisco)
IGT	Ispettorato Generale della Telcomunicazioni (Italy)
IGW	International Gateway Switch
IGY	International Geophysical Year
IHF	1. Institute of High Fidelity
	2. Inherited-Halt Flip-flop
IHL	Internet Header Link
IHO	International Hydrographic Organization
IHP	Internal High-order path Protection
IIA	1. Information Industry Association (U.S.A.)
	2. Irish Internet Association
	3. Iran Informatics Association
IICA	International Intellectual Capital Codes Association
IICCA	International Intellectual Capital Codes Association
IIH	Is-IS Hellow (routing protocol)
III	Information Industry Index
IIIA	1. Integrated Internet Information Architecture
	2. Internet Information Infrastructure Architecture
	3. International Internet Industrial Association
IIL	Integrated-Injection Logic (digital electronics)
IILC	Information Industry Liaison Committee
IIN	Issuer Identification Number
IIOP	Internet Inter-ORB Protocol
IIR	1. Interactive Information Response
	2. Infinite Impulse Response (filter)
	3. Interactive Information Response
IIS	Internet Information Server (Windows NT)
IISP	1. Information Infrastructure Standards Panel (ANSI)
	2. Interim Interface Signaling Protocol
	3. Interim Interswitch Signaling Protocol
IITF	Information Infrastructure Task Force
IIW	ISDN Implementers Workshop
IJCAI	International Joint Conference on Artificial Intelligence
IJT	Intrinsic-Junction Transistor (semiconductors)
IKBS	Intelligent Knowledge-Based System (ICT)
IKE	Internet Key Exchange (IPsec)

IKP	Internet Keyed-payments Protocol
IL	1. Insertion Loss (transmission)
	2. Intermediate Language (programming)
ILA	1. Intermediate Light Amplification
	2. Injection-Locked Amplifier
ILAN	1. Interactive LAN (protocol)
	2. Inline Amplifier Node
ILAS	Improved Limb Atmospheric Sounder (remote sensing)
ILCR	International Least Cost Route
ILD	Injection Laser Diode
ILEC	Incumbent Local Exchange Carrier
ILF	Infra low Frequency (300–3000 Hz)
ILLED	Integral Lens Light-Emitting Diode
ILLP	Inter Link-to-Link Protocol
ILMI	Interim Link Management Interface (ATM)
ILNP	Interim Line Number Portability
ILO	Injection-Locked Oscillator
ILOVEYOU	"I LOVE YOU" (VBScript worm)
ILP	Internal Low-order path Protection
ILS	1. International Launch Service (satellite company in U.S.A.)
	2. Instrument Landing System (avionics)
	3. Input-buffer Limiting Scheme
	4. Integrated Learning System (e-learning)
	5. Internet Locator Service
ILSR	IPX Link State Router (networking)
ILT	1. Idle Line Termination
	2. Instructor-Led Training (e-learning)
	3. Institute for Learning and Teaching in Higher Education (U.K.)
IM	1. Intermodulation
	2. Intensity Modulation
	3. Isochronous Modulation
	4. Information Model
	5. Instant Messaging
IMA	1. Interactive Multimedia Association (Malaysia)
	2. Integrated Modular Avionics
	3. Inverse Multiplexing for ATM
IMAC	Isochronous Media Access Control (FDDI Architecture)
IMAG	Induction Magnetometer
IMAP	Internet Messaging Access Protocol
IMAP4	Internet Messaging Access Protocol 4

I

IMAS	Intelligent Maintenance Administration System (software)
IMASS	Intelligent Multiple Access Spectrum Sharing
IMAX	"I"—eye Maximum (cinematic system)
IMC	1. Internet Mail Consortium (U.S.A.)
	2. Interagency Management Council
	3. Instrument Meteorological Condition (avionics)
IMD	Intermodulation Distortion
IMDN	Intelligent Mobile Data Network
IME	Internally Mounted Equipment
IMEI	International Mobile-station Equipment Identifier (GSM)
IMEISV	**IMEI** plus a Software Version (GSM)
IMFET	Internally Matched **FET** (semiconductors)
IMG	Interferometric Monitor for Greenhouse Gases (remote sensing)
IMN	Inmarsat Mobile Number
IMO	International Maritime Organization
IMCC	ISDN Management Coordinating Committee
IML	Incoming Mail Loss (telephone company)
IMP	1. Interface Message Processor (ARPANet)
	2. Intermodulation Products
IMPATT	**Imp**act Avalanche Transit Time (semiconductors)
IMPDU	Initial MAC Protocol Data Unit
IMPS	Infinite Monkey Protocol Suite
IMR	Intermodulation Ratio
IMS	1. IP Multimedia Subsystem
	2. Information Management System (ICT)
IMS/VS	Information Management System/Virtual System
IMSI	International Mobile Subscriber Identity (GSM)
IMSO	International Mobile Satellite Organization
IMT	1. International Mobile Telecommunications
	2. Inter-Machine Trunk
IMT-2000	International Mobile Telecommunications-**2000**
IMTA	International Mobile Telecommunications Association
IMTC	International Multimedia Teleconferencing Consortium (U.S.A.)
IMTS	1. Improved Mobile Telephone System
	2. International Mobile Telecommunications System
IMUIMG	ISDN Memorandum of Understanding Implementation Management Group
IMUX	Inverse Multiplexer
IMW	Intelligent Music Workstation
in	inch

in/s	**in**ch **per s**econd
in²	**square in**ch
IN	**I**ntelligent **N**etwork
INA	1. **I**nformation **N**etwork **A**rchitecture (Telcordia Technologies)
	2. **I**nteractive **N**etwork **A**dapter (broadcasting)
INAP	1. **I**ntelligent **N**etwork **A**ccess **P**rotocol (INs)
	2. **I**ntelligent **N**etwork **A**pplication **P**art (INs)
InAs	**In**dium–**Ar**senide (semiconductors)
INAT	**In**termode **A**larm **T**ransport
Inbox	**In**coming mails **box** (e-mail)
INC	**In**ternational **C**arrier
INCC	**I**nternal **N**etwork **C**ontrol **C**enter
INCM	**I**ntelligent **N**etworks **C**all **M**odel
IND-E	**Ind**ian Ocean Region **E**ast (Inmarsat)
IND-W	**Ind**ian Ocean Region **W**est (Inmarsat)
INDIGO	**In**tegrated **Dig**ital **O**verlay
INDIX	**I**nternational **N**etwork for **D**evelopment **I**nformation **Ex**change
INE	**I**ntelligent **N**etwork **E**lement
INet	**I**nstitutional **Net**work
INET	short form of **Inter**net
info	**info**rmation field (LANs)
Infobahn	**Info**rmation auto**bahn** (superhighway)
INFOSEC	**Info**rmation **Sec**urity system
InGaAs	**In**dium–**Ga**llium–**Ar**senide (semiconductors)
INGECEP	**I**ntegrated **N**ext **G**eneration **E**lectronic **C**ommerce **E**nvironmental **P**roject
INH	**Inh**ibiter (microchips)
.ini	Name extension for **Ini**tialization files (computer)
INIC	**I**SDN **N**etwork **I**dentification **C**ode
INIM	**I**SDN **N**etwork **I**nterface **M**odule
INIRC	**I**nternational **N**on-**I**onizing **R**adiation **C**ommittee
INL	1. **In**ter**n**ode **L**ink
	2. **I**ntegral **N**on-**L**inearity
INMAC	**I**nternational **N**etwork **M**anagement **C**enter
Inmarsat	**In**ternational **mar**itime **sat**ellite (organization) Now it is called "**In**ternational **m**obile **sat**ellite Company"
INMC	**I**nternational **N**etwork **M**anagement **C**enter
INMLM	**I**ntegrated **N**etwork **M**anager **L**ink **M**odule
INMS	1. **In**marsat **M**onitoring **S**ystem
	2. **I**ntegrated **N**etwork **M**anagement **S**ervices

INN	1. **In**ter**n**ode **N**etwork
	2. **In**ter**n**et **N**ews
INNO	**I**nternational **N**eural **N**etwork **O**perator
INNS	**I**nternational **N**eural **N**etwork **S**ociety (Berkeley University, U.S.A.)
INode	Integrated **Node**
INOS	**I**ntelligent **N**etwork **O**perating **S**ystem
InP	**In**dium–**P**hosphide (semiconductors)
INP	1. **I**ntelligent **N**etwork **P**rocessor
	2. **I**nternational **N**etwork **P**lanning
	3. **I**nterim **N**umber **P**ortability
INPA	**I**nterchangeable **N**etwork **P**lanning **A**rea
INPE	1. **I**nstituto **N**acional de **P**esquisas **E**spaciais (Brazil)
	2. Brazilian Space Agency
INPS	**I**ntelligent **N**etwork **P**roduct **S**upport
INs	**I**ntelligent **N**etwork**s**
INS	1. **I**nformation **N**etwork **S**ystem
	2. **I**on **N**eutralization **S**pectroscopy (electronics)
	3. **I**nertial **N**avigation **S**ystem
	4. **I**ntegrated **N**etwork **S**ystem
	5. **I**nternet **N**aming **S**ervice
INSAT	**I**ndian **N**ational **Sat**ellite
INSP	**IN** **S**ervice **P**rovider
INSS	**IN** **S**ervice **S**ubscriber
int	**int**ernal
INT	**I**nduction **N**eutralizing **T**ransformer
INTAIP	**In**teroperability **T**echnology **A**ssociation for **I**nformation **P**rocessing
INTEGRAL	**In**ternational **G**amma **R**ay **A**strophysics **L**aboratory
Intelsat	**In**ternational **Tel**ecommunication **Sat**ellite (organization)
intercom	**intercom**munication system
InterLATA	Telecommunication services that originate in one **LATA** and terminate in another
Internaut	**Inter**net astro**naut**
internet	**inter**connection of two or more data **net**works
Internet	**Inter**national **Net**work
Internet2	The **second** Generation **Internet**
InterNIC	**Inter**net **N**etwork **I**nformation **C**enter (U.S.A.)
InterPBX	The calls **coming** to a **PBX**
INTERSPUTNIK	International Organization of Space Communications (Russia)
INTFC	**Inter**fa**c**e

IntraLATA	Telecommunication services that originate and terminate in the same as **LATA**
INTS	**Int**ernational **T**ransit **S**witch
IntServ	**Int**egrated **Serv**ices
INTUG	**Int**ernational **T**elecommunications **U**ser **G**roup
INV	**inv**erter
InWATS	**In**ward **W**ide-**A**rea **T**elephone **S**ervices
I/O	Input/Output
IO	**I**nformation **O**utlet
IOA	**I**nterim **O**perating **A**uthority (cellular networks)
IOC	1. **I**n-route/**O**ut-route **C**ontroller (electronics)
	2. **I**nput/**O**utput **C**ontroller
	3. **I**ntegrated **O**ptical **C**ircuit (electronics)
	4. **I**ntelsat **O**perations **C**enter
	5. **I**ntermediate **O**rbit **C**ommunications
	6. **I**ndependent **O**perating **C**arrier
	7. **I**nter-**O**ffice **C**hannel
	8. **IS**DN **O**rdering **C**ode
IOCTF	**I**ntelsat **O**perations **C**enter **TDMA F**acility
IOD	**I**dentified **O**utward **D**ialing (PBX)
IOEF	**I**nstantaneous **O**verride **E**nergy **F**unction (AT&T)
IOF	**I**nter-**O**ffice **F**acility
IOL	**I**nter-**O**perability **L**ab (New Hampshire University)
ION	1. **I**nternetworking **O**ver **N**BMA
	2. **I**ntelligent **O**ptical **N**etwork
IONL	**I**nternal **O**rganization of the **N**etwork **L**ayer (OSI model)
IOP	1. **I**nput/**O**utput **P**rocessor
	2. **interop**erability
IOPS	**I**nternet **O**perators Group
IOR	1. **I**ndian **O**cean **R**egion (Inmarsat)
	2. **I**ndex **O**f **R**efraction (fiber optics)
IOS	1. **I**nternetwork **O**perating **S**ystem (Cisco)
	2. **I**nter-**O**ffice **S**ection
	3. **I**ndian **O**cean **S**atellite (Intelsat)
	4. **I**nternational **O**rganization for **S**tandardization (Switzerland)
IOSA	**I**nter-**O**ffice **S**ection **A**daptation
IOST	**I**nter-**O**ffice **S**ection **T**ermination
IOT	1. **I**n-**O**rbit **T**esting (satellites)
	2. **I**nter-**O**perators **T**ariffs (GSM)
	3. **I**nter**O**perability **T**esting
IOTP	**I**nternet **O**pen **T**rading **P**rotocol
IOU	**I**nput/**O**utput **U**nit
IOV	**I**n-**O**rbit **V**erification (space researches)

IP	1. Internet Protocol (TCP/IP)
	2. Information Provider (Internet)
	3. Interface Processor (electronics)
	4. Intelligent Peripheral (device)
	5. Ingress Protection (hardware box standard)
	6. Intercom Profile (Bluetooth)
IP PBX	Internet Protocol PBX
IPA	1. Intermediate Power Amplifier
	2. Intellectual Property Attorney
IPARS	International Passenger Airline Reservation System (IBM)
IPBO	Input Power Back-Off
IPC	1. Inter-Process Communication (networking)
	2. Inter-Personal Communications
	3. Information Processing Center
	4. ISDN to POTS Converter
IPCC	Internet Protocol Contact Center
IPCE	Inter-Process Communication Environment
IPCH	Initial Paging Channel (GSM)
IPCI	Integrated Personal Computer Interface
IPComp	IP Payload Compression Protocol
IPconfig	Internet Protocol configuration (utility program)
IPCP	Internet Protocol Control Protocol
IPDC	Internet Protocol Device Control
IPDR	Internet Protocol Data Record
IPDRs	IP Detail Records
IPDS	1. Inmarsat Packet Data Service
	2. Intelligent Printer Data Stream (IBM)
IPDU	Internet Protocol Data Unit
IPE	1. Intelligent Peripheral Equipment (Northern Telecom)
	2. Initial Pointing Error
IPEI	International Portable Equipment Identities (wireless)
IPENZ	Institute of Professional Engineers of New Zealand
IPL	Initial Program Load (of operating systems)
IPLC	International Private Line Circuits
ipm	1. impulses per minute
	2. interruptions per minute
IPM	1. Inter-Personal Messaging system (X.400)
	2. Interference Prediction Model (propagation)
	3. Incremental Phase Modulation
	4. Internal Polarization Modulation

IPMC	Industrial Process Measurement and Control
IPMS	Inter-Personal Messaging Service
IPN	Impulse Noise
IPNC	IP Network Controller
IPND	Integrated Public Number Database
IPNG	Internet Protocol Next Generation
IPNGWG	IP Next Generation Working Group
IPNS	ISDN PBX Networking Specification
IPO	1. Initial Public Offering (finance)
	2. Input-Process-Output
IPoATM	Internet Protocol over ATM
IPoD	Internet Protocol phone over Data
IPONZ	Intellectual Property Office of New Zealand
IPoS	IP over SONET (IETF)
IPP	Internet Printing Protocol
IPPV	Impulse Pay Per View (broadcasting)
IPR	Intellectual Property Right
IPRA	Internet Policy Registration Authority
IPRS	Internet Protocol Routing Service (Bell)
IPS	1. Internet Protocol Suite
	2. Integrated Power Systems
IPS7	Internet Protocol Signaling 7
IPSec	IP Security policy (VPNs)
IPSN	International Public-Switched Network
IPSO	IP Security Option
IPSS	International Packet-Switched System
IPT	IP Telephony
IPT Gateway	IP Telephony Gateway
IPTC	IP Telephony solution for Carriers (Ericsson)
IPTV	Internet Protocol TV (Cisco)
IPU	Intelligent Processing Unit
IPv4	Internet Protocol version 4
IPv5	Internet Protocol version 5
IPv6	Internet Protocol version 6
IPX	Internetwork Packet Exchange (networking)
IPXCP	IPX Control Protocol (networking)
IPXODI	Internetwork Packet Exchange Open Data-link Interface
IPX/SPX	Internet Packet Exchange Sequenced Packet Exchange
IPXWAN	IPX Wide-Area Network (Novell specification)
I/Q	In-phase/Quadrature
IQF	Intrinsic Quality Factor (transmission)

IR	1. In-Route
	2. Infrared
	3. Internet Registry
	4. Information Retrieval
IRA	1. Internet Resource Access
	2. Impulse Radiating Antenna
IRAC	1. International Radiocommunications Advisory Committee
	2. Interagency Radio Advisory Committee
	3. Interdepartmental Radio Advisory Council (U.S.A.)
	4. Internal Review and Audit Compliance
	5. Infrared Array Camera
IRAM	Intelligent RAM (memory)
IRAMS	Innovative Real-time Antenna Modeling System
iraser	infrared maser
IRC	1. Internet-Relay Chat (ICT)
	2. Integrated Receiver decoder
	3. International Record Carrier
	4. Interference Rejection Combining (cellular networks)
IrCOMM	Provides COM (serial or parallel) port emulation or connections using IrDA protocol
IRCN	Inter-Regional Control Node
IRD	1. Integrated Receiver/Descrambler
	2. Integrated Receiver/Decoder (DSB)
	3. Internal Resources Database
IrDA	Infrared Data Association (standards organization)
IRDB	Intelligent Roaming Data-Base (cellular networks)
irdome	infrared dome
IRE	Institute of Radio Engineers
IRED	Infrared-Emitting Diode (electronics)
IREQ	Interrupt Request (signal)
IRF	1. Industrial Radio Frequency
	2. Impulse Response Function
	3. Inherited Rights Filter (networking)
IRFBP	Inmarsat Request For Business Plan
IRFU	Integrated Radio Frequency Unit
IRG	Inter-Record Gap (storage disks or tapes)
IRGB	Intensity Red Green Blue (monitors color encoding)
IRIB	Islamic Republic of Iran Broadcasting
IRICA	Iran Informatics Companies Association
IRIG	Inter-Range Instrumentation Group
iRIP	iCalendar Real-time Interoperability Protocol
IRL	Inter-Repeater Link (networking)
IrLAP	Infrared Link Access Protocol (IrDA Network)

IrLMP	Infrared Link Management Protocol (IrDA Network)
IRM	Image Rejection Mixer
IrMC	Infrared Mobile Communications
IROB	In Range Of Building
IRP	I/O Request Packet
IRQ	Interrupt Requests (computer)
IRR	1. Internet Routing Registry
	2. Image Rejection Ratio (heterodyning)
IRS	1. Inertial Reference System
	2. Indian Remote Sensing System
IRSC	Iranian Remote Sensing Center
IRSG	Internet Research Steering Group
IRSU	ISDN Remote Subscriber Unit
IRTF	Internet Research Task Force (IAB subcommittee)
IRTU	Integrated Remote Test Unit
IRU	Indefeasible Right of Use or User (undersea fiber cable)
IS	1. Information Services (ICT)
	2. Information Systems (ICT)
	3. Information Superhighway
	4. Information Separator (control character)
	5. Interactive Services
	6. Interim Standard
	7. Internal Shield (tube-base diagrams)
IS-2000	Interim Standard 2000 (cellular network)
IS-41	Interim Standard 41 (cellular network)
IS-410	Interim Standard 41 Zero (cellular network)
IS-41A	Interim Standard 41A (cellular network)
IS-95	Interim Standard 95 or CDMAone (cellular network)
IS-IS	1. Intermediate System to Intermediate System (routing protocol)
	2. Intelligent Scheduling and Information System
ISA	1. Iranian Space Agency (Iran)
	2. Industry Standard Architecture (old computers)
	3. Interactive Service Association
	4. Instrumentation, Systems, and Automation society
	5. Instrumentation Society of America
ISACA	Information Systems Auditability and Control Association
ISAKMP	Internet Security Association and Key Management Protocol
ISAM	Indexed Sequential-Access Method (databases)
ISAP	Istanbul Action Plan (ITU)
ISAPI	Internet Server Application Programming Interface (Microsoft)

I

ISAR	Inverse Synthetic-Aperture Radar
ISAS	Institute of Space and Astronautical Science (Japan)
ISB	1. Inter-Shelf Bus
	2. Independent Side-Band (modulation)
ISBN	Integrated Satellite Business Network
ISC	1. International Switching Center
	2. Internet Software Consortium
ISCA	International Speech Communication Association
ISCCP	International Satellite Clouds Climatologic Project (remote sensing)
ISCSI	Internet Small Computer Systems Interface
ISCP	Integrated Services Control Point (Telcordia Technologies)
ISD	1. Incremental Service Delivery
	2. Information Systems Department
	3. Internet Standards Document
ISDB	Integrated Services Digital Broadcasting
ISDE	Integrated Services Digital Exchange
ISDL	Integrated Services Digital Line
ISDN	Integrated Services Digital Network (transmission)
ISDN2	**ISDN** with **two** BRI channels and one D channel
ISDN2e	European **ISDN** with **two** BRI channels and one D channel
ISDN30	**ISDN** service delivering **30** BRI lines over one line
ISE	Integrated Switching Element
ISEC	Information Security Exploratory Committee
ISG	Incoming Service Group
ISHM	International Society of Hybrid Microelectronics
ISI	1. Information Sciences Institute
	2. Inter-Symbol Interference (transmission)
	3. Informatics Society of Iran
ISIA	Information and Software Industry Association
ISIS	1. Internal Switch Interface System (packet service)
	2. Intelligent Scheduling and Information System
ISL	1. Inter-station Signaling Link (Inmarsat)
	2. Inter-Satellite Link (Intelsat)
	3. Inter- Switch Link (Cisco Ethernet)
	4. ISDN Signaling Link (Northern Telecom)
ISLU	Integrated Services Line Unit (AT&T)
ISM	Inter-Stellar Medium
ISM Band	Industrial, Scientific and Medical frequency **Band**
ISMC	International Switching Management Center
	International Society for Measurement and Control

ISMN	Integrated Switching Management Network
	International Standard Music Number
ISMS	Interactive SMS (GSM)
ISN	1. Inmarsat Serial Number
	2. Intelligent Service Node (AT&T)
	3. Information Systems Network (AT&T)
	4. Initial Sequence Number
ISNAP	Intelligent Services Network Applications Processor
ISNET	Inter-Islamic Network on Space Sciences and Technology
ISNI	Intermediate Signaling Network Identification
ISO	1. International Standards Organization (Switzerland)
	2. isolator
ISOC	Internet Society
ISoD	Interactive Service on Demand
ISODE	ISO Development Environment (OSI model)
IsoENET	Isochronous Ethernet
ISO/OSI model	International Organization for Standardization Open System Interconnection **model**
ISP	1. Internet Service Provider
	2. Internet Safety Policy
	3. Inmarsat Service Provider
	4. Integrated Service Provider
	5. In-System Programmable
	6. Information Service Platform
	7. ISDN Signal Processor
ISPBX	Integrated Services Private Branch Exchange
ISPC	1. International Sound Program Center
	2. International Signaling Point Code
ISPT	Instituto Superiore delle Poste e delle Telecomunicazioni (Italy)
ISR	1. Intermediate Session Routing (routing algorithm)
	2. Interrupt Service Routine (computer)
	3. International Simple Resale
ISRO	Indian Space Research Organization (India)
ISS	1. Inter-Satellite Service
	2. Intelligent Services Switch
	3. International Space Station (NASA)
	4. Integrated Satellite Solutions Corporation
ISSA	Information Systems Security Association
ISSI	Inter-Switching System Interface
ISSLL	Integrated Services over Specific Link Layers (QoS)
ISSS	Information Society Standardization System (ICT)
ISSN	Integrated Special Services Network

IST	1. Initial Service Term
	2. Independent Sideband Transmission
	3. Information Society Technologies
ISTF	Integrated Services Test Facility
ISU	1. Initial Signal Unit
	2. Iridium Subscriber Unit
ISUP	ISDN User Part (SS7)
ISV	Independent Software Vendor (ICT)
ISVC	Inmarsat Standard Voice Codec
IT	1. Information Technology (ICT)
	2. Inter-Toll Trunk
IT&T	Information Technology **and** Telecommunication
ITA	1. International Telegraph Alphabet
	2. Integrated Trunk Access
	3. Industrial Telecommunications Association
ITA1	International Telegraph Alphabet **No.1** (ITU)
ITA5	International Telegraph Alphabet **No.5** (ITU)
ITAA	Information Technology Association of America (standards)
ITAB	Information Technical Advisory Board
ITAC	1. Information Technology Association of Canada
	2. ISDN Terminal Adapter Circuit
Italsat	Italian Domestic Satellite System
ITANZ	Information Technology Association of New Zealand
ITAR	International Traffic in Arms Regulations (U.S. State Dept.)
ITB	Intermediate Block Character (transmission)
ITC	1. International Transit Center
	2. Independent Television Commission (broadcasting)
	3. Information To Controller
	4. International Teletraffic Congress
ITCA	International Tele-Conferencing Association (U.S.A.)
ITCC	International Telecommunications Clearing Corporation
ITCI-DC	Internet Training Centers Initiative for Developing Countries (ITU)
ITCO	Independent Telephone Company
ITDM	Intelligent Time-Division Multiplexer
ITESF	Internet Traffic Engineering Solution Forum (Telcordia Co.)
ITF	Interframe Time Fill
ITFS	1. Instructional Television Fixed Service
	2. International Toll Free Service
ITG	Integrated Telemarketing Gateway (AT&T)

ITI	1. Interactive Terminal Interface (X.25)
	2. Intelligent Transportation Infrastructure (U.S.A.)
	3. Iran Telecommunication Industries
	4. Idle Trunk Indicator
ITIC	Information Technology Industry Council (U.S.A.)
iTIP	iCalendar Transport-Independent Interoperability Protocol
ITM	Information Technology Management
ITMC	1. International Transmission Management Center
	2. Iran Telecommunications Manufacturing Company
ITMN	Integrated Transmission Management Network
ITN	1. Independent Telephone Network
	2. Inter-office Transport Network
ITO	International Telecommunication Organization
ITORP	IntraLATA Toll Originating Responsibility Plan
ITPD	Iran Telephone Planning and Development Company
ITR	1. International Telecommunication Regulations (ITU)
	2. Internet Talk Radio
ITRC	1. Information Technology Research Center
	2. Iran Telecommunication Research Center
ITS	1. Institute for Telecommunication Sciences (U.S.A.)
	2. Intelligent Transportation Systems
ITSec	Information Technology Security
ITSO	International Telecommunications Satellite Organization
ITSP	Internet Telephony Service Provider
ITT	1. International Telephone and Telegraph (company)
	2. Invitation To Tender (contracts)
ITU	International Telecommunications Union (standards organization)
ITU-D	ITU-Telecom Development Sector
ITU-R	ITU-Recommendation Sector
ITU-SB	ITU Standardization Bureau
ITU-TS	ITU-Telecommunication Standardization Sector
ITUR	Italy–Turkey–Ukraine–Russia (fiber optic cable consortium)
ITUSA	Information Technology User's Standards Association
ITV	Interactive Television (broadcasting)
ITXC	Internet Telephony Exchange Carrier
IU	Interface Unit
IUA	ISDN User Adaptation Layer
IUS	Interim Upper Stage
IUT	Implementation Under Test

IUW	ISDN Users Workshop
IV	Interactive Video (ICT)
IVAN	International Value-Added Network
IVC	Inmarsat Voice Codec
IVCP	Installation Verification Certification Program
IVD	1. International Volume Discount
	2. Inside Vapor Deposition (optical fibers)
	3. Integrated Voice/Data
IVDM	Integrated Voice and Data Multiplexer
IVDS	Interactive Video Data Service (FCC)
IVDT	Integrated Voice/Data Terminal
IVHS	Intelligent Vehicle Highway System
IVI	Intel Video Interactive
IVMS	In-Vehicle Multiplexing System (standard)
IVN	Intervening Network
IVoD	Interactive Video-on-Demand
IVPN	International Virtual Private Network
IVPO	Inside Vapor Phase Oxidation process (optical fibers)
IVR	Interactive Voice Response
IVS	1. Interactive Voice Service
	2. Interactive Video Services
IW	1. Inside Wiring (telephone wire)
	2. intraworking
	3. interworking
	4. Information Warfare
	5. Intelligent Workstation
IW3C2	International World Wide Web Conference Committee
IWA	Intelligent Workstation Architecture
IWF	1. Interworking Function (PLMN)
	2. Internet Watch Foundation
IWG	Intersessional Working Group (Inmarsat)
IWS	Intelligent Work-Station
IWTA	International Wireless Telecommunications Associations
IWU	1. Internet Working Unit
	2. Interworking Unit
IWV	Impulse Wahlverfahren (German pulse dialing)
IX	Internet Exchange
IXC	Inter-Exchange Carrier (long-distance telephone company)
IXM	Inter-Exchange Mileage
IXO	Information Exploitation Office (DARPA)

J j

J	Symbol for **J**oule (unit of work and energy)
J Box	Junction **Box**
J-Carrier	**J**apanese equivalent of T-**Carrier**
J-hook	**J**-shaped **hook** connector
J-SAC	**J**ournal on **S**elected **A**reas in **C**ommunications (IEEE)
J0	Regeneration section trace byte (MUX)
J1	Path trace byte (MUX)
J2	Path overhead byte (MUX)
J2ME	**J**ava **2** Platform, **M**icro **E**dition
JABWT	**J**ava **A**pplets for **B**luetooth **W**ireless **T**echnology
JAD	**J**oint **A**pplication **D**esign
JAE	**J**ava **A**pplication **E**nvironment
JAM	**J**ini Technology **A**ccess **M**odule
JAMSAT	**J**apanese affiliate of **AMSAT** (satcom)
JANet	**J**oint **A**cademic **Net**work (U.K.)
JAR	**J**ava **Ar**chive
JARL	**J**apan **A**mateur **R**adio League, Inc.
JAROS	**Ja**pan **R**esources **O**bservation **S**ystem (organization)
JASC	**Ja**pan **S**ea **C**able
JASREP	**Ja**pan **S**hip **Rep**orting system
JATE	**J**apan **A**pprovals Institute for **T**elecommunications **E**quipments
JAXA	**J**apan **A**erospace **Ex**ploration **A**gency (Japan)

JB7	Jam Bit 7
JBIG	Joint Bitonal Image Group
JBOD	Just a Bunch Of Disks (storage technology)
JCALS	Joint Computer-aided Acquisition and Logistics Support
JCC	Joint Call-back Center
JCL	Job Control Language (IBM mainframes)
JCSA	Japan Computer Security Association
JDAP	Java Directory Access Protocol
JDBC	Java DataBase Connectivity
JDC	Japanese Digital Cellular System
JDK	Java Development Kit
JDS	Japanese Digital Standard
JEC	Jahan Electronic Company (Iran)
JECF	Java Electronic Commerce Framework
JECS	Job-by-Email Control System
JEDEC	Joint Electronic Device Engineering Council (standards)
JEDI	Joint Electronic Document Interchange
JEIDA	Japan Electronic Industry Development Association
JEMA	Japan Electronic Messaging Association
JEPI	Joint Electronics Payments Initiative (W3C)
JERS	Japanese Earth Resource Satellite (remote sensing)
JES	Job Entry Subsystem
JESSI	Joint European Semiconductor Silicon Initiative
JET	Just-Enough-Time (protocol)
JF	Junction Frequency
JFET	Junction Field Effect Transistor (semiconductors)
.jfif	Name extension for JPEG File Interchange Format (computer)
JHTML	Java and HTML (programming)
JI	Junction Isolation (electronics)
JIOA	Japan Institute of Office Automation
JIPS	JANET Internet Protocol Service
JIS	Japanese Industrial Standard
JISA	Japan Information Service Association
JISC	1. Joint Information Systems Committee (U.K. ICT) 2. Japanese Industrial Standards Committee
JIT	Just-In-Time (product manufacturing)
JIVN	Joint Intelligence Virtual Network
JMA	Japanese Meteorological Agency
JMAPI	Java Management Application Programming Interface
JMOS	Junction MOS (semiconductors)
JMTS	Jordanian Mobile Telephone System

JMTSS	**J**oint **M**ultichannel **T**runking and **S**witching **S**ystem (military)
JNT	**J**oint **N**etwork **T**eam
JOFX	**J**ava **O**pen **F**inancial **Ex**change
JOLT	**J**ava **O**pen **L**anguage **T**oolkit (project)
JOSS	**J**ohnniac **O**pen **S**hop **S**ystem (programming language)
JOTP	**J**ava **O**pen **T**rading **P**rotocol
JOVIAL	**J**unior **O**fficer's **V**ersion of an **I**ncomprehensible **A**rithmetic **L**anguage (programming)
JPC	**J**oint **P**rocurement **C**onsortium
JPCD	**J**ust **P**erceptible **C**olor **D**ifference
.jpeg	Name extension for **J**oint **P**hotographic **E**xpert **G**roup files
JPEG	**J**oint **P**hotographic **E**xperts **G**roup (image format)
.jpg	Name extension for **JP**EG format Images files (computer)
JPIX	**Ja**pan **I**nternet **Ex**change
JPS	**J**oint **P**roduct **S**pecification
JPSA	**J**apan **P**ersonal Computer **S**oftware **A**ssociation
JRC	**J**apan **R**adio **C**ompany
JRCC	**J**oint (aeronautical and maritime) **RCC** (COMSAR)
JRG GII	**J**oint **R**apporteur **G**roup **G**lobal **I**nformation Infra-structure
JSAT	**J**apan **Sat**ellite Systems (broadcaster)
JSC	**J**ohnson **S**pace **C**enter (U.S.A)
JSD	**J**ustification **S**ervice **D**igit
JskyB	**J**apan **sky** **B**roadcasting (company)
JSP	1. **J**ackson **S**tructured **P**rogramming (specifications) 2. **J**ava **S**erver **P**ages (specifications)
JST	**J**apan **S**olderless **T**erminal
JT	**J**ordan **T**elecom (operator)
JTAC	**J**apanese **T**otal **A**ccess **C**ommunications
JTACS	**J**apanese **T**otal **A**ccess **C**ommunications **S**ystem
JTAG	**J**oint **T**est **A**ction **G**roup (IEEE)
JTAPI	**J**ava **T**elephony **API**
JTC	**J**oint **T**echnical **C**ommittee
JTC³A	**J**oint **T**echnical **C**ommand, **C**ontrol, and **C**ommunications **A**gency
JTEC	**J**apanese **T**echnology **E**valuation **C**enter
JTFA	**J**oint **T**ime-**F**requency **A**nalysis (DSP)
JTG	**J**ordan **T**elecom **G**roup
JTIDS	**J**oint **T**echnical **I**nformation **D**istribution **S**ystem
JTK	**J**ava **T**ool **K**it
JTM	**J**ob **T**ransfer and **M**anipulation (file transfer)

J

JTRB	Joint Telecommunications Resources Board
JTRS	Joint Tactical Radio Systems (U.S. Army)
JTSSG	Joint Telecommunications Standards Steering Group
JUG	Joint Users Group
JUNET	Japan Unix Network
JUNOS	Juniper Networks Operating System
JUTCPS	Joint Uniform Telephone Communications Precedence System
JVC	Victor Company of Japan
JVM	Java Virtual Machine (software)
JvNCnet	John von Neumann Center Network
JWICS	Joint Worldwide Intelligent Communications System
JXML	Java XML (programming language)

K k

k	1. Symbol for **k**ilo denoting 1000 or 10^3
	2. Symbol for **k**ilo or 1024 of data elements
	3. Symbol for a constant value (mathematics)
K	1. Symbol for **K**elvin degree (unit of temperature)
	2. Symbol for **C**athode (circuit diagrams)
	3. Symbol for **C**oefficient of absorption (fiber optics)
K&R C	Brian **K**ernighan and Dennis **R**itchie **C** programming Language
K-band	Radio frequency **band** ranging from 10.9 to 36 GHz
kA	**k**iloampere
Ka-band	Radio frequency **band** ranging from 18 to 22 GHz
KACST	**K**ing **A**bdulaziz **C**ity of **S**cience and **T**echnology (Saudi Arabia)
KADS	**K**nowledge **A**nalysis and **D**esign **S**ystem (ICT)
KAFOS	**KA**RA Daniz **F**iber **O**ptic **S**ystem (Turkey)
KAI	**K**orea **A**erospace **I**ndustries Ltd
KARI	**K**orean **A**erospace **R**esearch **I**nstitute (South Korea)
kb	**k**ilo**b**it, which is 1000 bits (transmission)
kB	**k**ilo**b**yte, which is 2^{10} or 1024 bytes (computer)
KBC	**K**orean **B**roadcasting **C**ommission (South Korea)
kBd	**k**ilo**b**au**d** (1024 baud)
KBHCA	**Thousand B**usy **H**our **C**all **A**ttempts
kb/s	**k**ilo**b**its **per s**econd

K

175

kbps	kilobits per second
kB/s	kilobytes per second
kBps	kilobytes per second
KBS	Knowledge-Based System (ICT)
kc	kilocycle
kCHAR	kilocharacter
kCi	kilocurie
KCI	Kerman Cable Industries (Iran)
KCL	Kirchhoff's Current Law (electronics)
KDC	Key Distribution Center
KDD	Kokusai Denshin Denwa Company, Ltd (Japan)
KDDI	Kokusai Denshin Denwa International (Japan)
KDE	Kool Desktop Environment project (Linux)
KDR	Keyboard Data Recorder (computer)
KDT	Keyboard Display Terminal
KEA	1. Key Exchange Algorithm
	2. Key Encryption Algorithm
keV	kiloelectronvolts
KEYB	keyboard
KFPA	K-band Full Performance Antenna
kft	kilofeet or one thousand feet
kg	kilogram
kG	kilogauss
kHz	kilohertz (10^3 Hz)
KIF	Knowledge Interchange Format
KIS	Knowbots Information Service
kLm	kilolumen
KLY	klystron (tube)
km	kilometer
km/s	kilometer per second
km/h	kilometer per hour
KM	Knowledge Management (ICT)
KMBA	Knowledge Management Benchmarking Association (ICT)
KMID	Key Material Identifier
KMS	Knowledge Management System (ICT)
KNet	Kangaroo Network product
KNI	Katmai New Instructions set (Intel)
KΩ	Kiloohm (electronics)
KOMPSAT	Korea Multipurpose Satellite program
KP	Key Pulsing
KPA	Klystron Power Amplifier
KPI	Key Performance Indicator

KQML	**K**nowledge **Q**uery and **M**anipulation **L**anguage (programming)
KS	**K**earney **S**ystem
KSIA	**K**orean **S**emiconductor **I**ndustry **A**ssociation
ksps	**k**ilo**s**ymbols **p**er **s**econd
KSR	**K**eyboard **S**end and printer **R**eceive (device)
KSU	**K**ey **S**ervice **U**nit
Kt	**k**no**t**
KT	1. Symbol for Boltzman's **Constant**-time **T**emperature 2. **K**orea **T**elecom
KTA	**K**ey **T**elephone **A**dapter
KTF	**K**orea **T**elecom **F**reeTel (South Korea)
KTI	**K**ey **T**elephone **I**nterface
KTILA	**D**evelopment **C**enter for **T**elecommunications (Greece)
KTN	**K**ernel **T**ransport **N**etwork
KTS	**K**ey **T**elephone **S**ystem
KTU	**K**ey **T**elephone **U**nit
Ku-band	Radio frequency **band** ranging from 11 to 14.5 GHz
kV	1. **k**ilo**v**olt 2. **K**ey **V**oice (telephone)
kVA	**k**ilo**v**olt**a**mpere
kVAC	**k**ilo**v**olt **A**lternating **C**urrent
kVDC	**k**ilo**v**olt **D**irect **C**urrent
KVL	**K**irchhoff's **V**oltage **L**aw (electronics)
KVM	**K** **V**irtual **M**achine (J2ME)
KVM Switch	**K**eyboard **V**ideo **M**ouse **Switch**
kVp	**k**ilo**v**olt **p**eak
kVp-p	**k**ilo**v**olt **p**eak-**to-p**eak
kVr.m.s	**k**ilo**v**olt **r.m.s**
kW	**k**ilo**w**att
kWh	**k**ilo**w**att-**h**our
KWIC	**K**ey **w**ord-**I**n-**C**ontext (index)
KWOC	**K**ey **w**ord **O**ut of **C**ontext (index)

K

L

L	1. Symbol for an **inductor** or **inductanc**e (coil)
	2. Symbol for **L**ambert (unit of luminance)
	3. Symbol for **L**iter (unit of volume)
	4. Symbol for **L**ength (mathematics)
L-ADM	**L**inear **A**dd/**D**rop **M**ultiplexer (MUX)
L-band	Radio frequency **band** in the spectral region of 1.5 GHz
L-to-C	**L**-band **to** **C**-band
L1 cache	**L**evel **1 cache** (memory)
L2F	**L**ayer **2 F**orwarding protocol (Cisco)
L² FET	**L**ogic **L**evel **FET** (semiconductors)
L2TP	**L**ayer **2 T**unneling **P**rotocol (VPNs)
L2TPext	**L**ayer **2 T**unneling **P**rotocol **Ext**ension (Cisco)
L3	**L**ayer **3** switching
LA	1. **L**ocation **A**rea
	2. **L**ine **A**dapter
	3. **L**isted **A**ddress (Internet)
LAA	**L**ocally **A**dministered **A**ddress
LAAS	**L**ocal **A**rea **A**ugmentation **S**ystem (GPS)
LAB	**L**ogic **A**rray **B**lock
LAC	1. **L**oop **A**ssignment **C**enter
	2. **L**ocation **A**rea **C**ode (GSM)
LACE	**L**ow-power **A**tmospheric **C**ompensation **E**xperiment
LACN	**L**ocal **A**rea **C**ontrol **N**etwork

LAD	1. **L**ATA **A**rchitecture **D**atabase
	2. **L**ife **A**fter **D**eath (satcom)
	3. **L**inear **A**mplitude **D**ispersion
ladar	1. **la**ser **d**etection **a**nd **r**anging
	2. **la**ser ra**dar**
LADS	**L**ocal **A**rea **D**ata **S**ervice
LADT	**L**ocal **A**rea and **D**ata **T**ransport (transmission)
LAE	**L**iquid **A**pogee **E**ngine (satellites)
LAGEOS	**La**ser **Geo**dynamics **S**atellite (U.S.A.)
LAI	**L**ocation **A**rea **I**dentity (GSM)
LAIIX	**L**os **A**ngeles **I**nternational **I**nternet E**x**change
LAM	1. **L**iquid **A**pogee **M**otor (satellites)
	2. **L**ine **A**dapter **M**odule
	3. **L**obe **A**ttachment **M**odule
LAMA	**L**ocal **A**utomatic **M**essage **A**ccounting
LAMAR	**L**arge-**A**rea **M**odular **A**rray of **R**eflectors (antennas)
LAMM	**L**arge-**A**ntenna **M**ini-**M** terminal (Inmarsat)
LAMMR	**L**arge-**A**ntenna **M**ultifrequency **MW** **R**adiometer
LAMP	**L**arge **A**dvanced **M**irror **P**rogram
LAN	**L**ocal **A**rea **N**etwork
LANAC	**La**minar **N**avigation **A**nti-**C**ollision (aviation)
LANalyzer	**LAN** **analyzer** (networking)
LANCE	**L**ocal **A**rea **N**etwork **C**ontroller for **E**thernet
LANDA	**L**ocal **A**rea **N**etwork **D**ealers **A**ssociation (Canada)
LANE	**LAN** **E**mulator (ATM)
LANNET	**L**arge **A**rtificial **N**euron **Net**work
LANS	**L**ocal **A**rea **N**etwork **S**ervices
LAP	1. **L**ink **A**ccess **P**rotocol
	2. **L**ink **A**ccess **P**rocedure
LAP-B	**L**ink **A**ccess **P**rocedure-**B**alanced (X.25)
LAP-D	**L**ink **A**ccess **P**rocedure for **D** channel (ISDN)
LAP-F	**L**ink **A**ccess **P**rocedure for **F**rame Relay
LAP-M	**L**ink **A**ccess **P**rocedure for **M**odems
LArc	1. **L**empel–Zive data **Arc**hiving program
	2. **L**ivermore **A**utomatic **r**esearch **c**omputer
LAS	**L**ow-**A**ltitude **O**bservation **S**atellite
LASA	**L**arge-**A**perture **S**eismic **A**rray (antennas)
LASCR	**L**ight-**A**ctivated **S**ilicon-**C**ontrolled **R**ectifier (electronics)
LASCS	**L**ight-**A**ctivated **S**ilicon-**C**ontrolled **S**witch (electronics)
laser	**L**ight **A**mplification by **S**imulated **E**mission of **R**adiation
LASINT	**Las**er **Int**elligence
LASS	**L**ocal **A**rea **S**ignaling **S**ervices (PBX)
LASSI	**L**ow-**A**ltitude **S**atellite **S**tudies of **I**onosphere

LAST	1. **L**ocal **A**rea **S**ystems **T**echnology
	2. **L**arge-**A**perture **S**canning **T**elescope
Lat	**lat**itude
LAT	1. **L**ocal **A**rea **T**ransport (transmission)
	2. **L**ocal **A**rea **T**erminal
	3. **L**ocal **A**rea **T**ransit (DEC protocol)
LATA	1. **L**ocal **A**ccess and **T**ransport **A**rea (U.S. telephone)
	2. **L**ocal **A**rea **T**ransport **A**rrangement
LATNET	**Lat**vian **Net**work
LATTIS	**L**ocal **A**rea **T**ransport **T**ariff **I**nformation **S**ystem
LAU	**L**obe **A**ccess **U**nit
LAVC	**L**ocal **A**rea **V**AX **C**luster
LAWN	**L**ocal **A**rea **W**ireless **N**etwork
LAWRS	**L**imited **A**irport **W**eather **S**tation
Lb	Symbol for pound (unit of weight)
LB	1. **L**oop-**B**ack
	2. **L**eaky **B**ucket (ATM)
LBA	**L**ogical **B**lock **A**ddressing (enhanced IDE)
LBC	1. **L**inear **B**lock **C**oding
	2. **L**inked **B**oundary **C**ondition
LBEN	**L**ow-**B**yte **En**able
LBI	**L**oad **B**alance **I**ndex (telephone company)
LBO	**L**ine **B**uild **O**ut (cables)
LBRV	**L**ow-**B**it-**R**ate **V**oice
LBS	1. **L**ocation-**B**ased **S**ervices (wireless and cellular networks)
	2. **L**oad-**B**alancing **S**ystem
LBT	**L**isten **B**efore **T**alk
LC	1. **L**ine **C**ard
	2. **L**ine **C**lock
	3. **L**eased **C**ircuit
	4. **L**ead **C**hannel
	5. **L**ocal **C**ompany
	6. **L**ocal **C**hannel
	7. **Inductance**–**C**apacitance (circuit diagrams)
LCA	1. **L**ogical **C**hannel **A**ssignment
	2. **L**ocal **C**alling **A**rea
LCAS	**L**ink **C**apacity **A**djustment **S**cheme
LCC	**L**eadless **C**hip **C**arrier (circuit board mounting)
LCD	1. **L**iquid **C**rystal **D**isplay (electronics)
	2. **L**oss of **C**ell **D**elineation (ATM)
	3. **L**inear **C**ollider **D**etector
LCE	**L**ine **C**onditioning **E**quipment

L

LCI	1. Logical Channel Identifier (X.25)
	2. Line-Conducted Interference
LCL	Longitudinal Conversion Loss
LCM	1. Line Control Module
	2. Least Common Multiple
	3. Line Concentrating Module
LCMS	Learning Content Management System (e-learning)
LCN	1. Logical Channel Number (ISDN)
	2. Local Communication Network
LCOS	Liquid Crystal On Silicon (semiconductors)
LCP	1. Link Control Protocol (OSI model)
	2. Left Circularly Polarized (antennas)
LCPGA	Low-Cost Pin-Grid Array (microchips)
LCR	1. Least Cost Routing (service)
	2. Line Concentrating Ratio
	3. Level Crossing Rate
	4. **Inductance**–Capacitance–**R**esistance (circuit diagrams)
LCS	1. Low-order Connection Supervision
	2. Live Call Screening (feature)
	3. Laboratory for Computer Science (MIT University)
LCT	Large Capacity Trunk
LCU	1. Lightweight Computer Unit
	2. Line Control Unit
LCV	Line Code Violation (data encoding)
LD	1. Long Distance
	2. Line Driver
	3. Laser Diode
	4. Loop Disconnect
LD-AAC	Low-Delay Advanced Audio Compression
LD-CELP	Low-Delay Code-Excited Linear Prediction (voice coding)
LDA	Long-Distance Alerting
LDAP	Lightweight Directory Access Protocol (network management)
LDB	1. Local Database
	2. Long-Distance Bypass (switching)
LDBS	Local Database Services
LDD	Limited Distance Dialing
LDDC	Long-Distance Direct Current (dialing system)
LDE	Long-Delayed Echo
LDF	1. Local Distribution Frame
	2. Lightwave Distribution Frame
LDGPS	Local Differential GPS

LDIP	Long-Distance Internet Provider
LDL	Long-Distance Line
LDM	Limited Distance Modem
LDMC	Loop Data Maintenance Center
LDMOS	Laterally Diffused MOS (semiconductors)
LDMS	Local Multipoint Distribution Service
LDN	Listed Directory Number (PBX)
LDP	Label Distribution Protocol (MPLS)
LDR	Light-Dependent Resistor (semiconductors)
LDS	Local Digital Switch
LDT	1. Line appearance on a Digital Trunk
	2. Lightning Data Transfer bus
LDTV	Low-Definition Television
LDU	1. Line Director Unit
	2. Local Distribution Utility
LE	1. Local Exchange
	2. LAN Emulator
	3. Link Encapsulation
LE ARP	LAN Emulator Address Resolution Protocol
LEA	1. Law Enforcement Authority
	2. Longitudinally Excited Atmosphere (laser)
LEAF	1. Law Enforcement Access Field
	2. Large Effective Area Fiber
LEAP	Lightweight Extensible Authentication Protocol
LEAPS	Long-term Equipment Anticipation Securities
LEAS	LATA Equal Access System
LEC	1. LAN Emulation Client (ATM)
	2. Local Exchange Carrier (company)
	3. Local Engineering Circuit
	4. Loop Electronic Coordinator
LECID	LAN Emulation Client Identifier (ATM)
LECS	LAN Emulation Configuration Server (ATM)
LED	Light-Emitting Diode (electronics)
LEED	Low-Energy Electron Diffraction
LEF	Library-Exchange Format
LEMS	Low-Emission Mobile System
LEN	1. Line Equipment Number
	2. Low-Entry Networking (SNA)
LEO	Low Earth Orbit (satellite)
LEOS	Low Earth Orbit Satellite
LEOW	Low-cost Exploration Operator Workstation (image processing)
LEP	1. Large Electron Positron Collider
	2. Light-Emitting Polymer

	3. Left Elliptically Polarized (antennas)
LER	Label Edge Router (MPLS)
LERG	Local Exchange Routing Guide (Telcordia Technologies)
LES	1. Land Earth Station (Inmarsat)
	2. LAN Emulation Server (ATM)
	3. Loop Emulation Service (ATM)
	4. Label Edge Switch
	5. Lincoln Experimental Satellite
	6. Line Errored Second
LES-SIG	Land Earth Station Signaling channel (Inmarsat)
LESA	Land Earth Station Assignment channel
LESD	Land Earth Station Data channel
LESI	Land Earth Station Interstation signaling channel
LESO	Land Earth Station Operator (Inmarsat)
LESV	Land Earth Station Voice channel
LETN	Law Enforcement Television Network (broadcaster)
LF	1. Low-Frequency (30–300 KHz range)
	2. "Linefeed" (character control code)
LFA	Loss of Frame Alignment
LFACS	Loop Facility Assignment and Control System
LFAP	Lightweight Flow Admission Protocol
LFB	Look-ahead-For-Busy (information)
LFI	Link Fragmentation and Interleave (VoIP)
LFC	Large Format Camera (remote sensing)
LFN	Long File Name (Microsoft standard)
LFOV	Limited Field-Of-View (radar)
LFSR	Linear Feedback Shift Register
LFSS	Link Failed Signal State
LG	Line Group
LGA	Low-Gain Antenna
LGC	Line Group Controller
LGCI	ISDN Line Group Controller
LGE	Loop and Ground start Exchange card
LGN	1. Logical Group Node (ATM)
	2. Logical Group Number
LGRS	Local Government Radio Service
LGS	Loop and Ground start Subscriber card
LHARC	Lempel–Zive and Huffman data Archiving program
LHC	Long-Haul Communications
LHCN	Long-Haul Communications Network
LHCP	Left-Hand Circular Polarization (microwave)
LHEP	Left-Hand Elliptical Polarization (microwave)
LHMC	Long-Haul Mileage Calculation

LHOTS	Long-Haul Optical Transmission Set
Li-Ion	Lithium-Ion (battery)
LIB	Label Information Base (MPLS)
LIC	1. Light guide Interconnect Cable
	2. Lowest Incoming Channel
LID	Line Identification/Information Database
LIDAR	1. Light Detection And Ranging (remote sensing)
	2. Laser Infrared radar
LIDB	Line Information Database
LIF	1. Location Interoperability Forum
	2. Logarithmic Intermediate Frequency (amplifier)
	3. Low Insertion Force (sockets)
LIFO	Last-In, First-Out (buffer memories)
LIGBT	Lateral Insulated-Gate Bipolar Transistor (semiconductors)
LightSAR	**Light S**ynthetic **A**perture **R**adar
LIJP	Leaf Initiated Joint Parameter (ATM)
LILO	Linux Loader
LIM	1. Line-access Interface Module (Ericsson)
	2. Link Interface Module
	3. Lotus Intel Microsoft (committee)
LIM-EMS	Lotus Intel Microsoft—Expanded Memory Specification
LIMS	1. Legal Interception Management System
	2. Laboratory Information Management System
LINC	Laboratory Instruments Computer (MIT University)
LINCOMPEX	**Lin**ked **Comp**ressor and **Ex**pander
LINCS	Leased Interfaculty National Airspace Communications System
LINE	Logistic Information Network Enterprise
LINX	London Internet Exchange
LIP	1. Loop Initialization Protocol
	2. Large Internet Packet (networking)
LIPS	1. Lightweight Internet Person Schema (information retrieval)
	2. Linear Inferences Per Second (AI machines)
	3. Language Independent Program Subtitling Group (India)
LIS	1. Link Interface Shelf
	2. Local Interconnection Service
	3. Label Interface Structure (ISDN)
	4. Logical IP Sub-network
	5. Lightning Imaging Sensor (remote sensing)

L

LISN	1. Line Impedance Stabilization Network
	2. Low-Incidence Support Network
LISP	List Processing (programming language)
LISS	1. Linear Imaging Self-Scanner (remote sensing)
	2. Low-Imaging Sensing Satellite (remote sensing)
LIST SERV	List Server (Internet)
LIT	1. Line Insulation Test
	2. Logic Integrity Test
LIU	Line Interface Unit
LIU7	Line Interface Unit for CCS 7
LL	1. Leased Line
	2. Link Level
	3. Long Lines
LLA	1. Logical Layered Architecture
	2. Low-Level Amplifier
LLB	1. Line Loop-Back (function)
	2. Local Loop-Back
LLC	1. Logical Link Control (OSI model)
	2. Leadless Chip Carrier method (chip mounting)
LLC2	Logical Link Control 2
LLD	Low-Level Dispatcher
LLDP	Local Loop Demarcation Point
LLF	1. Line Link Frame
	2. Low-Layer Function
LLIF	Linear LIF (amplifier)
LLM	L-band Land Mobile
LLP	Low-level Protocol
LLWAS	Low-Level Wind-shear Alarm System
lm	lumen (unit of luminous flux)
lm/ft²	lumen per square foot
lm/m²	lumen per square meter
lm/W	lumen per Watt
lms	lumen second
LM	1. Linear Modulation
	2. Line Monitor
	3. Long-distance Marketer
LMCS	Local Multipoint Communications System
LMDS	Local Multipoint Distribution Service (Telcordia Technologies)
LME	Layer Management Entity
LMEI	Layer Management Entity Identifier
LMES	Land Mobile Earth Station
LMF	Language Media Format

LMGT	**L**ockheed **M**artin **G**lobal **T**elecommunications (company)
LM Hosts	**L**AN **M**anager **Hosts** (Windows file)
LMI	1. **L**ocal **M**anagement **I**nterface (frame relay)
	2. **L**ogical **M**odem **I**nterface
LMOS	**L**oop **M**aintenance **O**perations **S**ystem
LMP	**L**ink **M**anager **P**rotocol
LMR	**L**and **M**obile **R**adio
LMS	1. **L**and **M**obile **S**ervice (ITU)
	2. **L**and **M**obile **S**tation
	3. **L**earning **M**anagement **S**ystem (e-learning)
	4. **L**ibrary **M**anagement **S**ystem
	5. **L**oop **M**onitoring **S**ystem
	6. **L**ocal **M**easured **S**ervice
	7. **L**ocal **M**essage **S**witch
	8. **L**ocation and **M**onitoring **S**ervice
	9. **L**east **M**ean **Squa**re Error (algorithm for filters)
LMSS	**L**and **M**obile **S**atellite **S**ervice
LMST	**L**ightweight **M**ultiband **S**atellite communications Terminal
LMU	**L**ine **M**onitor **U**nit
ln	1. **n**atural **l**ogarithm (mathematics)
	2. **N**eperian **l**ogarithm (mathematics)
	3. **N**apierian **l**ogarithm (mathematics)
LN	**L**inear **N**etwork
LNA	1. **L**ow-**N**oise **A**mplifier (satcom)
	2. **L**aunch **N**umerical **A**perture (fiber optics)
LNB	**L**ow-**N**oise **B**lock down-converter (satcom)
LNBF	**L**ow-**N**oise **B**lock **F**eed (satcom)
LNC	**L**ow-**N**oise **C**onverter (satcom)
LND	**L**ast **N**umber **D**ialed
LNMF	**L**ocal **N**etwork **M**anagement **F**unction
LNNI	**L**ANE **N**etwork-to-**N**etwork **I**nterface (ATM)
LNP	**L**ocal **N**umber **P**ortability (cellular networks)
LNPA	**L**ocal **N**umber **P**ortability **A**dministration
LNR	1. **L**ow-**N**oise **R**eceiver
	2. **lin**ear
LNRU	**L**ike **N**ew **R**epair and **U**pdate
LO	1. **L**ocal **O**scillator (electronics)
	2. **L**ocal **O**perator
LO-FERF	**L**ow-**O**rder **F**ar-**E**nd **R**eceive **F**ailure alarm (MUX)
LOA	**L**etter **O**f **A**gency
LOAD	**L**aser **O**pto-**A**coustic **D**etection
LOB	**L**ine **O**f **B**earing (radio direction finding)

LOC	1. Loss Of Cell Delineation (ATM)
	2. Large Optical Cavity (diode)
	3. Lead-On-Chip package (microchips)
LOCAL TV	Launching Our Communities Access to Local Television (Act)
LOD	Letter Of Disconnect
LODE	Large Optics Demonstration Experiment
LOF	1. Loss Of Frame (MUX)
	2. Lowest Operating Frequency
LOFAR	Low-Frequency Array (antennas)
log	natural logarithm (mathematics)
logFTC	logarithmic Fast Time Constant (algorithm)
logamp	logarithmic amplifier
LOH	Line Overhead (SONET)
LOI	Letter Of Intend
LOM	1. Loss Of Multiframe (MUX)
	2. List Of Materials (contracts)
LON	Local Operating Network
LONAL	Local Off Network Access Line
Long	Longitude
LOP	1. Loss Of Path
	2. Loss Of Pointer (MUX)
	3. Low-Order Path
	4. Line Of Position
LOPA	Lower Order Path Adaptation
LOPSN	Lower Order Path Sub-Network (MUX)
LOPT	Lower Order Path Termination (MUX)
Lorac	Long-range-accuracy (radio system)
Loran	Long-range aid to navigation system
LORG	Large-size Organizations (Microsoft)
LORO	Lobe On Receive Only (radar)
LOS	1. Line-Of-Sight (microwave path)
	2. Loss Of Signal
	3. Launch On Schedule
LOTOS	Language Of Temporal Ordering Specification
LOVC	Lower Order Virtual Container (MUX)
LOW	1. Launch On Warning
	2. Link Order Wire
LP	1. Linear Polarization (wave)
	2. Line Printer
	3. Low-Pass
	4. Low-Power
	5. Low-Pressure
	6. Long-Play (recording)
	7. Log-Periodic (antenna)

LPA	1. Lower order Path Adaptation (MUX)
	2. Linear Power Amplifier
LPAC	Lossless Predictive Audio Compression
LPC	1. Linear Predictive Coding
	2. Lower order Path Connection (MUX)
LPD	1. Line Pointer Daemon
	2. Link Protocol Discriminator (GSM)
	3. Low Probability of Detection (radio and optical signals)
LPDT	Low-Power Distress Transmitter
LPE	Liquid Phase Epitaxy (semiconductors)
LPES	Land Portable Earth Station
LPF	Low-Pass Filter
LPFM	Low-Power FM (radio station)
LPI	1. Low Probability of Interception
	2. Lines Per Inch (facsimile machine)
LPIC	IntraLATA Primary Interexchange Carrier
LPM	1. Lines Per Minute (printer speed parameter)
	2. Library of Parameterized Modules (standard)
LPMUD	Lars Pensjo Multiuser Dungeon
Lport	Logical **port**
LPP	1. Link Peripheral Processor
	2. Lightweight Presentation Protocol
LPRF	1. Low-Power Radio Frequency
	2. Low-Pulse-Repetition Frequency
LPS	1. Line Profile System
	2. Line Protection Switching (MUX)
	3. Lightning Protection Subsystem
LPT	1. Lower order Path Termination (MUX)
	2. Line Print Terminal (PC parallel port)
LPTV	Low-Power Television (service)
LQA	Link Quality Analysis
LQR	Linear Quadratic Regulator
LR	Location Register
LRB	Least Reliable Bit
LRC	Longitudinal Redundancy Check (transmission)
LRE	1. Low-bit-Rate Encoding
	2. Line Regenerating Equipment
	3. Lightwave Repeating Equipment
	4. Long Reach Ethernet (Cisco)
LRF	Laser Range Finder
LR FAX	Low-Resolution **Fac**simile
LRIC	Long-Run Incremental Cost

L

LRIT	Long-Range Identification and Tracking of ships (IMO)
LRL	Line-Reflect-Line (calibration)
LRM	1. Line-Reflect-Match (microwave)
	2. Line Route Map
LRN	Location Routing Number
LRR	Long-Range Radar
LRRM	Line-Reflect-Reflect-Match (calibration)
LRS	Line Repeater Station
LRU	1. Line Replaceable Unit
	2. Least Recently Used
LS	1. Local Switch
	2. Linear Systems
	3. Loop-Start (signal)
LS Trunk	Loop-Start **Trunk**
LSA	1. Leased Space Agreement
	2. Limited Space-charge Accumulation
	3. Link-State Advertisement (routing)
	4. Local Service Agency
LSA diode	Limited Space-charge Accumulation **diode**
LSAP	Link Service Access Point (ATM)
LSAS	Line Side Answer Supervision
LSB	1. Least Significant Bit (in a byte)
	2. Lower Sideband (modulation)
LSC	1. Least Significant Character (in a string)
	2. Lambda-Switched Capable (GMPLS)
	3. Local Switching Center
LSCH	Low-Speed Channel
LSCIE	Light guide Stranded-Cable Interconnect Equipment
LSCIM	Light guide Stranded-Cable Interconnect Module
LSCIT	Light guide Stranded-Cable Interconnect Terminal
LSD	Low-Speed Data (Inmarsat)
	Least Significant Digit (in a number)
LSDU	Link Layer Service Data Unit
LSE	1. Local Support Element
	2. Local System Environment (OSI model)
	3. Line Supervisory Equipment
LSG	Large Signal Gain (nonlinear devices)
LSI	1. Large-Scale Integration (microchips)
	2. Line Status Identification
LSIC	Large-Scale Integrated Circuit
LSL	Link Support Layer (networking)
LSMA	Large-Scale Multicast Applications
LSMS	Local Service Management System
LSN	Local Signal Network

LSNG	Large-Scale Networking Group
LSO	1. Local Service Office
	2. Laser Shut Off
LSOA	Local Service Order Administration system
LSOD	Laser Shut Off Disable
LSOG	Local Service Ordering Guidelines
LSP	1. Local Service Provider
	2. Learning Service Provider (e-learning)
	3. Link-State Packet (routing)
	4. Label-Switched Path (MPLS)
LSR	1. Label Switching Router (MPLS)
	2. Line Service Request
	3. Local Service Request
	4. Leaf Setup Request (ATM)
LSS	Local Switching System
LSSC	Lower Sideband Suppressed Carrier
LSSD	Level-Sensitive Scan Design
LSSGR	LATA Switching System General Requirements
LSSU	Link Status Signal Unit (SS7)
LSSW	Low-Speed Switch
LSTP	1. Linear Search & Track Processor
	2. Local Signal Transfer Point (mobile)
LSTTL	Low-power Schottky Transistor-to-Transistor Logic
	(digital electronics)
LSU	1. Link-State Update (routing)
	2. Logical Storage Unit
	3. Line Switching Unit
LSV	Line Status Verifier
LT	1. Line Terminal (MUX)
	2. Logical Terminal
	3. Lower Tester (ATM)
LTA	Line Turn-Around (time)
LTAC	Lossless Transform Audio Coding
LTB	Last Trunk Busy
LTC	1. Line Trunk Controller
	2. Line Traffic Coordinator
	3. Longitudinal Time Code (satcom)
	4. Longitudinal Transmission Check
LTCC	Low-Temperature Co-fired Ceramic (substrate
	technology)
LTCI	Line Trunk Controller of ISDN
LT CLASS	Logical Terminal Class (ISDN terminal)
LTCN	Long-Term Care Network (TV broadcaster)
LTCS	Label Traffic Control System

L

LTDP	Line Terminator type **DP**
LTE	1. Line Terminating Equipment (MUX)
	2. Long-Term Evolution (3GSM)
LTG	Line Timing Generator
LTGRP	Logical Terminal Group (ISDN)
LTI	Linear Time Invariant
LTID	Logical Terminal Identifier (ISDN)
LTN	Local Transport Network
LTNUM	Logical Terminal Number (ISDN)
LTO	Linear Tape-Open storage technology (HP and IBM)
LTP	1. Line Test Position
	2. List Transmission Path
LTPA	Line Terminator type **PA**
LTPB	Line Terminator type **PB**
LTR	1. Line Terminal Regenerator
	2. Local Transport Restructuring (FCC)
LTRS	Letters Shift
LTS	1. Loop Testing System
	2. Lightwave Transmission System
LTSA	Linear Taper Slot Antenna
LTSN	Learning and Teaching Support Network (e-learning)
LTSP	Linux Terminal Server Project
LTTB	Line Terminator Test Board
LU	1. Line Unit
	2. Local Unit (SNA access port)
	3. Logical Unit (networking)
	4. Local Use flag
	5. Location Update (GSM)
LU6.2	Logical Unit interface, Version **6.2** (IBM protocol)
LUF	Lowest Usable Frequency
LUHF	Lowest Usable High Frequency
LULT	Line-Unit-Line Termination (fiber optics)
LUM	Line Utilization Monitor
LUN	Logical Unit Number (computer)
LUNI	LANE User Network Interface (ATM)
LUNs	Logical Unit Numbers
LUNT	Line Unit Network Termination (fiber optics)
LUT	1. Local User Terminal (aviation)
	2. Look Up Table
LUXPAC	Luxemburg Packet-Switched Network
LVCMOS	Low-Voltage Complementary MOS (semiconductors)
LVD	Low-Voltage Differential (SCSI card)
LVDM	Low-Voltage Differential Multipoint
LVDS	Low-Voltage Differential Signal

LVDT	Linear Variable-Differential Transducer
LVTTL	Low-Voltage Transistor-to-Transistor Logic (digital electronics)
LW	Long-Wave (radio broadcast band)
LWER	Lightweight Encoding Rules (data encoding)
LWT	Listen While Talk
lx	lux (unit of illumination)
LX	Local Exchange
LXC	Long-**distance** Carrier
LZ77	Lempel–Ziv compression method developed in 1977
LZARI	Lempel–Ziv **Ari**thmetic (data compression method)
LZB	Lempel–Ziv **B**ell (data compression method)
LZFG	Lempel–Ziv–Fiala–Green (data compression method)
.lzh	Name extension for files compressed with Lempel–Ziv (computer)
LZHUF	Lempel–Ziv–**Huf**fman (data compression method)
LZJ	Lempel–Ziv–Jakobsson (data compression method)
LZJH	Lempel–Ziv–Jeff–Heath (data compression method)
LZMW	Lempel–Ziv–Miller–Wegman (data compression method)
LZP	Lempel–Ziv Prediction (data compression method)
LZR	Lempel–Ziv–Roden (data compression method)
LZS	Lempel–Ziv–Stac (data compression method)
LZSS	Lempel–Ziv–Sorer–Szymanski (data compression method)
LZW	Lempel–Ziv–Welsh (data compression method)

L

M m

m	1. Symbol for prefix **m**illi-, denoting one-thousandth or 10^{-3}
	2. Symbol for **m**eter
mA	**m**illi**a**mpere (electronics)
mbar	**m**illi**bar**
mcd	**m**illi**c**an**d**ela
mCi	**m**illi**c**ur**i**e
m²	**square m**eter
m³	**cubic m**eter
m³/s	**cubic m**eters **per s**econd
M	1. Symbol for prefix **m**ega-, denoting 2^{20} or 10^6
	2. **M**obile, radiodetermination, and amateur services (ITU-T)
M bauds	**M**ega**bauds**
M bit	**M**ark **bit**
M hop	**M**-shaped **hop** (satellite transmission)
M port	**M**aster **port** (FDDI Architecture)
M VTS	**M**arconi **V**ideo **T**elephone **S**tandard
M&C	**M**onitor **and C**ontrol
M-ES	**M**obile **E**nd **S**ystem
M-Learning	**M**obile **Learning** (e-learning)
M-PSK	**M**-ary **P**hase **S**hift **K**eying (modulation)
M-quad	**M**ini-**quad** (antenna)

M

M/V	Motor Vessel (navigation)
M/W	microwave
M/Yr	Millions **per Year**
M1	Management interface **1**
M2	Management interface **2**
M2M	Machine **To M**achine (interface)
M5	Management interface **5**
M12	Multiplex DS **1**-to-DS **2**
M13	Multiplex DS **1**-to-DS **3**
M23	Multiplex DS **2**-to-DS **3**
MA	1. **M**ultiple **A**ccess
	2. **M**emory **A**ddress
	3. **M**obile **A**llocation (GSM)
MAA	1. **MA**C **A**ccess **A**rbitration
	2. **M**aximum **A**cceptance **A**ngle (optical waveguides)
MAAP	**M**aintenance **A**nd **A**dministration **P**anel
MAC OS	**Mac**intosh **O**perating **S**ystem
MAC	1. **M**edia **A**ccess **C**ontrol (MPLS)
	2. **M**edia-specific **A**ccess **C**ontrol (protocol)
	3. **M**onitoring, **A**larm, and **C**ontrol
	4. **M**ultiplexed **A**nalog **C**omponents (TV broadcast format)
	5. **M**aritime **A**ir **C**ommunications
	6. **M**obile **A**dvisory **C**ouncil
	7. **M**ultiply **A**ccumulate function (computer operations)
	8. **Mac**intosh computer
	9. **M**essage **A**uthentication **C**ode
MACE	**M**acintosh **A**udio **C**ompression and **E**xpansion
MACs	**M**oves, **A**dds, and **C**harge**s** (phone system installation)
MACSAT	**M**ultiple **A**ccess **C**ommunications **Sat**ellite
MacTCP	A **Mac**intosh extension that allows Macintosh computers to use **TCP**/IP
MAD	1. **M**agnetic **A**nomaly **D**etector
	2. **M**ean **A**verage **D**ifference
	3. **M**onitoring **A**nd **D**iagnostic unit (Intelsat)
MADAR	**M**alfunction, **A**nalysis, **D**etection, **A**nd **R**ecording (avionics)
MAdCaP	**M**odel for **Ad**vanced **C**apital **P**lanning
MADN	**M**ultiple **A**ppearance **D**irectory **N**umber (ISDN)
MADRE	**M**agnetic **D**rum **R**eceiving **E**quipment (radar)
MAE	1. **M**etropolitan-**A**rea **E**xchange
	2. **M**ERIT **A**ccess **E**xchanges
	3. **M**acintosh **A**pplication **E**nvironment
MAF	**M**anagement **A**pplication **F**unction

MAFET	Microwave and Analog Front End Technology project
MAG	magnetometer
magamp	magnetic amplifier
MAGIC	Multidimensional Applications and Gigabit Internetwork Consortium
MAGLEV	Magnetically Levitated transportation system
magram	magnetic radar absorbing material
mAh	milliampere-hour
MAHO	Mobile-Assisted Hand-Off (wireless)
MAI	1. Mobile Allocation Index
	2. Multiple Access Interference
	3. Message Alignment Indicator
MAINT	maintenance
MAIO	Mobile Allocation Index Offset (GSM)
MAJ	major (alarm)
MALLOC	Multicast Address Allocation (IETF protocol)
MAM	Major Account Manager
MAMA	Multiple ALOHA Multiple Access
MAMI	Modified Alternate Mark Inversion (code or signal)
MAMS	Marine Automatic Meteorological Station
MAMSK	Multitone Auxiliary Manual Select Keyboard
MAN	1. Metropolitan-Area Network
	2. manual
MAO	Maintenance, Administration, and Operations
MAP	1. Manufacturing Automation Protocol (General Motors)
	2. Mobile Application Part (GSM)
	3. Maintenance and Administration Panel
	4. Media Access Project
	5. Maximum Power
MAPI	Messaging Application Programming Interface (Windows)
MAPOS	Multiple Access Protocol Over SONET
MAPS	Measurement of Atmospheric Pollution from Satellite
MAR	1. Memory Address Register (computer)
	2. Major Account Representative
MARECS	1. Maritime European Communication Satellite
	2. Maritime Experimental Comsat System
MARISAT	Maritime Satellite Communications System (U.S.A)
MARS	1. Multicast Address Resolution Server (ATM)
	2. Military Affiliated Radio System (U.S. Army)
MARSIS	Mars Advanced Radar for Subsurface and Ionospher Sounding
MARTIAN	Mis-Addressed or Routed Telepacket In A Network

M

MAS	1. **M**ultiple **A**ddress **S**ystems (microwave)
	2. **M**ulti-**a**gent **S**ystems Laboratory
	3. **M**ulti-**a**dministration **S**atellite System
	4. **M**obile **A**pplication **S**ubsystem
	5. **M**inimum **A**verage **S**urcharge
MAS Lab	**M**ulti-**a**gent **S**ystems **Lab**oratory (Massachusetts University)
MASC	1. **M**ajor **A**ccounts **S**ervice **C**enter
	2. **M**ulticast **A**ddress **S**et **C**laim (IETF)
maser	**m**icrowave **a**mplification by **s**imulated **e**mission of **r**adiation
MASG	**M**aster **A**ddress **S**treet **G**uide (database)
MASP	**M**ediated **A**ttribute **S**tore **P**rotocol
MASQ	**mas**querade (IP)
MASS	**M**ajor **A**ccount **S**upport **S**pecialist
MASTER	**M**inimal **A**ccess **S**urgery by **Te**lecommunication and **R**obotics
MAT	**M**eridian **A**dministration **T**ools (Northern Telecom)
MATC	**M**ajor **A**ccount **T**echnical **C**onsultant
MATD	1. **M**aximum **A**cceptable **T**ransit **D**elay
	2. **M**ultiple **A**ntenna **T**ransmit **D**iversity
MATE	**M**odular **A**utomatic **T**est **E**quipment (standard)
MATEL	**M**ultiplex **A**utomatic **Tel**ephone system
MATR	**M**inimum **A**verage **T**ime **R**equirement
MATSE	**M**obile **A**utomatic **T**elephone **S**ystem-**E**urope
MATV	**M**aster-**A**ntenna **TV** system
MAU	1. **M**edia **A**ccess **U**nit (LANs)
	2. **M**edium **A**ttachment **U**nit (transceiver)
	3. **M**ultistation **A**ccess **U**nit (networking)
	4. **M**ath **A**cceleration **U**nit
MAX	1. **M**edia **A**ccess **E**xchange
	2. **max**imum
MaxCR	**Max**imum **C**ell **R**ate
MAYPAC	X.25 **Pac**ket-switched network of **Ma**lay**c**ia
mb	**m**illi**b**arn
Mb	**M**ega**b**it, equal to 10^6 bits (transmission)
mbar	**m**illi**bar**
Mbar	**M**ega**bar**
MB	1. **M**ega**b**yte, which is 2^{20} or 1024 times kilobytes (computer)
	2. **M**acro **B**lock
MBA	**M**ulti**b**eam **A**ntenna
Mbaud	**M**ega**baud**

MBB	1. Magnetic Blow-out Breaker
	2. Moisture Barrier Bag
MBCO	Mobile Broadcasting Corporation (Japan)
MBE	Molecular-Beam Epitaxy (semiconductors)
MBG	Multilocation Business Group
Mb/s	Megabit per Second
MB/s	Megabyte per Second
MBGP	Multicast Border Gateway Protocol
Mbits	Megabits
MBM	Magnetic-Bubble Memory
Mbone	Multicast backbone (Internet)
MBPA	Multi beam Phased Array (antennas)
Mbps	Megabit per second
MBps	Megabyte per second
MBR	1. Master Boot Record (hard disk or CD)
	2. Memory Buffer Register (computer)
MBTS	Micro Base Transceiver Station (Motorola)
Mbus	1. Message bus
	2. Multiprocessor bus
MByte	Megabyte
MBR	1. Memory Buffer Register
	2. Minimum Bend Radius (optical waveguides)
MBS	1. Maximum Burst Size (ATM)
	2. Mobile Broadband System
	3. Marine Broadcast Station
MBSAT	Multimedia Broadcasting Satellite (Japan)
MBSR	Mini BTS Sub-Rack (GSM)
MBWA	Management By Walking Around (Hewlett–Packard)
MBZS	Maximum Bandwidth Zero Suppression
Mc	Megacycle (electronics)
MC	1. Maintenance Center
	2. multicarrier
	3. Matrix Controller
	4. Main Cross-connect
MC-MVIP	Multichassis MVIP
MCA	1. Microchannel Architecture (IBM PCs)
	2. Maximum Calling Area (access)
MCAS	Multichannel Access System
MCB	Media Control Board
MCC	1. Mission Control Center
	2. Maintenance Control Center
	3. Maintenance Control Circuit
	4. Mobile Control Channel
	5. Mobile Country Code (GSM)

M

	6. Mobile Communications Company (Iran)
	7. Miscellaneous Common Carrier
	8. Multiplex Control Computer
	9. Microelectronics and Computer Technology Corporation
MCCS	Mechanized Calling Card Service
mcd	millicandela (light intensity)
MCD	1. Magnetic Circular Dichroism
	2. Manipulative Communications Deception
MCDN	MicroCellular Data Network
MCDU	Master Control and Display Unit
MCDV	Maximum Call Delay Variance (ATM)
MCE	Manufacturing Cycle Efficiency
MCEB	Military Communications Electronics Board (U.S.A)
MCEF	Military Common Emergency Frequency
MCF	1. Multimedia Communication Forum
	2. Message Confirmation Frame
	3. Meta-Content Format
MCG	Magnetocardiograph (medicine)
MCGA	Multicolor Graphics Adapter (monitor resolution)
MCH	Multipacket Command Handler
MCHG	Mass Change
MCHI	Mobile Communications Holdings Inc.
MCHO	Mobile Controlled Hand-Off
mCi	millicurie
MCi	Megacurie
MCI	1. Media Control Interface (Windows application programming)
	2. Mobile Company of Iran
	3. Microwave Communications Inc. (U.S.A)
MCID	Malicious Call Identification
MCIF	Miniature Card Implementer's Forum
MCIT	Ministry of Communications and Information Technology (Iran)
MCL	1. Mercury Communications Ltd. (U.K.)
	2. Microstrip Constrained Lens
MCLD	Modifying Calling Line Disconnect
MCLR	Maximum Cell-Loss Ratio (ATM)
MCM	1. Multichip Module (IC design)
	2. Multicarrier Modulation
MCMI	Modified Coded Mark Inversion (data encoding)
MCN	1. Master Customer Number
	2. Monitoring and Control Network
MCNC	Microelectronic Center of North Carolina

MCNS	Multimedia Cable Network System
MCP	1. Microsoft-Certified Professional
	2. Master Control Program
MCPA	Multicarrier Power Amplifier
MCPC	Multiple Channel Per Carrier
MCPW	Microstrip Co-planar Waveguide (microwave)
MCR	Minimum Cell Rate (ATM)
MCRL	Multimedia Communications Research Laboratory (Bell)
MCS	1. Maritime Communication System
	2. Master Control System
MCSE	Microsoft-Certified System Engineer (Microsoft)
MCSP	Microsoft-Certified Solution Provider (Microsoft)
MCT	1. MOS-Controlled Thyristor (semiconductors)
	2. Mobile Communications Technologies company (Iran)
MCTD	Maximum Cell Transfer Delay (ATM)
MCU	1. Multipoint Control Unit
	2. Microcontroller Unit
	3. Microcomputer Unit
MCVF	Multichannel Voice-Frequency
MCVFT	Multichannel Voice-Frequency Telegraph
MCW	Modulated Continuous Wave
MCXO	Microcomputer-Compensated Crystal Oscillator
MD	1. Mini-Disk
	2. Mediation Device
	3. Message Display
	4. Message Digest
	5. Manufacturer Discontinued
	6. Management Domain (X.400)
MD-IS	Mobile Data Intermediate System
MD5	Message Digest Version 5 (algorithm)
MDA	Monochrome Display Adapter
MDAC	Multiplying Digital-to-Analog Converter (electronics)
MDAS	Multiple Database Access System
MDB	Modified Duo-Binary (data encoding)
MDBS	Mobile Database Station
MDC	1. Meridian Digital Centrex (Northern Telecom)
	2. Manipulation Detection Code (algorithm)
MDCT	Modified Discrete Cosine Transform (mathematics)
MDDB	Multidrop Data Bridge
MDF	Main Distribution Frame (telephone exchange)
MDHCP	Multicast DHCP (protocol)

M

MDI	1. Medium-Dependent Interface
	2. Multiple-Document Interface
MDIS	Meta Data Interchange Specification
MDK	Modem Developer's Kit
MDLP	Mobile Data Link Protocol (cellular networks)
MDM	1. Modulator-Demodulator (modems)
	2. Modulation Division Multiplexing
MDMF	Multiple Data Message Format
MDN	Managed Data Network
MDNS	Managed Data Network Service
MDPE	Medium Density Polyethylene
MDQ	Market-Driven Quality (IBM)
MDR	Mission Data Recorder (remote sensing)
MDRAM	Multibank Dynamic RAM (memory)
MDS	Multipoint Distribution System (Telcordia Technologies)
MDSI	1. Message Delivery Service Interoffice
	2. Mobile Data Solutions, Inc. (company)
MDSL	Moderate-speed DSL (access)
MDSP	Microwave Digital Signal Processor
MDT	1. Mobile Data Terminal
	2. Mean Down-Time
	3. Manufacturer Delegated Testing
MDTRS	Mobile Digital Trunked Radio System
MDU	1. Message Display Unit
	2. Multiple Dwelling Unit
MDUS	Medium-scale Data Utilization Station (remote sensing)
MDVC	Mobile Digital Voice Channel (cellular network)
MDWDM	Metropolitan Dense Wavelength-Division Multiplexing
ME	Mediation Equipment (TMN)
MEA	1. Metropolitan Economic Area
	2. Messaging Enabled Applications
MEB	Megabit Erlang Bit rate
MEBS	Middle East Broadcast and Satellite (magazine)
MEC	1. Mobile E-Commerce
	2. Middle East Communications (magazine)
MECAB	Multiple Exchange Carrier Access Billing
MECAL	Master Event Calendar
MECCA	Multiple Engineering Control Center Activity
MECL	Motorola Emitter-Coupled Logic (digital electronics)
MECOD	Multiple Exchange Carriers Ordering and Design
MED	Manipulative Electronic Deception
MEDIA	Missile Error Data Integration Analysis
MEE	Multiple Equipment Engineering

MEECN	**M**inimum **E**ssential **E**mergency **C**ommunications **N**etwork
MEF	**M**aintenance **E**ntity **F**unction (TMN)
Megapel	**Mega**pi**xel**
MELCAS	**M**ercury **E**xchange **L**imited **C**hannel **A**ssociated **S**ignaling
MEM	**M**acro**e**ncoded **M**essage (Inmarsat)
MEMO	**M**ultimedia **E**nvironment for **Mo**biles (broadcasting)
MEMS	**M**icro**e**lectro**m**echanical **S**ystems (microwave)
MENIDS	**M**anufacturer's **ENIDs** (Inmarsat)
MEO	**M**edium **E**arth **O**rbiting (satcom)
MEP	**M**ultiple **E**quipment **P**rovisioning
MERCAST	**Mer**chant ship broad**cast** system
MERIS	**Me**dium **R**esolution **I**maging **S**pectrometer (remote sensing)
MERS	**M**ost **E**conomic **R**oute **S**election
MES	1. **M**obile **E**arth **S**tation (Inmarsat)
	2. **M**aster **E**arth **S**tation
MES-SIG	**M**obile **E**arth **S**tation-**Sig**naling channel (Inmarsat)
MESA	**M**obility for **E**mergency and **S**afety **A**pplications
MESC	**M**olecular **E**quipment **S**ubcommittee for **C**ommunications
MESD	**MES** **D**ata channel (Inmarsat)
MESFET	**M**etal-**E**pitaxial **S**emiconductor **F**ield-**E**ffect **T**ransistor
MESHCO	**M**atn **E**rtebat **Sh**abake **Co**mpany (Iran)
MESRP	**MES** **R**es**p**onse channel (Inmarsat)
MESRQ	**MES** **R**e**q**uest channel (Inmarsat)
MESS	**mess**age
MESSR	**M**ultispectral **E**lectronic **S**elf-**S**canning **R**adiometer
MESV	**MES** **V**oice channel (Inmarsat)
METAREA	**Met**rological **Area** (IMO)
METEOSAT	**Meteo**rological **Sat**ellite
METOP	**Met**erological **O**perational **P**olar-orbit (weather satellite)
METRAN	**M**anaged **E**uropean **Tra**nsmission **N**etwork
MeV	**M**ega**e**lectron**v**olt
MEW	**M**icrowave **E**arly **W**arning
MEWS	**M**issile **E**arly **W**arning **S**ystem
MEXE	**M**obile **Ex**tension **E**nvironment (wireless)
MF	1. **M**edium **F**requency (300–3000 kHz range)
	2. **m**ulti**f**requency
	3. **m**ulti**f**rame
	4. **m**ulti**f**unction
MFAS	**M**ulti**f**rame **A**lignment **S**ignal

M

MFC	1. **M**ulti**f**requency **C**oding
	2. **M**icrosoft **F**oundation **C**lass
MFD	1. **M**icro**f**ara**d** (capacitance)
	2. **M**ode **F**ield **D**iameter (fiber optics)
MFJ	**M**odification of **F**inal **J**udgment (AT&T)
MFlop	**M**ega**flop** (computer speed)
MFLOPS	**M**illion **F**loating-point **O**perations **Per** **S**econd (computer speed)
MFM	1. **M**odified **F**requency **M**odulation
	2. **M**agnetic **F**ield **M**odulation
MFN	**M**etropolitan **F**iber **N**etwork
MFOS	**M**ulti**f**unction **O**perations **S**ystem (AT&T)
MFP	**M**ulti**f**unction **P**eripherals
MFS	1. **M**etropolitan **F**iber **S**ystems
	2. **M**acintosh **F**ile **S**ystem
	3. **M**odem **F**ault **S**imulator
MFSK	**M**ultiple **F**requency **S**hift **K**eying (modulation)
MFTP	**M**ulticast **F**ile **T**ransport **P**rotocol
MFV	**M**ehr **F**requenz **V**erfahren (German for Tone Dialing)
mg	**m**illi**g**ram
mG	**m**illi**g**auss
MG	1. **M**aster **G**roup (MUX)
	2. **M**ega**g**auss
MBG	**M**ultilocation **B**usiness **G**roup
MGC	1. **M**anual **G**ain **C**ontrol
	2. **M**edia **G**ateway **C**ontroller
MGCP	**M**edia **G**ateway **C**ontrol **P**rotocol (VoIP)
MGD	**M**OS **G**ate **D**river (electronics)
Mget	**M**ultiple **get** (Computer command)
MGIF	**M**obile **G**aming **I**nteroperability **F**orum
MGN	**M**ulti**g**rounded **N**eutral (conductors)
MGT	**M**ulti-**G**igabit **T**ransceiver
MGTS	**M**essage **G**enerator **T**raffic **S**imulator
mH	**m**illi**h**enry (electronics)
MH	**M**odified **H**uffman (data compression method)
MHD	**m**agneto**h**ydro**d**ynamics
MHDL	**M**imic **H**ardware **D**escription **L**anguage (microwave)
MHEC	**M**idwestern **H**igher **E**ducation **C**ommission
MHEG	**M**ultimedia and **H**ypermedia information coding **E**xpert **G**roup
MHEMT	**M**etamorphic **H**igh-**E**lectron **M**obility **T**ransistor (microwave)
MHF	**M**obile **H**ome **F**unction
MHI	**M**itsubishi **H**eavy **I**ndustries

MHIC	Microwave Hybrid Integrated Circuit
MHM	Multiplexed Hierarchical Modeling
MHP	Multimedia Home Platform (broadcasting)
MHS	1. Message Handling System/Service (e-mail)
	2. Microwave Humidity Sounder (remote sensing)
MHT	Mean Holding Time (telephone call)
MHTML	MIME encapsulation of aggregate HTML (programming)
MHW	Magnetohydrodynamic Wave
MHz	Megahertz (1,000,000 or 10^6 Hz)
mi	mile (unit of length)
mi²	square mile
mi/h	mile per hour
MI	1. Multiple Inheritance (object-oriented programming)
	2. Micromodge Industries (Iran)
MI-DFB	Modulator Integrated-Distributed Feed-Back (laser diode)
MIA	Media Interface Adapter
MIB	1. Management Information Base (ICT)
	2. Module Interface Bus
MIBD	Minimum In-Band Dispersion
MIBS	Minimum BER Strategy
MIC	1. Microwave Integrated Circuit (microwave)
	2. Monolithic Integrated Circuit
	3. Minimum Ignition Current
	4. Microphone (electronic diagrams)
	5. Mode Indicate Common
	6. Media Interface Connector
	7. Mutual Interface Chart
MICE	Mobile Interworking Control Element
MICR	1. Magnetic-Ink Character Recognition (scanners)
	2. Magnetic-Ink Character Reading (scanners)
micro	microcomputer
MID	1. Maritime Identification Digits (Inmarsat)
	2. Message Identifier (ATM)
	3. Minimum Dispersion
MIDI	Musical Instrument Digital Interface
MIDP	Mobile Information Device Profile (**J2ME**)
MIDR	MES Identification Record
MIDS	Multifunction Information Distribution System
MIF	1. Management Information File
	2. Minimum Interworking Functionality
MIFR	Master International Frequency Register (ITU-R)

M

MII	1. **M**edia **I**ndependent **I**nterface (Ethernet)
	2. **M**inistry of **I**nformation **I**ndustry (China)
Mike	Slang term for **Mic**rophone
MIL	**Mil**itary specification
MIL-STD	**Mil**itary **Stand**ard
MILNet	**Mil**itary **Net**work (ARPANet)
Milstar	**Mil**itary **s**trategic and **ta**ctical **r**elay (network)
MIM	1. **M**echanically **I**nduced **M**odulation
	2. **M**etal–**I**nsulator–**M**etal (capacitor)
MIMD	**M**ultiple-**I**nstruction, **M**ultiple-**D**ata stream (parallel processing)
MIME	**M**ultipurpose **I**nternet **M**ail **E**xtension (e-mail attachments)
MIMIC	**M**icrowave **M**onolithic **I**ntegrated **C**ircuit
MIMO	**M**ultiple-**I**nput **M**ultiple-**O**utput
MIN	1. **M**obile **I**dentification **N**umber (cellular networks)
	2. **min**or (alarm)
	3. **min**imum
MIP	**M**edium **I**nterface **P**oint
Mips	**M**illions of **i**nstructions **per s**econd (processor speed)
MIR	1. **M**ultiplex **I**nput **R**outer
	2. **M**icrowave **I**mpulse **R**adar
MIRAN	**M**issile **Ran**ging
MIS	1. **M**anagement **I**nformation **S**ystems (ICT)
	2. **M**anagement **I**nformation **S**ervices (ICT)
	3. **M**etal-**I**nsulator **S**emiconductor
	4. **M**obile **I**nternet **S**ervice
MISFET	**M**etal **I**nsulator **S**emiconductor **F**ield **E**ffect **T**ransistor
MISR	**M**ultiangle **I**maging **S**pectro**r**adiometer (remote sensing)
MISSI	**M**ultilevel **I**nformation **S**ystems **S**ecurity **I**nitiative
MITS	**M**icro**i**nstrumentation and **T**elemetry **S**ystems
MITT	**M**inistry of **I**nformation **T**echnology and **T**elecommunications
MIU	**M**odem **I**nterface **U**nit
MIX	1. **mix**er
	2. **M**ultinational **I**nternet **E**xchange
	3. **M**ulticasting **I**nternational **E**xchange
MJ	**M**odular **J**ack (telephone cable)
MJU	**M**ulti**j**unction **U**nit
MKDIR	**M**a**k**e a new **Dir**ectory (MS DOS Command)
MKS	**M**eter-**k**ilogram-**S**econd (parametric unit)
MKSA	**M**eter-**k**ilogram-**S**econd-**A**mpere (parametric unit)
mL	**m**illi**l**iter

MLA	1. **M**ail **L**ist **A**gent
	2. **M**icrowave **L**ink **A**nalyzer
MLB	**M**ulti**l**ayer **B**oard
MLC	1. **M**ulti**l**evel **C**oding
	2. **M**ono**l**ithic **M**ultilayer **C**apacitor (electronics)
MLCM	**M**ulti**l**evel **C**oded **M**odulation
MLD	1. **M**aximum **L**ikelihood **D**ecoding
	2. **M**ulti**l**evel **C**oded **M**odulation
	3. **M**edian **L**evel **D**epressions fading
MLE	**M**aximum **L**evel **E**rror algorithm
MLHG	**M**ulti**l**ine **H**unt **G**roup
MLI	**M**ultiple **L**ink **I**nterface (networking)
MLID	**M**ultiple **L**ink **I**nterface **D**river (networking)
MLK	**M**ail **L**ist **K**ey
MLL	1. **M**onthly **L**eased **L**ine
	2. **M**ultimedia **L**earning **L**ab (e-learning)
mlm	**m**illi**l**u**m**en
MLM	1. **M**eridian **L**ink **M**odule
	2. **M**ulti**l**ongitudinal **M**ode (laser)
MLP	1. **M**ulti**l**evel **P**recedence
	2. **M**icro **L**eadframe **P**ackage (microwave)
MLPP	**M**ulti**l**evel **P**recedence and **P**reemption
MLS	**M**icrowave **L**anding **S**ystem
MLSC	**M**ultiple **L**ow-**S**peed **C**SU (PBX)
MLST	**M**inimum **L**ine **S**can **T**ime
MLT	1. **M**echanized **L**ine **T**esting equipment
	2. **M**echanical **L**oop **T**est
MLT-3	**M**ulti**l**evel **T**ransmit-**3 levels**
MLTS	**M**ultiple **L**anguage **T**ext **S**ystem (TV subtitles)
MLU	**M**obile **L**ocator **U**nit (GSM)
mm	**m**illi**m**eter (10^{-3} meter)
MM	1. **M**an–**M**achine
	2. **M**obility **M**anagement (GSM)
	3. **M**uti**m**edia
MMAC	**M**ultimedia **M**obile **A**ccess **C**ommunication
MMB	1. **M**ulti**m**aster **B**us
	2. **M**illi**m**etric **B**and (ETSI)
	3. **M**ass **M**edia **B**ureau (FCC)
MMC	1. **M**aintenance, **M**onitoring, and **C**ontrol
	2. **M**an–**M**achine **C**ommunication
	3. **M**obile **M**ultimedia **C**ommunication project
	4. **M**obile **M**aritime **C**ommittee (U.S. Coast Guard)
	5. **M**ultimedia **M**ultiparty **C**onferencing

M

	6. **M**ulti**m**ode **C**ommunications
	7. **M**ulti**m**edia **C**ard (Flash memory)
MMCA	**M**ulti**m**edia **C**ard **A**ssociation
MMCD	**M**ulti**m**edia **C**ompact **D**isk
MMCF	**M**ulti**m**edia **C**ommunication **F**orum
MMCX	**M**ulti**m**edia **C**ommunication **Ex**change
MMDC	1. **M**assachusetts **M**icroprocessor **D**esign **C**enter (U.S.A)
	2. **M**ulti**m**edia and **D**igital **C**ommunications laboratories
	3. **M**ulti**m**edia **D**esktop **C**ollaboration (software)
	4. **M**ulti**m**edia **D**evelopment **C**enter
	5. **M**ulti**m**ode **D**ata **C**ompression
	6. **M**ultiple **M**odule **D**ata **C**omputer
	7. **M**ultiservice **M**ulticarrier **D**istributed **C**ommunications
MMDS	**M**ultichannel **M**ultipoint **D**istribution **S**ervice
MME	1. **M**obility **M**anagement **E**ntity
	2. **M**ulti**m**edia **M**essage **E**ntity
	3. **M**icrosoft **M**obile **E**xplorer
	4. **M**obile **M**eteorological **E**quipment
MMEG	**M**ulti**m**edia **E**xpert **G**roup (WAP Forum)
mmf	**m**agneto**m**otive **f**orce (voltage)
MMF	1. **M**obile **M**anagement **F**orum
	2. **M**ulti**m**ode **F**iber
	3. **M**aximum **M**odulating **F**requency
MMFD	**m**icro**m**icro**f**ara**d** (electronics)
MMFS	**M**anufacturing **M**essage **F**ormat **S**tandard
MMI	1. **M**an-to-**M**achine **I**nterface
	2. **M**ulti**m**edia **I**nterface
MMIC	**M**onolithic **M**icrowave **I**ntegrated **C**ircuit
MMITS	**M**odular **M**ultifunction **I**nformation **T**ransfer **S**ystem
MMJ	**M**odified **M**odular **J**ack (computer)
MML	**M**an–**M**achine **L**anguage (programming)
MMM	1. **M**ulti**m**edia **M**essaging
	2. **M**ulti**m**edia **M**ail
MMMS	**M**ulti**m**edia **M**ail **S**ervice
MMMSec	**M**ulti**m**edia **M**ail **Sec**urity
MMN	**M**ulti**m**edia **N**etwork
MMP	**M**an–**M**achine **P**rocessor
MMPS	**M**ilstar **M**essage **P**rocessing **S**ystem
MMR	1. **M**obile **M**arine **R**adio Inc.
	2. **M**odified **M**odified **R**ead (compression method)
MMS	1. **M**ulti**m**edia **M**essaging **S**ervice (GSM)
	2. **M**ultiplex **M**anagement **S**ystem (TV access)

	3. Maritime Mobile Service
	4. Modular Measurement System
	5. Meteorological Measurement System
	6. Manufacturing Message Specification (ISO)
MMSC	Multimedia Messaging Service Center (GSM)
MMSE	1. Minimum Mean Square Error
	2. Multimedia Messaging Service Environment
MMSER	Minimum Mean Square Error Reception
MMSI	1. Maritime Mobile Service Identity code (Inmarsat)
	2. Manchester Museum of Science & Industry
MMSK	Modified MSK (modulation)
MMSP	Modular Multisatellite Preprocessor
MMSS	Maritime Mobile Satellite Service
MMSU	Modular Metallic Service Unit
MMT	Multimedia Terminal
MMTA	Multimedia Telecommunications Association (standards)
MMTF	Mean Mission Time to Failure
MMU	1. Manned Maneuvering Unit (U.S. space shuttle)
	2. Memory Management Unit (processors)
MMUSC	Multiparty Multimedia User Session Control (IETF)
MMVF	Multimedia Video File (NEC)
MMW	millimeter Wave (microwave)
MMX	1. Multimedia Extension (processors)
	2. Matrix Math Extensions (processors)
MNA	Multimedia Network Applications
MNA7	Multiple CCS7 Network Addresses
MNC	Mobile Network Code (GSM)
MNLP	Mobile Network Location Protocol (cellular network)
MNO	Mobile Network Operator
MNOS	Metal-Nitride-Oxide Semiconductor
MNP	1. Microcom Networking Protocol (data compression)
	2. Mobile Number Portability (cellular network)
MNP10	Microcom Networking Protocol Class 10 (cellular networks)
MNRP	Mobile Network Registration Protocol (cellular network)
MNRU	Modulated Noise Reference Unit
MNRZ	Modified Non-Return-to-Zero level (data encoding)
MNS	Master of Network Science (3COM)
.mn.us	Internet domain name extension for the addresses located in Minnesota, United States
.mo	Internet domain name extension for the addresses located in Macao

M

MO	1. **M**agneto-**O**ptical
	2. **M**aster **O**scillator
MOA	**M**obitex **O**perators **A**ssociation (Korea)
MobMan	**Mob**ility **Man**agement system (Inmarsat)
Mobo	**Mo**ther**bo**ard (computer)
MOC	**M**obile **O**riginating **C**all (GSM)
MOCT	**M**ultiorder **C**ancellation **T**echnique (microwave)
MOCVD	**M**etal-**O**rganic **C**hemical **V**apor **D**ecompression (semiconductors)
MOD	1. **mod**ulator
	2. **M**agneto-**O**ptical **D**isk
modec	**mo**dem and co**dec** (device)
modem	**mo**dulator and **dem**odulator (device)
MODFET	**Mo**dulated-**D**oped **FET** (semiconductors)
MODIS	**Mod**erate-resolution **I**maging **S**pectrometer (remote sensing)
MOEMS	**M**icro-**O**ptical **E**lectro**m**echanical **S**ystems
MOF	**M**agneto-**O**ptical **F**ilter Development Group
MOH	**M**usic **O**n **H**old (PBX)
mΩ	**milliohm** (electronics)
MΩ	**Meg**a**ohm** (electronics)
MOL	**M**anned **O**rbiting **L**aboratory (satcom)
MOM	1. **M**aster **O**scillator **M**odule
	2. **M**essage **O**riginated **M**iddleware
MOMA	**M**essage **O**riginated **M**iddleware **A**ssociation (standards)
MOMBE	**M**etal-**O**rganic **M**olecular-**B**eam **E**pitaxy (semiconductors)
MONET	1. **M**ultiwavelength **O**ptical **Net**work
	2. **M**obile Inter**net**
MOO	**M**UD **O**bject-**O**riented
MOON	**M**agneto **O**ptical **O**n **N**etwork
.moov	Name extension for QuickTime **MooV** video files (computer)
MooV	The file format for QuickTime **Mov**ie files (computer)
MOP	1. **M**ethod **O**f **P**rocedure (SONET)
	2. **M**aintenance **O**perations **P**rotocol
MOPA	**M**aster-**O**scillator **P**ower-**A**mplifier (electronics)
MOPS	1. **M**inimum **O**perational **P**erformance **S**tandards
	2. **M**illions of **O**perations **P**er **S**econd (processors)
MOPTT	**M**inistry **O**f **P**ost **T**elegraph and **T**elecommunications (Lebanon)
MOR	**M**ultiwavelength **O**ptical **R**epeater
MORC	**M**ulti-**o**utput **R**esonant **C**onverter

morphing	meta**morph**osing (graphics)
MOS	1. **M**etal-**O**xide **S**emiconductor (integrated circuits)
	2. **M**ean **O**pinion **S**core (voice quality)
	3. **M**arine **O**bservation **S**atellite
MOSFET	**M**etal-**O**xide **S**emiconductor **FET** (semiconductors)
MOSIGT	**M**etal-**O**xide **S**emiconductor **I**nsulated-**G**ate **T**ransistor
MOSPF	**M**ulticast **OSPF** (routing protocol)
MOSS	**M**IME **O**bject **S**ecurity **S**ervices
MOST	**M**etal-**O**xide **S**emiconductor **T**ransistor (semiconductors)
MOTIS	**M**essage-**O**riented **T**ext **I**nterchange **S**ystem (ISO)
MOTO	**M**ail **O**rder **T**elephone **O**rder
MoU	1. **M**inutes **o**f **U**se
	2. **M**emorandum **o**f **U**nderstanding
.mov	Name extension for **Mov**ie file in Apple's QuickTime format
MOV	**M**etal-**O**xide **V**aristor (semiconductors)
MOVPE	**M**etal-**O**rganic **V**apor **P**hase **E**pitaxy (semiconductors)
MP	1. **M**ulti**p**ort
	2. **M**ultilink **P**rotocol
MP+	**M**ultilink **P**rotocol **Plus**
MP-MLQ	**M**ulti**p**ulse-**M**aximum **L**ikelihood **Q**uantization
MP/M	**M**ultitasking **P**rogram for **M**icrocomputers (operating system)
MP1	**M**PEG layer-**1** (audio encoding)
MP2	**M**PEG layer-**2** (video encoding)
MP3	**M**PEG layer-**3** (audio encoding)
MP4	**M**PEG layer-**4** (video encoding)
MPC	1. **M**ulti**p**ort **C**luster
	2. **M**ulti**p**ort **C**ard
	3. **M**ultimedia **P**ersonal **C**omputer
	4. **MP**OA **C**lient (ATM)
MPDS	**M**obile **P**acket **D**ata **S**ervice (Inmarsat)
.mpeg	Name extension for **MPEG** image files (computer)
MPEG	**M**otion-**P**ictures **E**xperts **G**roup (data compression standard)
MPFM	**M**otion-**P**icture **F**ile **M**anager
mpg	File extension for encoded data streams containing compressed information using **MPEG** format
MPG	1. **M**icrowave **P**ulse **G**enerator
	2. **M**ultiple-image **P**ortable **G**raphics
MPI	1. **M**ulti **P**ath **I**nterface
	2. **M**edia **P**latform **I**nterface
MPLS	**M**ulti**p**rotocol **L**abel **S**witching (IETF)
MPM	**M**aster **P**rocessor **M**odule

M

MPML	Main Profile Main Level (MPEG-2)
MPMP	Microwave Point to Multipoint
MPN	1. Mode-Partition Noise
	2. Manufacturer's Part Number
MPoA	Multiprotocol over ATM
MPOE	Minimum Point Of Entry
MPOP	1. Minimum Point Of Presence
	2. Metropolitan Point Of Presence
MPP	1. Multichannel Point-to-point Protocol
	2. Massively Parallel Processor
MPPP	Multilink Point-to-Point Protocol
MPPS	Million Packets Per Second (LANs and WANs)
MPR	Maximum Pulse Rate
MPS	1. Multipage Signal
	2. Mobile Positioning Service
MPSC	Multiprotocol Serial Controller
MPSK	Multiple Phase-Shift Keying (modulation)
MPT	Ministry of Post and Telecommunications (Lebanon)
MPTN	Multiprotocol Transport Networking (IBM)
MPTy	multiparty (GSM supplementary service)
MPU	1. Main Processor Unit (computer)
	2. Microprocessor Unit (computer)
	3. Message Processor Unit
MPX	multiplex
MQA	Multiple Queue Assignment (PBX)
MQFP	Metric-pitch Quad Flat-Pack (microchips)
MQW	Multiple-Quantum Well structure (semiconductors)
mR	milliroentgen
MR	1. Message Router
	2. Modem Ready (modem light)
	3. Magnetic Resonance
mrad	milliradian
MRAM	Magnetic RAM (memory)
MRC	1. MF Receiver Card
	2. Microwave Radio Communications (company)
MRCC	Maritime Rescue Coordination Center (COMSAR)
mrd	millirad
mrem	millirem
mrep	millirep
MRFR	Master Reference Frequency Rack
MRI	Magnetic Resonance Imaging (medicine)
MRN	1. Mobile Roaming Number
	2. Maritime Radio Navigation (ITU)
MRNS	Maritime Radio Navigation Service (ITU)

MRNSS	Maritime Radio Navigation Satellite Service (ITU)
MROP	Marine Radio Operator Permit (FCC)
MRP	1. Multiservice Route Processor (Cisco)
	2. Materials Resource Planning
MRR	Monthly Recurring Revenue
MRSC	Maritime Rescue Coordination Sub-Center (COMSAR)
MRSII	Minimum Request Sequence Initiation Interval
MRTIE	Maximum Relative Time Interval Error
MRU	1. Maximum Receive Unit
	2. Mountain Rescue Unit (COMSAR)
ms	millisecond (10^{-3} of a second)
mS	millisiemens
MS	1. Mobile Station (GSM)
	2. Mobile Service
	3. Message Storage
	4. Message Switch
	5. Multiplex Section (MUX)
	6. Microprocessor System
	7. Microsoft (company)
	8. Military Standard
MS-AIS	Multiplex Section, Alarm Indication Signal (MUX)
MS-DOS	Microsoft Disk Operating System
MS-FF	Master-Slave Flip-Flop (digital electronics)
MS-SPRing	Multiplex Section-Shared Protection Ring (MUX)
MSA	1. Metropolitan Statistic Area (cellular networks)
	2. Message Service Application
	3. Multiplexer Section Adaptation
MSAN	MultiService Access Node
MSAS	MTSAT Satellite-based Augmentation System (Japan)
MSat	Mobile Satellite (AMSC of Canada)
MSATCOM	Mobile Satellite Communications
MSAU	Multistation Access Unit (networking)
MSB	Most Significant Bit (in a byte)
MSBVW	Magneto-Static Backward Volume Waves (microwave)
MSC	1. Most Significant Character (in a string)
	2. Mobile Services Switching Center (GSM)
	3. Management Services Contractor
	4. Meteorological Satellite Center (remote sensing)
	5. Maritime Safety Committee (IMO)
	6. Malaysian Multimedia Super Corridor
MSCS	Microsoft Cluster Server (Windows NT)
MSD	1. Medium Speed Data (Inmarsat)
	2. Most-Significant Digit (in a number)
MSDN	Multiservice Data Network

M

MSDSL	Multirate Symmetrical **DSL** (access)
MSE	1. Mobile Subscriber Equipment
	2. Mean Square Error (MUX)
msec	millisecond
MSED	Minimum Squared Euclidean Distance
MSF	1. Mobile Serving Function
	2. Multiservice Switching Forum
MSFVW	Magneto-Static Forward Volume Waves (microwave)
MSG	Message
MSI	1. Maritime Safety Information (COMSAR)
	2. Medium-Scale Integration (microchips)
	3. Modular Station Interface
MSIN	Mobile Station Identification Number
MSISDN	Mobile Subscriber **ISDN** number
MSIX	Metered Service Information Exchange
MSK	Minimum-Shift Keying (modulation)
MSL	1. Mirror Server Link (networking)
	2. Multisatellite Link (satcom)
MSME	Multiplexing and Signaling Management Equipment
MSMQ	Microsoft Message Queuing
MSMT	Micro Surface-Mount Technology (electronics)
MSN	1. Microsoft Network
	2. Mobile Station Number (GSM)
	3. Multiple Subscriber Numbering (ISDN)
MSNF	Multisystem Networking Facility
MSO	1. Multiple System Operator (cable TV)
	2. Marine Safety Office (U.S.A.)
MSOH	Multiplex Section Over head (MUX)
MSOP	Micro Small Outline Package
MSP	1. Message Security Protocol (Internet)
	2. Mixed Signal Processor
	3. Multiplexer Section Protection switching (MUX)
	4. Metropolitan Service Provider
MSPP	Mobile Solutions Partner Program (Microsoft)
MSPS	Million of Samples **Per** Second
MSR	1. MES Status Record
	2. Microwave Scanning Radiometer (remote sensing)
	3. Magnetic-inductance Super Resolution
	4. Maximum Stuffing Rate (data communications)
MSRN	Mobile Station Roaming Number (GSM)
MSS	1. Mobile Satellite Services (ITU)
	2. Multispectral Scanner (remote sensing)
	3. MAN Switching System
MSSC	Maritime Satellite Switching Center

MSSE	Master of Science in Software Engineering
MSSW	Magneto-Static Surface Wave
MST	1. Multiplexer Section Termination function (MUX)
	2. Maintenance Support Terminal
MST-RR	Multiplex Section Termination for Sub-STM-1 Radio-Relay (MUX)
MSTS	Mobile Subscriber Test System
MSTU	Main Signal Transmission Unit
MSTV	Association for Maximum Service Television
MSU	1. Message Signaling Unit (SS7)
	2. Microwave Sounding Unit (remote sensing)
	3. Mobile Subscriber Unit
	4. Mobile Satellite Unit
	5. Modem Sharing Unit
MSUA	Mobile Satellite Users Association
MSV	Master Supervisory Equipment
MSW	Magneto-Static Wave
MSX	Mobile Switching Exchange (cellular networks)
MSymbol/s	Mega Symbols per second (TV)
MT	1. Message Transfer
	2. Message Type (ATM)
	3. Machine Translation (computer)
	4. Maroc Telecom (Moroccan operator)
	5. Mobile Termination
MTA	1. Message Transfer Agent (X.400)
	2. Macintosh Telephony Architecture
	3. Major Trading Area
	4. Microwave Transition Analyzer
MTBF	Mean Time Between Failures (production rating)
MTBM	Mean Time Between Maintenances (production rating)
MTBO	Mean Time Between Outages (production rating)
MTBPM	Mean Time Between Preventive Maintenances
MTC	1. Mobile Terminating Call (GSM)
	2. Mobile Telecommunications Company (Kuwait GSM operator)
MTD	1. Memory Technology Driver
	2. Moving-Target Detection (radar)
MTF	Modulation Transfer Function (optical fiber)
MTFT	Multilayer Thick-Film Technology
MTG	Multiplexer Timing Generator
MTI	Moving-Target Indicator (radar)
MTIE	Maximum Timing Interval Error
MTL	1. Merged-Transistor Logic (digital electronics)
	2. Message Transfer Layer

M

MTM	1. **M**aintenance **T**runk **M**onitor
	2. **M**ean **T**ime of **M**aintenance (production rating)
	3. **M**cLintock **T**heater **M**odel
MTP	1. **M**essage **T**ransfer **P**art (SS7)
	2. **M**edia **T**ermination **P**oint (VoIP)
	3. **M**any-**T**ime **P**rogrammable memory
MTPI	**M**ultiplexer **T**iming **P**hysical **I**nterface (MUX)
MTPT	**M**inistry of **T**ransportation, **P**ost and **T**elecommunications
MTR	**M**aster **T**ime **R**eference
MTS	1. **M**essage **T**elecommunications **S**ervice (regulation)
	2. **M**essage **T**ransfer **S**ervice
	3. **M**essage **T**elephone **S**ystem
	4. **M**ultichannel **T**elevision **S**ound
	5. **M**icrosoft **T**erminal **S**erver
	6. **M**ember of **T**echnical **S**taff
MTSAT	**M**ultifunctional **T**ransport **Sat**ellite
MTSO	**M**obile **T**elephone **S**witching **O**ffice
MTSR	**M**ean **T**ime to **S**ervice **R**estoral (production rating)
MTTF	**M**ean **T**ime **T**o **F**ailure (production rating)
MTTR	**M**ean **T**ime **T**o **R**epair (production rating)
MTTSR	**M**ean **T**ime **T**o **S**ervice **R**estoral (production rating)
MTU	1. **M**aximum **T**ransmission **U**nit (data packets)
	2. **M**ultiple **T**enant **U**nit
	3. **M**aintenance **T**ermination **U**nit
MTV	**M**agyar **Tel**evision (broadcaster)
MTX	**M**obile **T**elephone **Ex**change (Northern Telecom)
MU	**M**onitoring **U**nit (wireless)
MUA	**M**ulti**u**ser **A**gent
MUD	1. **M**ulti**u**ser **D**ungeon (Internet games)
	2. **M**ulti**u**ser **D**omain (Internet games)
MUF	**M**aximum **U**sable **F**requency
MULDEM	**Mul**tiplexer-**Dem**ultiplexer
MULTICS	**Mult**iplexed **I**nformation and **C**omputing **S**ystem
MUMPS	**M**assachusetts **U**tility **M**ulti**p**rogramming **S**ystem
MUNDIS	**M**ultiplexed **N**etwork for **D**istributed and **I**nteractive **S**ervices
MUP	**M**anagement of **U**ser **P**art
MUS	**M**arine **U**tility **S**tation
MUSA	**M**ultiple-**U**nit **S**teerable **A**ntenna
MUSE	1. **Mu**ltiple-**S**ubNyquist **S**ampling **E**ncoding (compression)
	2. **M**ulti-**U**ser **S**imulation **E**nvironment

MUSICAM	**M**asking **P**attern **A**dapted **U**niversal **S**ub-band **I**ntegrated **C**oding **A**nd **M**ultiplexing (Audio broadcasting)
MUSR	**M**aximum **U**ser **S**ignaling **R**ate
MUT	**M**ulti**u**ser **T**alk (chat program)
MUX	**mu**ltiple**x**er
MUXER	**mu**ltiple**xer**
MUXing	**mu**ltiple**xing**
mV	**m**illivolt
mVAC	**m**illivolt **A**lternating **C**urrent
mVDC	**m**illivolt **D**irect **C**urrent
MV	1. **M**ega**v**olt
	2. **m**acro**v**ision
MVA	**M**ega**v**olt**a**mpere
MVBR	**m**ulti**v**i**br**ator (digital circuit)
MVDDS	**M**ultichannel **V**ideo **D**istribution and **D**ata **S**ervice
MVDS	**M**ultipoint **V**ideo **D**istribution **S**ystem
MVIP	**M**ulti**v**endor **I**ntegration **P**rotocol (Telcordia Technologies)
MVL	1. **M**ultiple **V**irtual **L**ines (access technology)
	2. **M**ultimedia **V**irtual **L**aboratory (Japan)
	3. **M**ultimedia and **V**isualization **L**aboratory (Original of MLL)
MVNE	**M**obile **V**irtual **N**etwork **E**nabler
MVNO	**M**obile **V**irtual **N**etwork **O**perator
MVP	1. **M**ultichannel **V**ideo **P**rogramming
	2. **M**ultimedia **V**ideo **P**rocessor
MVPD	**M**ultichannel **V**ideo **P**rogramming **D**istribution (cable TV)
MVPN	**M**obile **V**irtual **P**rivate **N**umber
MVS	**M**ultiple **V**irtual **S**torage (IBM mainframes)
mW	**m**illi**w**att (10^{-3} Watts)
MW	1. **M**ega**w**att (10^6 Watts)
	2. **M**edium **W**ave (radio broadcast band)
	3. **m**icro**w**ave
MWh	**M**ega**w**att-**h**our
MWI	**M**essage **W**aiting **I**ndicator (telephone set)
MWIF	**M**obile **W**ireless **I**nternet **F**orum
MWR	**M**icro**w**ave **R**adiometer (test equipment)
MWV	**M**aximum **W**orking **V**oltage
Mx	**M**a**x**well (unit of magnetic flux)
MXB	**M**atri**x** **B**oard
MXR	1. **m**ultiple**xer**
	2. **m**i**xer**
MZI	**M**ach–**Z**ehnder **I**nterferometer

M

N n

n	1. Symbol for prefix **n**ano-, denoting one-billionth or 10^{-9}
	2. Symbol for an integer **n**umber (mathematics)
nA	**n**ano**a**mpere (electronics)
N	1. Symbol for **N**ewton
	2. Symbol for any integer **N**umber bigger than 1 (mathematics)
	3. Symbol for **N**oise
	4. Symbol for **N**egative (semiconductors)
	5. Symbol for **N**arrow-band
N&SS	**N**etwork **and S**ystems **S**upport
N-AMPS	**N**arrow-band **AMPS**
N-ISDN	**N**arrow-band **ISDN**
N-MOS	**N**-channel **MOS** (semiconductors)
N-Port	**N**ode **Port**
N-TACS	**N**arrow-band **TACS** (Motorola)
N/A	1. **N**ot **A**vailable (documents)
	2. **N**ot **A**pplicable (documents)
N/I	**N**oise-**to**-**I**nterference (ratio)
N/m²	**N**ewton **per square m**eter
NA	1. **N**etwork **A**dministration
	2. **N**umerical **A**perture (antenna)
NA-TDMA	**N**orth **A**merican **TDMA**

N

NAB	1. National Association of Broadcasters (U.S.A.)
	2. National Alliance of Business
NABTS	North American Basic Teletext Specification
NAC	1. Network Access Center
	2. Network Access Control
	3. Network Applications Consortium
	4. Numbering Advisory Committee
	5. Null Attachment Concentrator
NACC	North American Calibration Cooperative
NACD	Network Autonomous Commanding Display
NACIC	National Counterintelligence Information Center
NACISA	NATO Communications and Information Systems Agency
NACK	Negative Acknowledged (control character)
NACME	National Action Council for Minorities in Engineering
NACN	North American Cellular Network
NACS	Network Analysis Control Surveillance
NACSEM	National Communications Security Emanation Memorandum
NACSIM	National Communications Security Information Memorandum
NAD	Network Access Domain
NADC	North American Dual-mode Cellular system
NADF	North American Directory Forum
NADN	Nearest Active Downstream Neighbor (LANs)
NADP	North American Directory Plan
NADTP	National Association of Desktop Publishers
NAEC	Novell Authorized Education Center (Novel, Inc.)
NAK	Negative Acknowledgement (control character)
NAL	1. National Aerospace Laboratory (Japan)
	2. Network Access Line
NAM	1. Number Assignment Module (cellular networks)
	2. National Account Manager
	3. National Association of Manufacturers (U.S.A.)
NAMAS	National Measurement Accreditation Service
NAMPS	Narrow-band Advanced Mobile Phone System (cellular networks)
NAMTS	Nippon Advanced Mobile Telephone System (Japan)
NANC	North American Numbering Council
NAND	A combination of a NOT and an AND gate (digital electronics)
NANOG	North American Network Operator's Group (U.S.A.)
NANP	North American Numbering Plan

NANPA	North American Numbering Plan Administration
NAOM	National Accounts Operations Manager
NAP	1. Network Access Point (Internet backbone)
	2. Network Access Provider
	3. National Attachment Point (Internet)
NAPI	Numbering/Addressing Plan Identifier
NAPLPS	North American Presentation-Level Protocol Syntax
NAPP	National Aerial Photography Program
NAPT	Native American Public Telecommunications, Inc.
NAR	Network Access Register (Centrex)
NARS	Network Audio Response System
NARSAP	National Advanced Remote Sensing Application Program
NARTE	National Association of Radio and Telecommunications Engineers (U.S.A.)
NARUC	National Association of Regulatory Utility Commissioners (U.S.A.)
NAS	1. NetWare Access Server
	2. Network Access Server
	3. Network-Attached Storage (storage device)
	4. National Academy of Science (U.S.A.)
NASA	National Aeronautics and Space Administration (U.S.A.)
NASC	1. National Aeronautics and Space Council (U.S.A.)
	2. Number Administration and Service Center
NASCOM	NASA Communications Network (U.S.A.)
NASDA	National Space Development Agency (Japan)
NASDAQ	National Association of Security Dealers Automated Quotation
NASTD	National Association of State Telecommunications Directors
NAT	Network Address Translation (VPNs)
NATA	1. North American Telephone Association
	2. National Association of Testing Authorities
NATD	National Association of Telecommunications Dealers
NATI	Not A Telecom Issue
NATOA	National Association of Telecommunications Officers and Advisors (U.S.A.)
NAU	Network Addressable Unit (IBM SNA)
NAUCS	Network Access Usage and Cost System
NAUN	Nearest Active Upstream Neighbor (LANs)
NAV	1. Network Applications Vehicle
	2. Network Allocation Vector
	3. Net Asset Value
NAVAR	Navigation And Ranging (radar)

N

NAVAREA	**Nav**igational **Area** (IMO)
NAVSAT	**Nav**y **Sat**ellite system (U.S.A.)
NAVSTAR	**Nav**igation **S**ystem using **T**ime **A**nd **R**anging (satellite)
Nax	**N**etwork **Fax**ing
NB	1. **N**arrow-**B**and
	2. **N**ormal **B**urst
NBA	**N**ominated **B**arring **A**uthority (Inmarsat)
NBC	1. **N**ational **B**roadcasting **C**ompany
	2. **N**etwork **B**us **C**ontroller
	3. **N**on-**B**ackward **C**ompatible (softwares)
NBDP	**N**arrow-**B**and **D**irect-**P**rinting
NBF	**N**et **B**IOS **F**rame
NBFCP	**N**et **B**IOS **F**rames **C**ontrol **P**rotocol
NBFM	**N**arrow-**B**and **F**requency **M**odulation
NBFP	**N**et **B**IOS **F**rame **F**ormat **P**rotocol
NBH	**N**etwork **B**usy **H**our
NBMA	**N**on-**B**roadcast **M**ultiple **A**ccess
NBP	1. **N**ame **B**inding **P**rotocol (networking)
	2. **N**et **B**IOS **P**rotocol
NBPM	**N**arrow-**B**and **P**hase **M**odulation
NBRVB	**N**arrow-**B**and **R**adio **V**oice **B**andwidth
NBRVF	**N**arrow-**B**and **R**adio **V**oice **F**requency
NBS	**N**ational **B**ureau of **S**tandards (now NIST)
NBSP	**N**on-**B**reaking **Sp**ace (HTML command)
NBSV	**N**arrow-**B**and **S**ecure **V**oice (system)
NC	1. **N**etwork **C**omputer (Sun Microsystems)
	2. **N**etwork **C**ontrol
	3. **N**umerical **C**ontrol system
	4. **N**ormally **C**losed (relay or switch contacts)
	5. **N**o **C**onnection (diagrams)
NC&A	**N**etwork **C**onstruction **and A**dministration
NCA	1. **N**umber of **C**alls **A**bandoned
	2. **N**ational **C**onvergence **A**lliance (U.S.A.)
NCACHE	**N**egative **Cache**
NCAR	**N**ational **C**enter for **A**tmospheric **R**esearch
NCAS	**N**on-**C**all-path **A**ssociated **S**ignaling
NCB	**N**etwork **C**ontrol **B**lock (networking)
NCBS	**N**ew **C**ommercial **B**illing **S**ystem
NCC	1. **N**etwork **C**ontrol **C**enter (Inmarsat)
	2. **N**etwork **C**olor **C**ode (GSM)
	3. **N**ational **C**oordination **C**ommittee
	4. **N**ational **C**ommunications **C**ommittee
	5. **N**ational **C**omputer **C**onference
NCCF	**N**etwork **C**ommunications **C**ontrol **F**acility (IBM SNA)

NCCS	Network Control Center System
NCD	Non-linear Control Design
NCE	Nomadic Computing Environment (PC standard)
NCEO	Non-Compliant End Office
NCH	Number of Calls Handled
NCHD	National Customer Help Desk
NCHPC	National Consortium for High-Performance Computing
NCI	Network Channel Interface
NCIA	Native Client Interface Architecture (Cisco)
NCITS	National Committee for Information Technology Standardization
NCL	Network Control Language (programming)
NCMS	National Center for Manufacturing Sciences
NCO	1. National Coordination Office
	2. Neutral Central Office
	3. Numerically Controlled Oscillator
NCOP	Network Code Of Practice
NCP	1. NetWare Core Protocol (networking)
	2. Network Control Protocols (Internet)
	3. Network Control Point (AT&T)
	4. Network Control Program (IBM SNA)
	5. Network Configuration Process (Northern Telecom)
NCPS	National Customer Product Support
NCRA	National Cellular Resellers Association (U.S.A.)
NCS	1. Network Coordination Station (Inmarsat)
	2. Network Control System
	3. Network-based Call Signaling
	4. Network Computing System
	5. National Communications System (U.S.A.)
NCSA	1. National Center for Supercomputing Applications (U.S.A.)
	2. National Computer Security Associations
NCSANet	National Center for Supercomputing Applications Network
NCSC	1. National Computer Security Center (U.S.A.)
	2. National Communications Security Center (U.S.A.)
NCTA	National Cable Television Association (U.S.A.)
NCTE	Network Channel-Terminating Equipment
NCU	Node Control Unit
NCUG	National Centrex Users Group
NDA	1. Non-Disclosure Agreement (contracts)
	2. National Directory Assistance
NDAC	National Database Administration Center
NDC	National Destination Code (ISDN number)

N

NDCDB	National Digital Cartographic Database
NDCS	1. Network Data Control Station
	2. Network Data Control System
NDD	NetWare Directory Database (networking)
NDDS	Network Data-Delivery Service
NDE	Network Design Engineering
NDF	New Data Flag (MUX)
NDIR	Non-Dispersive Infrared
NDIS	1. Network Driver Interface Specification (networking)
	2. Network Design Interface Specification (networking)
NDLO-CIS	Norwegian Defense Logistic Organization—Communications and Information Systems
NDM	Network Data Mover
NDMP	Network Data Management Protocol
NDN	Non-Delivery Notification (Inmarsat)
NDO	Network Design Order
NDOD	Near-Data-On-Demand (service)
NDP	Neighbor Discovery Protocol (new Internet)
NDR	1. Non-Destructive Readout (memory)
	2. News Datacom Research, Ltd
	3. Negative Differential Resistance (transistor)
NDRO	Non-Destructive Read-Out (memory)
NDS	1. Network Desired State
	2. NetWare Directory Services (networking)
NDSF	Non-Dispersion Shifted Fiber
NDT	No Dial Tone
NDVD	**Non-Determination in Deductive Databases**
NDVI	Normalized Difference Vegetation Index (remote sensing)
Nd-YAG	Neodymium-doped Yttrium–Aluminum Garnet (laser)
NE	Network Element (MUX)
NE&I	Network Engineering **and** Implementation
NEAP	Novell Education Academic Partner (e-learning)
NEAR	Near-Earth Asteroid Rendezvous Spacecraft
NEARNet	New England Academic and Research Network
NEBE	Near-End Block Error
NEBS	Network Equipment Building Standard (Telcordia Technologies)
NEC	1. Nippon Electric Company (Japan)
	2. National Electric Code (U.S.A.)
NECA	National Exchange Carrier Association (standards organization)
NECP	Network Element Control Protocol

NECTAR	Network for Effective Collaboration Through Advanced Research
NECTAS	Network Equipment Craft Terminal Application Software
NEDA	National Electronics Distributors Association
NEDT	Noise-Equivalent Differential Temperature (electronics)
NEF	Network Element Function (MUX)
Negatron	**Nega**tive elec**tron**
NEI	Network Entity Identifier (cellular network)
NEL	Network Element Layer
NEM	Network Element Management (MUX)
NEMA	National Electrical Manufacturers Association
NEMAS	Network Management and Administration System
NEMKO	Norges Elektriske Material Kontroll (Norway)
NEMO	Not Emanating from Main Office (broadcasting)
NEMP	Nuclear Electromagnetic Pulse
NEMS	Network Element Managment Server
NENA	National Emergency Numbering Association (U.S.A.)
NEP	Noise Equivalent Power (electronics)
NEPA	National Environment Policy Act
NES	Noise Equivalent Signal
NESC	National Electrical Safety Code (IEEE)
NESDIS	National Environmental Satellite, Data, and Information Service
NET	1. Inter**net**
	2. **net**work
	3. Network Equipment Technologies, Inc.
	4. Normes Européenne de Télécommunications
NetPOP	**Net**work **P**oint **O**f **P**resence
NetBEUI	**NetB**IOS **E**xtended **U**ser **I**nterface (networking)
NetBIOS	**Net**work **BIOS** (networking)
NetBSD	A free version of **BSD** UNIX operating system
NetBT	**NetB**IOS over **T**PC/IP
netcast	A broad**cast** by using the Inter**net**
NETCOM	1. **Net**work Enterprise Technology **Com**mand (U.S. Army)
	2. **Net**work **Com**mand (information systems)
NETCONSTA	**Net**work **Con**trol **Sta**tion
NETDDE	**Net**work **D**ynamic **D**ata **E**xchange (Windows feature)
NETI	**Net**work **I**nformation Table
NETIPC	**Net**work **I**nter**p**rocess **C**ommunications
netiquette	**net**work et**iquette**
netizen	**net**work cit**izen**
NETOPS	**Net**work **O**perations **S**ystem

N

NetWare SFT	**NetWare** **S**ystem **F**ault **T**olerance (networking)
NEU	Network Extension Unit (AT&T)
.newsr	Name extension for UNIX-based **news r**eaders files (computer)
NEXT	Near-End **Cross-T**alk (cables)
nF	**n**anofarad (electronics)
NF	Noise Figure (satcom)
NFA	Noiseless Feedback Amplifier
NFAR	Network File Access Routine
NFAS	Non-Facility Associated Signaling (ISDN)
NFB	No-Fuse Circuit-Breaker
NFCB	National Federation of Community Broadcasters (U.S.A.)
NFM	1. Network Fault Management (Motorola)
	2. Noise Figure Meter
NFOEC	National Fiber Optic Engineers Conference
NFPA	National File Protection Association (U.S.A.)
NFR	Normal Frequency Reference unit
NFS	1. Network File Server
	2. Network File System (Unix standard)
NFSP	NetWare File Service Protocol
NFT	Network File Transfer
NGA	Noise Gain Analyzer
NGDC	National Geographical Data Center
NGDLC	Next Generation Digital Loop Carrier
NGFL	National Grid For Learning (e-learning)
NGI	Next Generation Internet
NGIN	Next Generation Intelligent Network
NGN	Next Generation Network
NGNP	Non-Geographic Number Portability
NGOSS	New Generation Operating System and Software
NGP	Network Group
NGSO	Non-Geostationary Orbit (satellites)
NGSO-FSS	Non-Geostationary Orbit—Fixed Satellite Service (ITU)
NGT	New Generation Transceiver
nH	**n**anohenry (electronics)
NH	National Host
NHAP	National High-Altitude Photography
NHK	Nippon Hose Kyokai (Japanese broadcaster)
NHRP	Next Hop Routing Protocol (IETF)
NHS	Next Hop Server
NI	1. Network Interface
	2. Network Implementation
	3. National ISDN

NI-1	National ISDN-**1**
NI-2	National ISDN-**2**
NI-3	National ISDN-**3**
NIA	Network Interface Adapter (IBM)
NIC	1. Network Interface Card (networking)
	2. Network Information Center
	3. Network Interface Controller
	4. National ISDN Council
	5. Near-Instantaneous Companding
NiCAD	Nickel–Cadmium (battery)
NICAM	Near-Instantaneous Companded Audio Multiplex
NiCd	Nickel–Cadmium (battery)
NICE	Network Information and Control Exchange
NICI	National Information and Communications Infrastructure (Africa)
NICS	NATO Integrated Communications System
NICT	National Institute of Information and Communications Technology (Japan)
NICTA	National Iranian ICT Agenda (locally known as TAKFA)
NID	1. Network Inward Dialing (feature)
	2. Network Interface Device
	3. Network Information Database
NIDS	Network Intrusion Detection System
NIF	Network Interface Function
NIH	National Information Highway
NII	National Information Infrastructure
NIM	1. Network Interface Module (standard)
	2. National Instrumentation Module (standard)
NIMA	National Imagery Mapping Agency
NIMBY	"Not In My Back Yard" (cellular networks)
NiMH	Nickel-Metal-Hydride (battery)
NIMS	Network Information Management System
NIOD	Network Inward/Outward Dialing
NIPO	Negative-Input Positive-Output (digital electronics)
NIS	1. Network Information Services (UNIX systems)
	2. Network Imaging Server
	3. NATO Identification System
NISP	National ISP (Internet Service Provider)
NIST	National Institute of Standards and Technology (U.S.A.)
NIT	Network Information Table (broadcasting)
NITF	National Image Transfer Format
NIU	Network Interface Unit
NIUF	North American ISDN Users' Forum (standards organization)

NIVT	Network Integration Verification Test
NJE	Network Job Entry (IBM)
NL	1. New Line (control character)
	2. Network Layer
NLA	Network Layer Address (LAN packet)
NLANR	National Laboratory for Applied Network Research (U.S.A.)
NLDC	Non-Linear Directional Coupler
NLDM	Network Logical Data Manager (IBM)
NLES	Navigational Land Earth Station
NLETS	National Law Enforcement Telecommunication System
NLM	NetWare Loadable Module (networking)
NLO	Non-Linear Optical material
NLOS	Non-Line-Of-Sight (transmission)
NLP	1. Neuro-Linguistic Programming
	2. Natural Language Processing (language recognition)
	3. Network Layer Packet
NLPI	Network Layer Protocol Identifier (LANs)
NLPID	Network Layer Protocol ID
NLQ	Near-Letter-Quality (printers)
NLS	1. Name Lookup Service
	2. Natural-Language Support (voice recognition)
	3. Non-Linear Schrodinger (equation)
NLSP	NetWare Link Services Protocol (networking)
nm	nanometer
NM	Network Management
NMA	1. Network Management Architecture
	2. Network Monitoring and Analysis system
NMAA	National Multimedia Association of America
NMACS	Network Monitor And Control System
NMAR	Network Management and Access Routine
NMB	Network Memory Bus
NMC	1. Network Management Center
	2. Network Monitoring and Control
	3. Non-Metallic Sheathed Cable
NMCC	Network Management Control Center (networking)
NMCCC	Network Management Command and Control Center
NMCM	Network Management Communications Manager
NMCN	Network Management Communications Network
NMEA	National Maritime Electronics Association
NMF	1. Network Management Forum (standards organization)
	2. Network Management Function

NMI	1. **N**on-**M**askable **I**nterrupt (PC chat)
	2. **N**etwork **M**anagement **I**nterface
NML	**N**etwork **M**anagement **L**ayer (TMN)
NMM	**N**etwork **M**anagement **M**odule
NMOS	**N**-channel **MOS** (semiconductors)
NMP	1. **N**etwork **M**anagement **P**rocessor (AT&T)
	2. **N**etwork **M**anagement **P**rotocol (networking)
NMPA	**N**ational **M**usic **P**ublishers' **A**ssociation
NMR	**N**uclear **M**agnetic **R**esonance (microwave)
NMRR	**N**ormal-**M**ode **R**ejection **R**atio
NMS	1. **N**etWare **M**anagement **S**ystem (networking)
	2. **N**etwork **M**anagement **S**ystem
NMSE	1. **N**etwork **M**anagement **S**ystem **E**ngineering
	2. **N**ormalized **M**ean **S**quare **E**rror
NMSI	**N**ational **M**obile **S**tation **I**dentification number
NMT	**N**ordic **M**obile **T**elephone system (Nordic Countries)
NMTC	**N**ational **M**obile **T**elecommunications **C**ompany (Kuwait)
NMU	1. **N**etwork **M**anagement **U**nit
	2. **N**oise **M**easurement **U**nit
NMVT	**N**etwork **M**anagement **V**ector **T**ransport
NN	1. **N**ational **N**etwork
	2. **N**ational **N**umber
	3. **N**etwork **N**ode
NNBS	**N**etView **N**etwork **B**illing **S**ystem
NNC	**N**eural **N**etwork **C**ouncil (IEEE)
NNCDE	**N**ortel **N**etworks **C**ertified **D**esign **E**xpert
NNCNA	**N**ortel **N**etworks **C**ertified **N**etwork **A**rchitecture
NNCSE	**N**ortel **N**etworks **C**ertified **S**upport **E**xpert
NNCSS	**N**ortel **N**etworks **C**ertified **S**upport **S**pecialist
NNG	**N**ational **N**umber **G**roup
NNI	1. **N**etwork **N**ode **I**nterface (MUX)
	2. **N**etwork-to-**N**etwork **I**nterface (ATM)
	3. **N**eural **N**otwork **I**nterface
	4. **N**earest **N**eighbor **I**nterpolation (remote sensing)
NNM	**N**ational **N**etwork **M**anagement
NNMC	**N**ational **N**etwork **M**anagement **C**enter (TPC/IP)
NNS	1. **N**ational **N**etwork **S**upport
	2. **N**etwork **N**ode **S**ystem
NNSC	**N**ational **N**etwork **S**urveillance **C**enter
NNSS	**N**avy **N**avigational **S**atellite **S**ystem
NNTM	**N**ational **N**etwork **T**rouble **M**anagement
NNTP	**N**etwork **N**ews **T**ransfer **P**rotocol (TPC/IP)
NNX	**N**etwork **N**umbering **E**xchange

N

NO	1. **N**etwork **O**bject
	2. **N**ormally **O**pen (relay or switch contacts)
NOAA	**N**ational **O**ceanic and **A**tmospheric **A**dministration (satellite)
NOBIS	**N**etwork **O**perations **B**usiness **I**nformation **S**ystem
NOC	**N**etwork **O**perations **C**enter (Inmarsat)
NOCC	**N**etwork **O**perations **C**ontrol **C**enter
NOCTAS	**N**etwork **O**ffice **C**raft **T**erminal **A**pplication **S**oftware
NOD	**N**etwork **O**utward **D**ialing (feature)
NODS	**N**ear **O**bstacle **D**etection **S**ystem
NOH	**N**etwork **O**perations **H**andbook (Inmarsat)
NOI	**N**otice **O**f **I**nquiry (FCC)
NOIS	**N**etscape, **O**racle, **I**BM, and **S**un Microsystems (Microsoft)
NOF	**N**etwork **O**perator **F**acility
NOMDA	**N**ational **O**ffice **M**achine **D**ealers **A**ssociation
Non-Div	**Non-Div**ersity
Non-GSO	**Non-G**eo**s**tationary **O**rbit (satellites)
NO-OP	**N**o-**Op**eration (instruction)
NOP	**N**etwork **O**perations **P**rotocol
NOPS	**N**ew **O**rder **P**rocessing **S**ystem
NOR Gate	A combination of a **N**O**T** and an **OR Gate** (digital electronics)
NORC	**N**etwork **O**perations **R**esearch **C**ommittee
NRE	**N**on-**R**ecurring **E**ngineering
NORM	**N**ack-**O**riented **R**eliable **M**ulticast
NOS	**N**etwork **O**perating **S**ystem (LAN software)
NOSC	**N**aval **O**cean **S**urveillance **C**enter
NOST	**N**ASA **O**ffice of **S**tandards and **T**echnology
NOTAM	**Not**ice to **Air**men (COMSAR)
NOTHS	**N**etwork **O**perations **T**rouble **H**andling **S**ystem
NOV	**N**ews **O**verview
Novin 3Com	**Novin Com**prehensive **Com**munication **Com**pany (Iran)
Np	**Ne**per (unit of ratio of two voltages, two currents, or two power values in a logarithmic manner)
NP	1. **N**umber **P**ortability
	2. **N**egative–**P**ositive (semiconductors)
NP&D	**N**etwork **P**lanning **and** **D**esign
NP0	**N**egative–**P**ositive-**Zero** (materials)
NPA	1. **N**umbering **P**lan **A**rea (Bell)
	2. **N**etwork **P**erformance **A**nalyzer (IBM)
	3. **N**etwork **P**rofessionals **A**ssociation (U.S.A.)
	4. **N**ational **P**ricing **A**greement (AT&T)

NPAC	1. Number Portability Administration Center
	2. Number Portability Access Center
NPBA	National Public Broadcasting Archives (Maryland University)
NPC	1. Network Parameter Control
	2. Network Processing Card
NPD	1. Network Protection Device (PBX)
	2. Noise Power Density
NPDA	Network Problem Determination Application (IBM)
NPDN	Nordic Public Data Network
NPDU	Network Protocol Data Unit
NPI	1. Network Product Implementation
	2. Not Plugged In
	3. Null Pointer Indicator (MUX)
NPL	Network Planning
NPM	1. Non-Penetrating Mount
	2. Network Process Monitor
NPN	1. Negative–Positive–Negative (semiconductors)
	2. Network Provider Network
NPNP	Negative–Positive–Negative–Positive (semiconductors)
NPO	Network Performance Objectives
NPR	1. National Public Radio
	2. Noise Power Ratio
	3. New Product Return (number)
NPRM	Notice of Proposed Rule Making (regulatory)
NPS	1. Network Product Support
	2. Number Portability Surcharge
NPSI	NCP Packet Switching Interface (X.25)
NPSP	Network Products and Systems Planning
NPSTC	National Public Safety Telecommunications Council
NPU	Numeric Processing Unit
NQR	Nuclear Quadruple Resonance
NR	1. Number (Morse code transmissions)
	2. Nyquist Rate
NRAM	Non-volatile RAM (memory)
NRC	1. Non-Recurring Charge
	2. National Research Council (U.S.A.)
	3. Network Reliability Council
NRCE	Non-Redundant Common Equipment
NRD	Network Reconciliation Descriptor
NRDA	National Routing and Database Administration
NREN	National Research and Education Network (e-learning)
NRI	Net Radio Interface

N

NRIC	Network Reliability and Interoperability Council (FCC)
NRM	1. Normal Response Mode (HDLC)
	2. Network Resource Manager
NRMS	Network Reference and Monitor Station
NRS	1. Network Routing System
	2. National Relay Service
	3. Network Reconfiguration System
NRSC	National Radio Systems Committee
NRT	Near-Real-Time
NRTC	National Rural Telecommunication Cooperative
NRTL	Nationally Recognized Test Laboratory
NRUG	National Rolm User Group
NRZ	Non-Return-to-Zero level (data encoding)
NRZ-1	Non-Return-to-Zero level, **one** (data encoding)
NRZ-C	Non-Return-to-Zero level, Change (data encoding)
NRZ-I	Non-Return-to-Zero level, Inverted (data encoding)
NRZ-L	Non-Return-to-Zero, Level (data encoding)
NRZ-M	Non-Return-to-Zero, Mark (data encoding)
NRZ-S	Non-Return-to-Zero, Space (data encoding)
NS-CPE	Non-Standard Customer Premise Equipment
NSA	1. National Security Agency
	2. Network Service Address
	3. Normalized Site Attenuation (microwave)
NSAI	National Standards Authority of Ireland
NSAP	Network Service Access Point (OSI model)
NSAPI	Netscape Server Application Programming Interface
NSB	National Science Board
NSC	1. Network Service Center
	2. Non-Standard Command
NSCAT	NASA Scatterometer (remote sensing)
NSCP	National Scalable Cluster Project
NSDETP	Network System Development Electronic Transaction Processing
NSDI	National Spatial Data Infrastructure (U.S.A.)
NSDP	Non-Signaling Data Protocol
NSDP-B&S	Non-Signaling Data Protocol for Billing **and** Settlements
NSDP-F	Non-Signaling Data Protocol for Fraud
NSDU	Network Service Data Unit (OSI model)
NSE	1. Network Support Element (networking)
	2. Network Surveillance Engineer
	3. Network System Engineer
NSE/OES	Network Support Element **and** Operations Evaluation System

nsec	**n**ano**sec**ond
NSESS	**N**etwork **S**ystems **E**ngineering **S**ystems **S**upport
NSF	1. **N**ational **S**cience **F**oundation
	2. **N**etwork **S**pecific **F**acilities
NSFNet	**N**ational **S**cience **F**oundation's **Net**work
NSG	**N**ational **S**ystems **G**roup
NSI	**N**etwork **S**olutions, **I**nc.
NSIC	**N**ational **S**torage **I**ndustry **C**onsortium
NSIE	**N**etwork **S**ecurity **I**nformation **E**xchange
NSIF	**N**etwork and **S**ervices **I**ntegration **F**orum (ATIS)
NSInet	**N**A**S**A **S**cience **Internet**
NSLOOKUP	**N**ame **S**erver **Lookup**
NSMD	**N**on-**S**older **M**ask **D**efined
NSN	**N**etwork **S**ervice **N**ode
NSO	1. **N**ational **S**tandards **O**rganization
	2. **N**ational **S**ervices **O**rganization
	3. **N**ational **S**witch **O**perations
NSOM	**N**ear-field **S**canning **O**ptical **M**icroscopy
NSP	1. **N**etwork **S**ervice **P**rovider
	2. **N**etwork **S**ervice **P**rotocol
	3. **N**etwork **S**ervice **P**oint
	4. **N**etwork **S**upervisory **P**rocessor
	5. **N**ative **S**ignal **P**rocessing
NSPE	**N**ational **S**ociety of **P**rofessional **E**ngineers
NSPIXP	**N**etwork **S**ervice **P**rovider **I**nternet **Ex**change **P**oint
NSR	1. **N**etwork **S**tatus **R**ecord
	2. **N**on-**S**ource **R**outed
	3. **N**yquist **S**ampling **R**ate
NSS	1. **N**etwork and **S**witching **S**ubsystem (PLMN)
	2. **N**etwork **S**tatus **S**ubsystem
	3. **N**etwork **S**upport **S**ystem
	4. **N**etwork **S**oftware **S**upport
	5. **N**ational **S**tandards **S**ystem (Canada)
	6. **N**on-**S**tandard **S**etup (command)
NSSN	**N**ational **S**tandard **S**ystem **N**etwork
NSTAC	**N**ational **S**ecurity **T**elecommunications **A**dvisory **C**ommittee
NSTS	**N**etwork **S**ystems **T**echnical **S**upport
NSU	**N**etwork **S**ynchronization **U**nit
NT	1. **N**orthern **T**elecom
	2. **N**etwork **T**erminal (ISDN)
	3. **N**ew **T**echnology (Windows NT)
	4. Windows **N**ew **T**echnology
NT1	**N**etwork **T**erminator type **1** (ISDN)

N

NT2	Network Terminator type **2**
NTACS	Narrow **TACS** (cellular networks)
NTAS	1. Network Traffic Analysis System
	2. Windows **NT A**dvanced **S**erver
NTB	Net-Top-Box (PC)
NTC	1. National Telecommunications Commission (Thailand)
	2. National Telecoms Corporation (Sudan)
	3. Negative Temperature Coefficient (thermistors)
NTCA	National Telephone Cooperative Association
NTDPC	National Telecommunications Damage Prevention Council (U.S.A.)
NTDS	1. Naval Tactical Data System
	2. National Television Design Standards
NTE	1. Network Termination Equipment (BT)
	2. Network Transmission Elements
NTELOS	NetView Traffic Engineering Line Optimization System
NTF	No Trouble Found
NTFS	New Technology File System (Windows NT)
NTI	Network Terminating Interface
NTIA	Network Telecommunications and Information Administration
NTIS	National Technical Information Service Technology Administration (U.S.A.)
NTL	National Transcommunications, Ltd.
NTM	1. Network Traffic Management (Telcordia Technologies)
	2. Notice To Mariners (shipping)
NTN	Network Terminal Number (X.25)
NTO	Network Terminal Option (IBM)
NTP	1. Network Time Protocol (TCP/IP)
	2. Network Termination Point
NTS	1. Network Test System
	2. Number Translation Services
	3. Non-Traffic Sensitive
NTSC	National Television Standards Committee (TV broadcast format)
NTSC STD	NTSC Standard (TV broadcasting)
NTT	Nippon Telegraph and Telephone Corporation (Japan)
NTTC	National Transportable Telecommunications Capability (comsat)
NTU	Network Terminating Unit
NTV	Non-linear Thickness Variation (IC wafer)

NU	1. **N**etwork **U**nit
	2. **N**etwork **U**navailable
NUA	**N**etwork **U**ser **A**ddress
NuBus	**New Bus**
NUDET	**Nu**clear **D**etonation **E**valuation **T**echnique
nudome	**nu**clear ra**dome** (nuclear radiations)
NUI	1. **N**et**W**are **U**sers **I**nternational (networking)
	2. **N**etwork **U**ser **I**dentifier
NUID	**N**etwork **U**ser **Id**entification
NUL	"**NULL**" (character control code)
NUMA	**N**on-**U**niform **M**emory **A**ccess (architecture)
nV	**n**ano**v**olt (electronics)
NV	**N**etwork **V**ideo
NVI	**N**ormalized **V**egetation **I**ndex (remote sensing)
NVIS	**N**ear-**V**ertical **I**ncident **S**kywave
NVLAP	**N**ational **V**oluntary **L**aboratory **A**ccreditation **P**rogram
NVM	**N**on-**V**olatile **M**emory
NVoD	**N**ear-**V**ideo-**o**n-**D**emand (service)
NVP	**N**etwork **V**oice **P**rotocol (ARPANet)
NVRAM	**N**on-**V**olatile **RAM** (memory)
NVS	**N**on-**V**olatile **S**torage
NVT	**N**ovell **V**irtual **T**erminals (Novell NetWare)
nW	**n**ano**w**att (electronics)
NWADMIN	**N**et**W**are **Admin**istrator (Novell NetWare)
NWAT	**N**ew **W**ireline **A**ccess **T**echnology
NWDM	**N**arrow-**W**avelength-**D**ivision **M**ultiplexing
NWL	**N**on-**W**ire**l**ine (cellular networks)
NWRA	**N**ational **W**ireless **R**esellers **A**ssociation (U.S.A.)
NWS	**N**et**w**ork **S**ervices
NXX	identifies the central office exchange allocated within the NPAs
NYIIX	**N**ew **Y**ork **I**nternational **I**nternet E**x**change
NYNEX	**N**ew **Y**ork **N**ew **E**ngland E**x**change (AT&T)
NZCS	**N**ew **Z**ealand **C**omputer **S**ociety
NZDSF	**N**on-**Z**ero **D**ispersion **S**hifted **F**iber
NZSA	**N**ew **Z**ealand **S**oftware **A**ssociation
NZWDF	**N**ew **Z**ealand **W**ireless **D**ata **F**orum

N

O o

O&M	**O**perations **and M**aintenance
O-ADM	**O**ptical **A**dd-**D**rop **M**ultiplexer (MUX)
O-band	**O**ptical Frequency wavelength ranging from 1260 to 1310 nm
O-QPSK	**Off**set **Q**uadrature **P**hase **S**hift **K**eying (modulation)
O/E	**O**ptical-**to-E**lectric converter
O/P	**o**ut**p**ut
O/R	**O**riginator **and R**ecipient (X.400)
OA	1. **O**ffice **A**utomation
	2. **O**ptical **A**mplifier
OA&M	**O**perations, **A**dministration **and M**aintenance
OAA	**O**pen **A**vionics **A**rchitecture
OACSU	**O**ff-**A**ir **C**all **S**et**u**p (GSM)
OAD	**O**ptical **A**dmittance **D**iagram
OADM	**O**ptical **A**dd-**D**rop **M**ultiplexer (MUX)
OAI	**O**pen **A**pplication **I**nterface
OAM	1. **O**ff-**A**ir **M**onitoring
	2. **O**perations, **A**dministration, and **M**aintenance
	3. **O**perations **and M**aintenance
OAM&P	**O**perations, **A**dministration, **M**aintenance **and** **P**rovisioning
OAMS	**O**peration **A**larms and **M**easurement **S**ystem
OAMT	**O**perations, **A**dministration and **M**aintenance **T**erminal

OAN	**O**ptical **A**ccess **N**ode
OAO	**O**rbiting **A**stronomical **O**bservatory
OAs	**O**perations **A**pplication**s** (AINs)
OATS	**O**pen **A**rea **T**est **S**ite
OAU	**O**perator **A**ccess **U**nit
OB	**O**utside **B**roadcasting (Intelsat)
OBC	**O**n-**B**oard **C**omputer
OBCS	**O**n-**B**oard **C**ommunications **S**tation
OBE	**O**ut-of-**B**and **E**mission (Intelsat)
OBF	**O**rdering and **B**illing **F**orum
OBI	1. **O**pen **B**uying on the **I**nternet Consortium (standards)
	2. **O**mni-**B**earing **I**ndicator (instrument)
OBL	Local Loop **O**perator of France
OBN	**O**ut-of-**B**and **N**oise (Intelsat)
OBO	**O**utput **B**ack-**O**ff
OBP	**O**n-**B**oard **P**rocessing (Intelsat)
OBS	1. **O**utdoor **B**ase **S**tation
	2. **O**ut-of-**B**and **S**ignaling
	3. **O**ptical **B**urst **S**witching
	4. **O**mni-**B**earing **S**elector (instrument)
OBSAI	**O**pen **B**ase **S**tation **A**rchitecture **I**nitiative
OBU	**O**n-**B**oard **U**nit
OBWR	**O**n-**B**oard **W**aveform **R**egenerator (Intelsat)
OC	1. **O**ptical **C**arrier
	2. **O**pen **C**ircuit
	3. **O**pen **C**ollector (circuits)
OC-1	**O**ptical **C**arrier level-**1**
OC-3	**O**ptical **C**arrier level-**3**
OC-12	**O**ptical **C**arrier level-**12**
OCA	**O**ptical **C**hannel **A**nalyzer
OCB	**O**il **C**ircuit-**B**reaker
OCC	1. **O**perations **C**ontrol **C**enter (Inmarsat and Intelsat)
	2. **O**ther **C**ommon **C**arrier (Intelsat)
	3. **O**ptical **C**ross-**C**onnect (MUX)
OCDMA	**O**ptical **CDMA** (access)
OCE	1. **O**pen **C**ollaboration **E**nvironment (Macintosh)
	2. **O**n-**C**all **E**ngineer (Intelsat)
OCG	**O**perational **C**oordination **G**roup (ETSI)
OCH	**O**ptical **C**hannel
OCIR	**O**vercoat **I**ncident **R**ecording
OCL	**O**ffice **C**ode **L**ocation
OCN	**O**perating **C**ompany **N**umber
OCO	**O**ven **C**ontrolled **O**scillator

OCP	1. **O**perator **C**ontrol **P**anel
	2. **O**ptical **C**ontrol **P**lane (Cisco)
OCPAR	**O**ptically-**C**ontrolled **P**hase-**A**rray **R**adar
OCR	1. **O**ptical **C**haracter **R**ecognition (scanners)
	2. **O**ptical **C**haracter **R**eader (scanners)
	3. **O**utgoing **C**all **R**estriction
OCRI	**O**ttawa **C**enter for **R**esearch and **I**nnovation (Canada)
OCS	**O**n-board **C**ommunications **S**tation
OCT	**O**ffice **C**raft **T**erminal
OCTS	**O**cean **C**olor and **T**emperature **S**canner (remote sensing)
OCU	**O**rderwire **C**ontrol **U**nit
OCUDP	**O**ffice **C**hannel **U**nit **D**ata **P**ort
OCVCXO	**O**ven-**C**ontrolled/**V**oltage-**C**ontrolled **Crystal Oscillator**
OCW	1. **O**pen**C**ourse**W**are (ICT)
	2. **O**scillator **C**ontrol **W**ord
OCXO	**O**ven-**C**ontrolled **Crystal O**scillator
OCX	**OLE C**ustom **C**ontrol
OD	1. **O**ptical **D**ensity
	2. **O**ptical **D**emultiplexer
ODA	1. **O**ffice **D**ocument **A**rchitecture (ISO)
	2. **O**bject **D**efinition **A**lliance (Oracle)
ODAPS	**O**ceanic **D**isplay **a**nd **P**lanning **S**ystem
ODAS	**O**cean **D**ata **A**cquisition **S**ystems (UNESCO)
ODB	1. **O**perator **D**etermined **B**arring (GSM)
	2. **O**pen **D**ual **B**us
ODBC	**O**pen **D**ata-**B**ase **C**onnectivity
ODBMG	**O**bject **D**ata-**B**ase **M**anagement **G**roup
ODC	**O**pen **D**evelopment **C**onsortium
ODF	**O**ptical **D**istribution **F**rame
ODI	1. **O**pen **D**ata-**L**ink **I**nterface (networking)
	2. **O**perational **D**ata **I**ntegrator (MCI)
ODIF	**O**ffice **D**ocument **I**nterchange **F**ormat
ODIN	**O**ptical **D**igital **I**ntegrated **N**etwork
ODINSUP	**O**pen **D**ata-link **I**nterface/**N**etwork **Sup**port (networking)
ODIS	**O**ceanographic **D**ata **I**nterrogating **S**tation
ODLC	**O**ptimum **D**ata-**L**ink **C**ontrol (protocol)
ODMG	**O**bject **D**atabase **M**anagement **G**roup
ODP	1. **O**pen **D**istributed **P**rocessing
	2. **O**ptical **D**istribution **P**oint
ODR	**O**ptical **D**ocument **R**ecognition
ODSI	**O**pen **D**irectory **S**ervices **I**nterface

O

ODTR	**O**ffice of the **D**irector of **T**elecommunications **R**egulation (Ireland)
ODU	**O**ut-**D**oor **U**nit (VSAT)
O/E	**O**ptical-**to**-**E**lectrical (converter)
Oe	**Oe**rsted (magnetic field strength)
OE	**O**ffice **E**quipment
OEIC	**O**pto-**E**lectronic **I**ntegrated **C**ircuit
OEL	**O**rganic **E**lectro **l**uminescent (display technology)
OELR	**O**verall **E**cho **L**oudness **R**ating
OEM	**O**riginal **E**quipment **M**anufacturer (products)
OEO	**O**ptical-to-**E**lectrical-to-**O**ptical regeneration
OES	**O**perations **E**valuation **S**ystem (MCI)
OF	**O**ptical **F**iber
OFA	**O**ptical **F**iber **A**mplifier
OFBG	**O**ptical **F**iber **B**ragg **G**rating
OFC	1. **O**pen **F**inancial **C**onnectivity (interface)
	2. **O**ptical **F**iber **C**able
OFCP	**O**ptical **F**iber **C**onductive **P**lenum
OFCR	**O**ptical **F**iber **C**onductive **R**iser
OFDM	**O**rthogonal **F**requency **D**ivision **M**ultiplexing (transmission)
OFDMA	**O**rthogonal **F**requency **D**ivision **M**ultiple **A**ccess
OFHC	**O**xygen-**F**ree **H**igh-**C**onductivity (copper wire)
OFM	**O**perational **F**ixed **M**icrowave
OFN	1. **O**ptical **F**iber **N**etwork
	2. **O**ptical **F**iber **N**on-conductive
OFNP	1. **O**ptical **F**iber **N**on-conductive **P**lenum
	2. **O**ptical **F**iber **N**etwork **P**lenum
OFNR	1. **O**ptical **F**iber **N**on-conductive **R**iser
	2. **O**ptical **F**iber **N**etwork **R**iser
OFR	**O**ffice of the **F**ederal **R**egister (U.S.A.)
OFS	1. **O**perational **F**ixed **S**ervice
	2. **O**ut-of-**F**rame **S**econds (MUX)
OFSCFC	**O**ptical **F**iber **C**able and **S**olar **C**ell **F**abrication **C**ompany (Iran)
OFTA	**O**ffice of the **T**elecommunications **A**uthority (Hong Kong)
OFTEL	The **O**ffice of **Tel**ecommunications (U.K.)
OFX	**O**pen **F**inancial **E**xchange
OGM	**O**ut-**G**oing **M**essage
OGT	**O**ut-**G**oing **T**runk
OH	**o**ver**h**ead (MUX)
OHA	**O**ver**h**ead **A**ccess (MUX)
OHB	**O**ver**h**ead **B**yte (MUX)

OHD	**O**ptical **H**ard **D**rive
OHG	**O**perators **H**armonization **G**roup (cellular networks)
OHI	1. **O**ver**h**ead **I**nterface
	2. **O**ne-**H**orn **I**nterferometer
OHQ	**O**ff-**H**ook **Q**ueue
OHR	1. **O**ptical **H**andwriting **R**ecognition
	2. **O**ver**h**ead **R**outer
OHVA	**O**ff-**H**ook **V**oice **A**nnounce
OIC	**O**ptical **I**ntegrated **C**ircuit
OICETS	**O**ptical **I**nter-orbit **C**ommunications **E**ngineering **T**est **S**atellite
OID	1. **O**riginating **ID**
	2. **O**bject **Id**entifier
OIDA	**O**ptoelectronics **I**ndustry **D**evelopment **A**ssociation
OIF	**O**ptical **I**nterworking **F**orum
OIR	**O**nline **I**nsertion and **R**emoval (equipment components)
OJT	**O**n-the-**J**ob **T**raining
OK-QPSK	**O**ffset-**K**eyed **QPSK** (modulation)
OLA	1. **O**pen-**L**ibrary **A**rchitecture
	2. **O**ptical **L**ine **A**mplifier
OLAP	**O**n-**L**ine **A**nalytical **P**rocessing (ICT)
OLC	**O**ptical **L**oop **C**arrier
OLE	**O**bject **L**inking and **E**mbedding (software technology)
OLEC	**O**riginating **L**ocal **E**xchange **C**arrier
OLED	**O**rganic **L**ight-**E**mitting **D**iode (display technology)
OLI	**O**ptical **L**oop **I**nterface
OLIU	**O**ptical **L**ine **I**nterface **U**nit
OLNS	**O**riginating **L**ine **N**umber **S**creening
OLO	**O**ther **L**icensed **O**perators
OLOS	**O**bstructed-**L**ine-**O**f-**S**ight (transmission)
OLR	**O**utgoing **L**ongwave **R**adiance (remote sensing)
OLRU	**O**ptical **L**ine **R**egenerating **U**nit
OLS	1. **O**riginating **L**ine **S**creening
	2. **O**n-**L**ine **S**ervices
	3. **O**ptical-**L**abel **S**witching (networking)
OLT	**O**ptical **L**ine **T**erminal
OLTC	**O**ptical **L**ine **T**erminal **C**ontroller
OLTE	**O**ptical **L**ine **T**erminal **E**quipment
OLTP	**O**n-**L**ine **T**ransaction **P**rocessing (ICT)
OLTS	**O**ptical **L**oss **T**est **S**et
OLTU	**O**ptical **L**ine **T**erminal **U**nit

O

OM	1. **O**perational **M**easurement
	2. **O**ld **M**an or radio officer of ship (Morse code transmissions)
	3. **O**ptical **M**ultiplexer
OM-1	**O**pen **M**PEG Consortium
OMA	1. **O**bject **M**anagement **A**rchitecture
	2. **O**pen **M**obile **A**lliance
OMAB	**O**bject **M**anagement **A**rchitecture **B**oard
OMAN	**O**ptical **M**etropolitan **A**rea **N**etwork
OmanTel	**Oman Tel**ecommunications Company (operator)
OMAP	1. **O**pen **M**ultimedia **A**pplications **P**rotocol (Texas Instruments)
	2. **O**perations and **M**aintenance **A**pplications **P**art
OMAT	**O**perational **M**easurement and **A**nalysis **T**ool
OMC	**O**perations and **M**aintenance **C**enter
OMC-Env	**O**perations and **M**aintenance **C**enter for **Env**ironment
OMC-IN	**O**perations and **M**aintenance **C**enter for **I**ntelligent **N**etwork
OMC-M	**O**perations and **M**aintenance **C**enter for **M**obile
OMC-Misc	**O**perations and **M**aintenance **C**enter for **Misc**ellaneous purposes
OMC-R	**O**perations and **M**aintenance **C**enter for **R**adio
OMC-S	**O**perations and **M**aintenance **C**enter for **S**witching
OMC-T	**O**perations and **M**aintenance **C**enter for **T**ransmission
OMC-WAN	**O**perations and **M**aintenance **C**enter for **W**ide **A**rea **N**etwork
OMF	**O**pen **M**odel **F**orum
OMG	**O**bject **M**anagement **G**roup (U.S. software standards)
OMI	**O**pen **M**icroprocessor **I**nitiative (European)
OMJ	**O**rtho-**M**ode **J**unction
OML	**O**perations and **M**aintenance **L**ink (GSM)
OMP	**O**ptimized **M**ulti**p**ath (routing service)
OMR	**O**ptical **M**ark **R**ecognition (scanners)
OMS	**O**perations and **M**aintenance **S**tation
OMT	**O**rthogonal-**M**ode **T**ransducer
OMUX	**O**utput **M**ulti**x**er
ONA	**O**pen **N**etwork **A**rchitecture (Bell packet networks)
ONAC	**O**perations **N**etwork **A**dministration **C**enter
ONAL	**O**ff-**N**etwork **A**ccess **L**ine
ONC	**O**pen **N**etwork **C**omputing
ONI	1. **O**ptical **N**etwork **I**nterface
	2. **O**perator **N**umber **I**dentification
ONMS	**O**pen **N**etwork **M**anagement **S**ystem
ONP	**O**pen **N**etwork **P**rovision

ONPT	**O**ffice **N**ationale des **P**ostes des **T**elecommunications (Morocco)
ONT	1. **O**pen **N**etwork **T**erminal
	2. **O**ptical **N**etwork **T**ester
	3. **O**ffice **N**ational de **T**élédiffusion (Morocco)
ONTC	**O**pen **N**etworks **T**echnology **C**onsortium
ONU	**O**ptical **N**etwork **U**nit
OO	**O**bject-**O**riented
OOCM	**O**bject-**O**riented **C**all **M**odel
OODB	**O**bject-**O**riented **D**ata-**B**ase
OOF	1. **O**ut-**O**f-**F**rame (transmission)
	2. **O**ut-**O**f-**F**ranchise
OOK	**O**n–**O**ff **K**eying (modulation)
OOO	1. **O**ptical-to-**O**ptical-to-**O**ptical regeneration
	2. **O**ptical-**O**riented **O**peration (TMN)
OOP	**O**bject-**O**riented **P**rogramming
OOPS	**O**pen **O**utsourcing **P**olicy **S**ervices (IETF)
OOR	**O**ut **O**f **R**egion
OORE	**O**perations **O**rder **Re**view
OOS	**O**ut **O**f **S**ervice
OOT	**O**bject-**O**riented **T**echnology (software)
OP	**op**erator (Morse code transmissions)
opamp	**op**erational **amp**lifier
OPAC	1. **O**utside **P**lant **A**ccess **C**abinet
	2. **O**nline **P**ublic **A**ccess **C**atalog (ICT)
OPB	**O**ptical **P**ower **B**udget
OPBO	**O**utput **P**ower **B**ack-**O**ff
OPC	1. **O**riginating **P**oint **C**ode (SS7)
	2. **O**n-**P**remises **C**abling
	3. **O**ptical **P**hase **C**onjugator
opcode	**op**eration **code**
opcom	**op**tical **com**munication
OPDAR	**Op**tical ra**dar**
OpenGL	**Open G**raphics **L**anguage (Silicon Graphics)
Opeval	**Op**erational **eval**uation
Opex	**Op**erations **ex**penditures
OPGW	**O**ptical **G**round **W**ire
OPIP	**O**utput **P**ower **I**ntercept **P**oint (microwave)
OPL	**O**ptical **P**ath **L**ength
OPLL	**O**ptical **P**hase-**L**ocked **L**oop
opm	**op**erations **p**er **m**inute
OP MODEL	**O**perations **Model**
OPP	**O**bject **P**ush **P**rofile (Bluetooth)
OPPV	**O**rder-ahead **P**ay **P**er **V**iew (pay TV)

O

Ops	operations
OPS	1. Off-Premises Station
	2. Open Profiling Standard
	3. Optical Sensor
	4. Operator Services
OPSEC	Operational Security
OPSEC-A	Operational Security Assessment
OPSEC-S	Operational Security Survey
Opt	optional
OPSK	Octal PSK (modulation)
OPTIS	Overlapped Phase Trellis-code Interlocked Spectrum
Optisolator	optically coupled isolator
OPW	On-Premises Wiring
OPX	Off-Premises Extension (PBX adapter)
OQPSK	Offset Quadrature PSK (modulation)
OR	1. Off-Route (aeronautical mobile service)
	2. Ocean Region
	3. Operations Research
	4. Optical Repeater
	5. Outage Ratio
ORB	1. Object Request Broker (CORBA interface)
	2. Office Repeater Bay
ORBCOMM	Orbital Communications Corporation (company)
ORDIG	Open Regional Dialog on Internet Governance
ORM	Optically Remote Module (AT&T)
ORPTR	Optical Repeater
ORR	Ocean Region Registration signal
ORS	Omni-directional Range Station
ORSA	Operations Research and Systems Analysis
ORU	Orbital Replacement Unit (remote sensing)
OS	1. Operating System (computer)
	2. Operator Services (telephony)
	3. Optical Section
OS/2	Operating System version 2 (IBM)
OSA	1. Open System Architecture
	2. Optical Spectrum Analyzer
	3. Optical Society of America
OSC	1. oscillator
	2. Out-band Signaling Channel
	3. Optical Supervisory Channel
OSCAR	Orbiting Satellite Carrying Amateur Radio
OSD	On-Screen Display
OSDM	1. Optical Spatial-Division Multiplexing
	2. Optical Space-Division Multiplexing

OSF	1. Open Software Foundation
	2. Operation Support Function
OSF1	Open Software Foundation version 1
OSI	Open Systems Interconnection (network architecture model)
OSI-TP	Open Systems Interconnection-Transport Protocol
OSI-RM	Open Systems Interconnection-Reference Model (ISO)
OSI-SD	Open Systems Interconnection-Service Definition
OSI-SM	Open Systems Interconnection-System Management
OSIE	Open Systems Interconnection-Environment (ISO)
OSINet	OSI Test Network
OSJ	Optical Society of Japan
OSMINE	Operations Systems Modifications for the Integration of Network Elements
OSN	Open Systems Networking
OSNS	Open Systems Network Support
OSP	Operator Service Provider
OSPF	Open Shortest Path First (routing protocol)
OSPFIGP	Open Shortest Path First Internet Gateway Protocol
OSPR	Optical Shared Protection Ring
OSPS	Operator Services Position System (AT&T)
OSR	1. Optical Solar Reflector
	2. Over Sampling Ratio (ADC)
OSRI	Originating Station Routing Indicator
OSS	1. Operations Support Systems (TMN)
	2. Operations Sub-Systems (GSM)
	3. Operational Service State
OSS7	Operator Services Signaling System Number 7
OSSI	Operations Support System Interface
OSSN	Originating Station Serial Number
OSSP	Object Serialization Stream Protocol
OST	Office of Science and Technology
OSTA	Optical Storage Technology Association (U.S.A.)
OSTP	Office of Science and Technology Policy
OSW	Out-band Service Word
OT	1. Office of Telecommunications
	2. Orascom Telecom (operator)
OTA	1. Office of Technology Assessment (U.S.A.)
	2. Orascom Telecom of Algeria
	3. Over-The-Air activation of services
OTAM	Over-The-Air Management (HF network nodes)
OTAR	Over-The-Air Rekeying
OTASP	Over-The-Air Service Provisioning
OTC	Operating Telephone Company

O

OTDOA	Observed Time Difference of Arrival (GSM)
OTDR	Optical Time-Domain Reflectometer (test equipment)
OTF	Optimum Transmission Frequency
OTGR	Operations Technology Generic Requirements
OTH	Over-The-Horizon (microwave)
OTIA	Office of Telecommunications and Information Applications
OTM	Optical Translation Measurement
OTN	1. Optical Transport Network
	2. Omam TradeNet
OTO	One-Time-Only (memory)
OTP	1. Office of Telecommunications Policy
	2. Open Trading Protocol
	3. One Time Programmable (memories)
OTQ	Outgoing Trunk Queuing
OTS	1. Operations Technical Support
	2. Off-The-Shelf
OUI	Organizationally Unique Identifier (Ethernet)
Outbox	Outgoing mails box (e-mail)
OutWATS	Outward Wide-Area Telephone Service
OVD	1. Optical Video Disk
	2. Outside Vapor Deposition process (optical fibers)
OVP	Over-Voltage Protection (power)
OVPO	Outside Vapor Phase Oxidation process (optical fibers)
OVS	Open Video System
OVSF	Orthogonal Variable Spreading Factor
OVTG	Optical Virtual Tributary Group (SONET)
OW	Order-Wire (signaling circuit)
OWM	Order-Wire Message
OWS	1. Operator Work station
	2. Optical Wavelength Switch
OWT	Operator Work Time
OXC	Optical Cross-Connect

P p

p	Symbol for prefix **p**ico-, denoting one-trillionth or 10^{-12}
P	1. Symbol for prefix **p**eta-, denoting 2^{50} or 10^{15}
	2. Symbol for prefix **p**seudo-, denoting false or deceptive
	3. Symbol for electrical **p**ermeance (electronics)
	4. Symbol for **p**ower
	5. Symbol for **p**ound (unit of weight)
	6. Symbol for **p**ositive (semiconductors)
	7. Symbol for **p**rimary winding (circuit diagrams)
	8. Symbol for **p**late (anode of electron tubes)
P-band	Radar Frequency **band** ranging from 0.22 to 0.39 GHz (SAR)
P-channel	Packet-mode **channel**
P-Code	1. Protected **Code**
	2. Precise **Code**
	3. Pseudo-**Code**
P-Frame	Predictive **Frame**
P-MAC	**P**acket **M**edia **A**ccess **C**ontrol
P-machine	Pseudo-**machine**
P-MOS	**P**-channel **MOS** (semiconductors)
P-P	**P**eak-**to**-**P**eak (alternating quantity)
P-picture	Predictive-coded **picture**
P-rating	Performance **rating**

P

P region	**P**ositive **region** or holes (semiconductors)
P-TMSI	**P**acket-**T**emporary **M**obile **S**ubscriber **I**dentity
P/A	**P**eak-**to**-**A**verage (ratio)
P/E Ratio	**P**rice-**to**-**E**arnings **Ratio**
P/P	**P**oint-**to**-**P**oint
P/S	**P**arallel-**to**-**S**erial (converter)
P1dB	**P**ower at **one dB** gain compression
P1dBin	**P1dB** referenced to the **in**put
P1dBout	**P1dB** referenced to the **out**put
P2P	1. **P**eer-**to**-**P**eer (networking)
	2. **P**oint-**to**-**P**oint (networking)
	3. **P**erson-**to**-**P**erson communication
P2MP	**P**oint-**to**-**M**ulti**p**oint (networking)
P2T	**P**ush-**to**-**T**alk
P3P	**P**latform for **P**rivacy **P**references **P**roject
P4	**P**hotolithographic **P**attern **P**lated **P**robe
P5	**5**86 **P**entium Microprocessor (Intel)
P54C	Official part number for **P**entium processor (Intel)
P55C	Official part number for **P**entium processor with MMX (Intel)
P6	**P**entium **Pro** Processor (Intel)
pA	**p**ico**a**mpere (electronics)
Pa	Symbol for **Pa**scal unit ($1\ Pa = 1\ N/m^2$)
PA	1. **P**re-**A**ssigned (access)
	2. **P**re**a**mble (FDDI)
	3. **P**ower **A**mplifier (electronics)
	4. **P**ooling **A**dministrator
PA-RISC	**P**recision **A**rchitecture-**R**educed **I**nstruction **S**et **C**omputing (Hewlett-Packard)
PA system	**P**ublic **A**ddress **system**
PAA	**P**hased-**A**rray **A**ntenna (microwave)
PABX	**P**rivate **A**utomatic **B**ranch **Ex**change (switching)
PAC	1. **P**latform **A**zimuth **C**ontrol
	2. **P**erceptual **A**udio **C**oder
PAC-E	**Pac**ific ocean region **E**ast (Inmarsat)
PACA	1. **P**riority **A**ccess and **C**hannel **A**ssignment
	2. **P**acific and **A**sian **C**ommunication **A**ssociation
	3. **P**icture **A**gency **C**ouncil of **A**merica
PACS	**P**ersonal **A**ccess **C**ommunications **S**ystem (U.S. version of PHS)
PACT	1. **P**ublic **A**ccess **C**ordless **T**elephony
	2. **P**BX **A**nd **C**omputer **T**esting (Siemens)
	3. **P**ersonal **A**ir **C**ommunications **T**echnology

PACVD	**P**lasma **A**ctivated **C**hemical **V**apor **D**eposition (optical fibers)
PACX	**P**rivate **A**utomatic **C**omputer **Ex**change
PAD	1. **P**acket **A**ssembler/**D**isassembler (X.25)
	2. **P**ersonal **A**uthentication **D**evice
	3. **P**rogram **A**ssociated **D**ata (broadcasting)
	4. **P**rogrammable **A**ddress **D**ecoder
PAE	**P**ower-**A**dded **E**fficiency
PAG	**P**rocedures **A**dvisory **G**roup
PAGEOS	**Pa**ssive **G**eodetic **E**arth-**O**rbiting **S**atellite (NASA)
PagP	**P**ort **ag**gregation **P**rotocol (LANs)
PAIS	**P**ath **A**larm **I**ndication **S**ignal
PakTel	**Pak**istan **Tel**ecommunications company Ltd
PAL	1. **P**hase **A**lternation by **L**ine (TV broadcast format)
	2. **P**ublic **A**ccess **L**ine
	3. **P**aradox **A**pplication **L**anguage (programming)
	4. **P**rogrammable **A**rray **L**ogic (digital electronics)
	5. **P**roprietary **Al**gorithm
PAL-M	**P**hase **A**lternation by **L**ine-**M**odified version (TV format)
PALC	**P**lasma-**A**ddressed **L**iquid **C**rystal (display)
PALI	**P**seudo-**A**utomatic **L**ocation **I**dentification
PALR	**P**recision **A**pproach and **L**anding **R**adar (aviation)
PALSAR	**P**hased **A**rray type **L**-band **S**ynthetic **A**perture **R**adar
PalTel	**Pal**estine **Tel**ecommunications Company
PAM	1. **P**ulse-**A**mplitude **M**odulation
	2. **P**ayload **A**ssist **M**odule (Intelsat)
	3. **P**ort **A**dapter **M**odule (Cisco)
PAM/FM	**P**ulse-**A**mplitude **M**odulation **on** **F**requency-**M**odulation
PAMA	1. **P**re-**A**ssigned **M**ultiple **A**ccess (satcom)
	2. **P**ulse **A**ddress **M**ultiple **A**ccess (satcom)
PAMR	**P**ublic **A**ccess **M**obile **R**adio (service)
PAMS	**P**erceptual **A**nalysis **M**easurement **S**ystem
PAN	1. **P**ersonal-**A**rea **N**etwork
	2. **P**ublic **A**ccess **N**etwork
Panamsat	**Pan**-**Am**erican **Sat**ellite
PANI	**P**seudo-**A**utomatic **N**umber **I**dentification
PANS	**P**retty **A**mazing **N**ew **S**ervice
PAP	1. **P**ublic **A**ccess **P**rovider (Internet)
	2. **P**ublic **A**ccess **P**rofile (DECT)
	3. **P**lug-**A**nd-**P**lay (computer hardware)
	4. **P**re-assignment **A**ccess **P**lan
	5. **P**assword **A**uthentication **P**rotocol
	6. **P**rinter **A**ccess **P**rotocol (AppleTalk)

P

PAR	1. **P**ositive **A**cknowledgement plus **R**etransmission
	2. **P**recision **A**pproach **R**adar
	3. **P**acket-to-**A**verage **R**atio
	4. **P**erformance **A**nalysis and **R**eview
paramp	**par**ametric **amp**lifier
PARC	**P**alo **A**lto **R**esearch **C**enter (Xerox)
PARD	**P**eriodic **A**nd **R**andom **D**eviation (power)
PARs	**P**urchase of **A**ccounts **R**eceivable**s**
PARS	**P**layback **A**nd **R**ecording **S**ystem
ParsTel	**Pars Tel**ephone company (Iran)
PAS	1. **P**ublic **A**ddress **S**ystem
	2. **P**ersonal **A**ccountability **S**ystem
	3. **P**riority **A**ccess **S**ervice
	4. **P**rivate **A**ircraft **S**tation
	5. **P**rofile **A**lignment **S**ystem
PASC	**P**erceptual **A**udio **S**ub-band **C**oding
PASL	**P**rimary **A**rea **S**witch **L**ocator
PAT	1. **P**ort **A**ddress **T**ranslation (LANs)
	2. **P**rogram **A**ssociation **T**able
PATU	**P**an-**A**frican **T**elecommunication **U**nion
PAX	**P**rivate **A**utomatic **E**xchange
Payphone	**Pay** Tele**phone** system
Pb	**P**eta**b**it, equal to 10^{15} or 2^{50} bits (transmission)
Pbps	**P**eta**b**its **p**er **s**econd
PB	1. **P**eta**b**yte, which is 2^{50} or 1024 times terabytes (computer)
	2. **P**lesiochronous **B**uffer
	3. **P**rimary **B**ody (orbits)
PBA	1. **P**ersonalized **B**asic **S**ervice
	2. **P**rinted **B**oard **A**ssembly
PBC	**P**olarization **B**eam **C**ombiner (laser)
PBER	**P**seudo-**B**it **E**rror **R**atio
PBGA	**P**lastic **B**all-**G**rid **A**rray (microchips)
PBGAM	**P**lastic **B**all-**G**rid **A**rray **M**ultilayer (microchips)
PBN	**P**acket **B**ackbone **N**etwork
PBNM	**P**olicy-**B**ased **N**etwork **M**anagement
PBO	**P**olymorphic **B**uffer **O**verflow (Internet)
PBP	**P**acket **B**urst **P**rotocol (networking)
PBS	1. **P**ersonal **B**ase **S**tation (PCS)
	2. **P**ublic **B**roadcasting **S**ervice
	3. **P**olarization **B**eam **S**plitter
PBT	1. **P**ush-**B**utton **T**elephone
	2. **P**oly**b**utylene **T**erephthalate (optical fibers)

PBX	**P**rivate **B**ranch **Ex**change (switching equipment)
pC	**p**ico-**C**oulomb (electronics)
P_{**C**}	**C**arrier **P**ower (transmission)
PC	1. **P**ersonal **C**omputer (IBM)
	2. **P**ersonal **C**ommunications
	3. **P**rogrammable **C**ontroller
	4. **P**ower **C**ontrol
	5. **P**rotocol **C**ontrol
	6. **P**rinted **C**ircuit
	7. **P**rimary **C**enter (switching)
	8. **P**rimary **C**hannel
	9. **P**roject **C**oordinator (AT&T)
PC-AT	**P**ersonal **C**omputer **AT** (IBM)
PC-QFP	**P**rinted-**C**ircuit **Q**uad **F**lat **P**ack (microchips)
PC-XT	**P**ersonal **C**omputer **XT** (IBM)
PCA	1. **P**oint of **C**losest **A**pproach (satcom)
	2. **P**rotective **C**onnecting **A**rrangement
	3. **P**rotective **C**oupling **A**rrangement
	4. **P**rotection **C**hannel **A**ccess
	5. **P**remises **C**abling **A**ssociation
PCAS	**P**ersonal **C**ommunications **A**ccess **S**ystem (cellular)
PCB	1. **P**rinted **C**ircuit **B**oard (electronics)
	2. **P**ower **C**ontrol **B**ox
	3. **P**ower **C**ircuit **B**reaker
	4. **P**rocess **C**ontrol **B**lock
	5. **P**rotocol **C**ontrol **B**lock
PCC	1. **P**roduct **C**ontrol **C**enter
	2. **P**ersonal **C**ompanion **C**omputer
	3. **P**repaid **C**alling **C**ard (INs)
	4. **P**lastic **C**hip **C**arrier
PCCA	**P**ortable **C**omputer and **C**ommunications **A**ssociation (U.S.A.)
PCCH	**P**hysical **C**ontrol **Ch**annel
.pcd	Name extension for kodak **P**hoto **CD** format files (graphics)
PCD	**P**ersonal **C**ommunications **D**evice
PCF	1. **P**hysical **C**ontrol **F**ield (LANs)
	2. **P**rint **C**oordination **F**unction
PCH	**P**aging **Ch**annel (GSM)
PCI	1. **P**eripheral **C**omponent **I**nterconnect (computer bus)
	2. **P**eripheral **C**omponent **I**nterface (computer bus)
	3. **P**rotocol **C**ontrol **I**nformation (OSI model)
	4. **P**re-**c**onnection **I**nspection
	5. **P**aya **C**ommunication **I**ndustries (Iran)

P

PCI SIG	**PCI S**pecial **I**nterest **G**roup
PCI-X	**P**eripheral **C**omponent **I**nterconnect-**Ex**tended (computer)
PCIA	**P**ersonal **C**ommunications **I**ndustry **A**ssociation (U.S. standards)
PCL	1. **P**rinter **C**ontrol **L**anguage (Hewlett-Packard)
	2. **P**roduct **C**omputer-module **L**oad
PCLEC	**P**acket **C**ompetitive **L**ocal **E**xchange **C**arrier
PCM	1. **P**ulse-**C**ode **M**odulation
	2. **P**hase **C**onjugation **M**irror
	3. **P**rocess **C**ontrol **M**onitor
PCM-IM	**P**ulse-**C**ode **M**odulation—**I**ntensity-**M**odulation
PCM/FM	**P**ulse-**C**ode **M**odulation **on F**requency-**M**odulation
PCM-30	2.048 Mbps T-1 (E-1)
PCMC	**PCM C**ontroller
PCMCIA	**P**ersonal **C**omputer **M**emory-**C**ard **I**nternational **A**ssociation
PCMDI	**PCM D**rop **I**nsert
PCMIA	**P**ersonal **C**omputer **M**anufacturer **I**nterface **A**dapter
PCN	**P**ersonal **C**ommunications **N**etwork
PCO	1. **P**ublic **C**all **O**ffice
	2. **P**oint of **C**ontrol and **O**bservation
	3. **P**rivate **C**able **O**perator
PCOM	**P**rivate **Com**munications
PCP	1. **P**redictor **C**ompression **P**rotocol
	2. **P**rivate **C**arrier **P**aging
	3. **P**ost **C**all **P**rocessing
PCR	1. **P**eak **C**ell **R**ate (ATM)
	2. **P**rogram **C**lock **R**eference (TV broadcast)
	3. **P**hase **C**hange **R**ewritable (memory)
	4. **P**rocessor **C**onfiguration **R**egister
PCRA	**P**roportional **C**ontrol **R**ate **A**lgorithm (ATM)
PCs	**P**ersonal **C**omputer**s**
PCS	**P**ersonal **C**ommunications **S**ystem (wireless and cellular)
PCS-1900	**P**ersonal **C**ommunications **S**ystems operating at **1900** MHz
PCSA	**P**ersonal **C**omputing **S**ystem **A**rchitecture
PCSC	**P**alestine **C**ommercial **S**ervices **C**ompany
PCSN	**P**rivate **C**ircuit **S**witching **N**etwork
PCSR	**P**arallel **C**hannel **S**ignaling **R**ate
.pct	Name extension for macintosh **PICT** format files (graphics)

PCT	1. **P**ersonal **C**ommunications **T**echnology
	2. **P**ortable **C**ontrol **T**erminal (Mux tester)
	3. **P**rogram **C**omprehension **T**ool (software)
PCTA	**P**ersonal **C**omputer **T**erminal **A**dapter (ISDN)
PCTE	**P**ortable **C**ommon **T**ool **E**nvironment
PCTS	**P**ublic **C**ordless **T**elephone **S**ervice (Canada)
PCU	1. **P**acket **C**ontrol **U**nit (GPRS)
	2. **P**eripheral **C**ontrol **U**nit
PCWG	**P**ersonal **C**onferencing **W**ork **G**roup
.pcx	Name extension for **PC** Paintbrush image files (graphics)
PD	1. **P**hase **D**etector
	2. **P**hoto-**D**etector
	3. **P**hoto-**D**iode
	4. **P**ulsed **D**oppler (radar)
	5. **P**re-**D**istortion
	6. **P**rotocol **D**iscriminator (GSM)
	7. **P**roduct **D**istributor (Inmarsat)
PD-channel	**P**acket-mode **D**ata **channel**
PDA	**P**ersonal **D**igital **A**ssistant
PDAIA	**PDA** **I**ndustry **A**ssociation (U.S. standards organization)
PDAS	**P**roduct **D**ocument **A**rchiving **S**ystem
PDAU	**P**hysical **D**elivery **A**ccess **U**nit
PDBM	**P**ulse-**D**elay **B**inary **M**odulation
PDC	1. **P**ersonal **D**igital **C**ellular system (Japan)
	2. **P**ersonal **D**igital **C**ellular (standard)
	3. **P**ersonal **D**igital **C**ommunications
	4. **P**rogram **D**elivery **C**ontrol (broadcasting)
	5. **P**rimary **D**omain **C**ontroller (Windows NT)
PDC-P	**P**ersonal **D**igital **C**ellular-**P**acket technology
PDCH	**P**hysical **D**ata **Ch**annel
PDD	1. **P**ost **D**ial **D**elay (phone calls)
	2. **P**ortable **D**igital **D**ocument (graphic file)
PDE	**P**roduct **D**ocument **E**xchange
PDEF	**P**hysical **D**esign **E**xchange **F**ormat
.pdf	Name extension for **PDF** format Adobe files (computer)
PDF	1. **P**ortable **D**ocument **F**ormat (files format)
	2. **P**ulse **D**uty **F**actor (parameter)
	3. **P**robability **D**istribution **F**unction
	4. **P**ower **D**istribution **F**rame
PDG	**P**rogram **D**ata **G**enerator (TV broadcast)
PDH	**P**lesiochronous **D**igital **H**ierarchy (MUX)

P

PDI	1. **P**icture **D**escription **I**nstructions
	2. **P**ersonal **D**ata **I**nterchange (format)
	3. **P**erspective **D**ark-field **I**maging (semiconductor wafer)
PDIAL	**P**ublic-access **Dial**-up (Internet)
PDIF	**P**-CAD **D**ata **I**nterchange **F**ormat
PDIP	**P**lastic **D**ual **I**nline **P**ackage (microchips)
PDL	1. **P**age-**D**escription **L**anguage (programming)
	2. **P**ositioning **D**ata **L**ink
PDLC	**P**olymer-**D**ispersed **L**iquid **C**rystal (display)
PDM	1. **P**ulse-**D**uration **M**odulation
	2. **P**lain-**D**ress **M**essage
	3. **P**ersonal **D**ata **M**odule
	4. **P**roduct **D**ata **M**anagement
PDM/FM	**P**ulse-**D**uration **M**odulation **on** **F**requency-**M**odulation
PDM/PM	**P**ulse-**D**uration **M**odulation **on** **P**ulse-**M**odulation
PDN	1. **P**acket **D**ata **N**etwork
	2. **P**ublic **D**ata **N**etwork
	3. **P**remises **D**istribution **N**etwork
	4. **P**rimary **D**irectory **N**umber (ISDN)
PDO	1. **P**acket **D**ata **O**ptimization
	2. **P**ortable **D**istributable **O**bjects (computer)
PDP	1. **P**ower **D**istribution **P**anel
	2. **P**ublic **D**ispersion **P**arameter (optical fibers)
	3. **P**lasma **D**isplay **P**anel (television set)
	4. **P**rogrammed **D**ata **P**rocessor (Digital Equipment Corporation)
	5. **P**acket **D**ata **P**rotocol (GPRS)
PDPS	**P**hotographic **D**ata **P**rocessing **S**ystem
PDR	**P**reliminary **D**esign **R**eview
PDS	1. **P**remises **D**istribution **S**ystem (Lucent Technologies)
	2. **P**rimary **D**istribution **S**ystem
	3. **P**ower **D**istribution **S**ystem
	4. **P**rogram **D**ata **S**ource
	5. **P**rotected **D**istribution **S**ystem (U.S. federal government)
	6. **P**ayload **D**ata **S**egment (satcom)
	7. **P**rocessor **D**irect **S**lot (Macintosh)
	8. **P**arallel **D**ata **S**tructure (computer)
PDT	1. **P**osition **D**etermination **T**echnology
	2. **P**rogrammable **D**ata **T**erminal
	3. **P**rototype **D**ebug **T**ool (Tektronics)
PDTS	**P**ublic **D**ata **T**ransmission **S**ervice

PDU	**P**rotocol **D**ata **U**nit (OSI model)
PDUS	**P**rimary **D**ata **U**ser **S**tation
PDV	**P**ath **D**elay **V**alue
PE	1. **P**eripheral **E**quipment
	2. **P**hase-**E**ncoded (recording)
	3. **P**rocessing **E**lement
	4. **P**rotocol **E**mulator
	5. **P**oly**e**thylene (cable)
PEAP	**P**rotected **E**xtensible **A**uthentication **P**rotocol
PEB	**P**CM **E**xpansion **B**us
PEBB	**P**ower **E**lectronic **B**uilding **B**lock
PEC	1. **P**hoto-**E**lectric **C**ell (electronics)
	2. **P**erfect **E**lectrical **C**onductor
PECC	**P**artially **E**rror-**C**ontrolled **C**onnections
PECL	**P**ositive **E**mitter-**C**oupled **L**ogic (digital electronics)
PECVD	**P**lasma-**E**nhanced **C**hemical **V**apor **D**eposition
PED	1. **ped**estal (antenna)
	2. **P**hase **E**rror **D**etector
PEDC	**P**an **E**uropean **D**igital **C**ommunications (GSM)
PEEC	**P**artial **E**lement **E**quivalent **C**ircuit
PEEK	**P**artners **E**arly **E**xperience **K**it (Taligent)
PEG	**P**ublic **E**ducational or **G**overnment access (cable TV)
PEG Ratio	**P**rice to **E**arning **G**rowth **Ratio**
PEL	**P**icture **El**ement (pixel)
PEM	1. **P**rivacy **E**nhanced **M**ail (Internet)
	2. **P**eripheral **E**quipment **M**odule
PEP	1. **P**acket **E**rror **P**robability
	2. **P**acket **E**nsemble **P**rotocol
	3. **P**acket **E**xchange **P**rotocol
	4. **P**ayment **E**xtension **P**rotocol
	5. **P**eak **E**nvelope **P**ower (electronics)
	6. **P**olicy **E**nforcement **P**oint
	7. **P**rimary **E**ntry **P**oint
	8. **P**artitioned **E**mulation **P**rogramming (IBM)
PEPCI	**P**rotocol for **E**xchange of **P**olicy **I**nformation (IETF)
PER	1. **P**acket **E**rror **R**ate
	2. **P**acket **E**ncoding **R**ules
Perl	**P**ractical **E**xtraction and **R**eport **L**anguage (programming)
PERT	**P**roject **E**valuation and **R**eview **T**echnique (ICT)
PES	1. **P**ersonal **E**arth **S**tation
	2. **P**acketized **E**lementary **S**tream (MPEG)
	3. **P**acket over **E**thernet over **S**ONET
	4. **P**rogram **E**lementary **S**tream

P

PESIS	**P**hoto **E**lectron **S**pectroscopy of **I**nner-**S**hell electrons
PESOS	**P**hoto **E**lectron **S**pectroscopy of **O**uter-**S**hell electrons
PESQ	**P**erceptual **E**valuation of **S**peech **Q**uality
PET	1. **P**ortable **E**arth **T**erminal
	2. **P**ositron-**E**mission **T**omography (medicine)
PEW	**P**acket over **E**thernet over **W**DM
pF	**p**ico**f**arad (electronics)
PF	1. **P**resentation **F**unction (TMN)
	2. **P**acking **F**raction (fiber optics)
	3. **P**BX **F**unction
	4. **P**ower **F**actor (electronics)
PF Xfer	**P**ower **F**ailure **Transfer** (electronics)
PFC	**P**ower **F**actor **C**orrection (PC power supply)
PFD	1. **P**ower **F**lux-**D**ensity
	2. **P**hase/**F**requency **D**iscriminator
PFM	1. **P**ulse-**F**requency **M**odulation
	2. **P**roto-**F**light **M**odel
PFN	**P**ulse-**F**orming **N**etwork
PFOC	**P**ending the **F**irm **O**rder **C**ommitment
PG	1. **P**re-**G**roup (MUX)
	2. **P**ower **G**ain
PGA	1. **P**in-**G**rid **A**rray (microchips)
	2. **P**rogrammable **G**ain **A**mplifier
	3. **P**rofessional **G**raphics **A**dapter (IBM)
PGBM	**P**ulse-**G**ated **B**inary **M**odulation
PGD	**P**rofessional **G**raphics **D**isplay (IBM)
PgDn	**P**age **D**own key (PC keyboard)
PGI	**P**arameter **G**roup **I**dentifier (ISO)
PGL	**P**eer **G**roup **L**eader
PGM	**P**ragmatic **G**eneral **M**ulticast (Cisco)
PGP	**P**retty **G**ood **P**rivacy (encryption software)
PGP fone	**PGP** Tele**phone** Protocol
PgUp	**P**age **U**p key (PC keyboard)
PGS	**P**rimary **G**uard **S**tation
PG Telecom	**P**ayam **G**ostar **Telecom** Company (Iran)
ph	**ph**ot (unit of illumination)
pH	**p**ico**h**enry (electronics)
PH	**P**acket **H**andler (ISDN)
PHB	**P**er-**H**op-**B**ehavior
PHEMT	**P**seudomorphic **H**igh-**E**lectron **M**obility **T**ransistor (semiconductors)
PHF	**P**acket **H**andling **F**unction
PHI	**P**BX-to-**H**ost **I**nterface
PHILCOM	**Phil**ippine Global **Com**munications Inc.

phone	1. tele**phone**
	2. head**phone**
PHOTINT	**Phot**ographic **Int**elligence
photocell	**photo**electric **cell**
phototube	**photo**electric **tube**
PHP	**P**ersonal **H**andy-**P**hone (Japan)
PHS	**P**ersonal **H**andyphone **S**ystem (Japan)
PHY	**Phy**sical layer (OSI model)
PHz	**P**eta**h**ertz (10^{15} Hertz)
pi filter	π-shaped **filter**
PI	1. **P**resentation **I**ndicator
	2. **P**arameter **I**dentifier (ISO)
	3. **P**rotection **I**nterval
PIA	1. **P**ersonal **I**nformation **A**ppliance
	2. **P**eripheral **I**nterface **A**dapter
PIAFS	**PH**S **I**nternet **A**ccess **F**orum **S**tandard (Japan)
PIC	1. **P**rimary **I**nterexchange **C**arrier
	2. **P**hotonic **I**ntegrated **C**ircuit
	3. **P**lastic **I**nsulated **C**onductor
	4. **P**icture **I**mage **C**ompression (graphics)
	5. **P**ersonal **I**ntelligent **C**ommunicator
	6. **P**rogrammable **I**nterrupt **C**ontroller (computer)
	7. **P**rogram **I**ntegration **C**ontrol (TV broadcast)
	8. **P**rogrammable **I**ntegrated **C**ircuit (ICs)
PICC	**P**rimary **I**nterexchange **C**arrier **C**harge
PICH	**Pi**lot **Ch**annel
PICMG	**PCI I**ndustrial **C**omputer **M**anufacturers **G**roup (U.S.A.)
PICS	1. **P**rotocol **I**mplementation **C**onformance **S**tatement
	2. **P**lug-in **I**nventory **C**ontrol **S**ystem
	3. **P**roduct **I**nventory **C**ontrol **S**ystem
	4. **P**latform for **I**nternet **C**ontent **S**election
.pict	Name extension for **PICT** format image files (computer)
PICT	**Pict**ure Format
PID	1. **P**rotocol **Id**entifier
	2. **P**acket **Id**entifier
PIDB	**P**eripheral **I**nterface **D**ata **B**us (AT&T)
PIE	**P**ersonal **I**nteractive **E**lectronics
PIECE	**P**roductivity, **I**nformation, **E**ducation, **C**reativity, **E**ntertainment
PIF	1. **P**ersonal **C**ommunications **S**ervices **I**ndustry **F**orum
	2. **P**ublic **I**nspection **F**ile
	3. **P**rogram **I**nformation **F**ile (MS Windows)
PILC	**P**erformance **I**mplication of **L**ink **C**haracteristics

P

PILOT	Programmed Inquiry, Learning Or Teaching (e-learning)
PIM	1. Passive Intermodulation
	2. Personal Information Manager (software)
	3. Personal Information Management
	4. Plug-in ISDN Module
	5. Protocol-Independent Multicast (IETF)
	6. Presence and Instant Messaging
PiMF	Parts in Metal Foil
PIMP	Passive Inter-Modulation Product
PIMS	Program Information Management System (TV broadcast)
PIN	1. Personal Identification Number (GSM)
	2. Positive-doped/Intrinsic/Negative-doped diode (semiconductors)
	3. Public Infrastructure Network
	4. Procedure Interrupt Negative (fax machine)
	5. PN junction with Isolated region (sold state)
PIN(BB)	PIN used on the Base-Band Level (bluetooth)
PIN-FET	PIN (diode) Field Effect Transistor (semiconductors)
PIN(UI)	PIN used on the User Interface Level (bluetooth)
PIND	Particle Impact Noise Detection
PING	1. Packet Internet Gopher (software utility)
	2. Process-Improvement Networking Group
PINT	PSTN and Internet Interworking
PINX	Private Integrated Network Exchange
PIO	Programmed Input/Output (hard drives)
PIP	Procedural Interface Protocol
PIRS	Positioning Inertial Reference System
PISD	Planned In-Service Data
PIP	1. Picture In Picture
	2. Path Independent Protocol
.pit	Name extension for PackIT files (computer)
PIU	Path Information Unit
PIV	Peak Inverse Voltage (electronics)
PIX	Private Internet Exchange (Cisco)
Pixel	Picture element
PIXIT	Protocol Implementation extra Information for Testing
pJ	picojoule (physics)
PJ/NF	Projection-Join Normal Form
PJCM	Pointer Justification Count Minus (MUX)
PJCP	Pointer Justification Count Plus (MUX)
PJE	Pointer Justification Event (MUX)
PJS	Pointer Justification Seconds (MUX)

PKCS	Public-Key Cryptography Standards
PKE	Public-Key Encryption
PKES	Public-Key Encryption System
PKI	Public-Key Infrastructure (ICT)
PKM	Perigee Kick Motor (satellite)
PKZIP	PKWARE, Inc.'s ZIP (data compression)
PL	1. Private Line
	2. Programming Language
PL/1	Programming Language 1 (IBM)
PL/M	1. Programming Language for Microcomputers
	2. Programming Language for Microprocessors
PLA	Programmable Logic Array (digital electronics)
PLAR	Private Line Automatic Ringdown
PLB	Personal Locator Beacon (COMSAR)
PLC	1. Programmable Logic Controller
	2. Power Line Carrier
	3. Problem Logging Control
	4. Planar Lightwave Circuit
PLCC	Plastic Leaded-Chip Carrier method (chip mounting)
PLCP	Physical-Layer Convergence Procedure (networking)
PLD	1. Programmable Logic Device (digital electronics)
	2. Phase Lock Detector
	3. Pulsed Laser Deposition
PLDT	Philippine Long-Distance Telephone
PLF	Polarization Loss Factor (microwave)
PLI	Programmable Language Interface
PLL	Phase-Locked Loop (electronics)
PLLC	Professional Limited Liability Corporation
PLLN	Public Leased Lines Network
PLM	1. Public Land Mobile (network)
	2. Pulse-Length Modulation
	3. Power-Line Modulation
	4. Perfectly Matched Layer
PLMN	Public Land Mobile Network (cellular networks)
PLMR	Private Land Mobile Radio
PLN	Private Line Network
PLO	Phase-Locked Oscillator
PLP	1. Packet Layer Protocol
	2. Packet Loss Probability
	3. Packet Level Procedure (X.25)
PLR	Pulse Link Repeater (signaling)
PLS	1. Physical Layer Signaling (network architecture)
	2. Premises Lightwave System (fiber optic)
	3. Programmable Logic Sequencer (digital electronics)

P

PLSC	Private Line Service Center
PLT	Power Line Telephony
PLTS	Private Line Transport Service
PLU	Percent Local Usage
PM	1. Phase Modulation
	2. Pulse Modulation
	3. Processor Module
	4. Packet Mode
	5. Polarization Maintaining (optical fiber)
	6. Preventive Maintenance
	7. Performance Monitoring (MUX)
	8. Permanent Magnet
	9. Presentation Manager (IBM)
	10. Physical Medium sublayer (ATM)
PMA	1. Prompt Maintenance Alarm
	2. Physical Medium Attachment
PMARS	Police Mutual Aid Radio System
PMB	Pilot Make-Busy (circuit)
PMBS	Packet Mode Bearer Service
PMC	1. Public Mobile Carrier
	2. Perfect Magnetic Conductor
PMCM	Plastic Multichip Module
PMD	1. Physical Medium Dependent (protocol sublayer)
	2. Polarization Mode Dispersion (fiber optics)
PMI	1. Project Management Institute
	2. Property Management Interface
PMMU	Paged Memory Management Unit (PC chip)
PMOS	P-channel MOS (semiconductors)
PMP	1. Point-to-Multipoint (radio system)
	2. Project Management Professional
PMR	1. Private Mobile Radio
	2. Poor Man's Routing (packet switching)
	3. Professional Mobile Radio
PMRS	Private Mobile Radio System
PMS	1. Pantone Matching System (programming language)
	2. Picturephone Meeting Service (AT&T)
	3. Property Management System (software)
PMT	1. Photo Multiplier Tube
	2. Photo Mechanical Transfer
	3. Personal Mobile Telecommunications
	4. Program Map Table (TV broadcast)
PMU	Parametric Measurement Unit
PN	1. Personal Number
	2. Positive-Negative Junction (semiconductors)

PNC	Police National Computer
PNdB	Perceived Noise level in **dB**
PNF	1. Portable Network Frame
	2. Planar Near-Field (range)
.png	Name extension for **PNG** format files (graphics)
PNG	Portable Network Graphics (file format)
PNI	Permit Next Increase (ATM)
PNM	Public Network Management
PNN	Probabilistic Neural Network
PNNI	Private Network-to-Network Interface (ATM)
PNO	Public Network Operator
PnP	Plug **and** Play (computer standard)
PNP	1. Positive-Negative-Positive (semiconductors)
	2. Permanent Number Portability
PNPN	Positive-Negative-Positive-Negative junction (semiconductors)
PNS	Personal Number Service
PNT	Private Network Termination
PO	Point of Origin
PoC	Push-to-talk over Cellular
POC	Point Of Contact
POCSAG	Post Office Code Standardization Advisory Group (protocol)
POD	1. Piece Of Data
	2. Processing Of Data
	3. Point Of Deployment (cable TV)
	4. Personal Operable Device
PODA	Priority Oriented Demand Assignment
PODP	Public Office Dialing Plan
POEM	Polar-Orbit Earth Observing Mission (remote sensing)
POF	Plastic Optical Fiber
POFS	Private Operational Fixed Service (microwave)
POGO	Post Office Goes Obsolete
POH	1. Path Overhead (MUX)
	2. Power-On Hours (production rating)
POI	1. Point Of Interconnection
	2. Point Of Interface (LATA)
	3. Parallel Optic Interfaces
POL	1. polarizer
	2. polarization
POLDER	Polarization and Directionality of the Earth's Reflectance
POLSK	Polarization Shift Keying (modulation)
PON	Passive Optical Network

P

PoP	Point of Presence (networking)
POP	Post Office Protocol (e-mail)
POP3	Post Office Protocol version 3 (e-mail)
POPS	Post Office Protocol-Secure
POR	Pacific Ocean Region (Inmarsat)
PoS	Packet over SONET
POS	1. Pacific Ocean Satellite
	2. Point Of Service
	3. Point Of Sale (electronic transactions)
	4. Personal Operating Space (IEEE)
POS-PHY	Packet Over SONET-Physical layer
POSI	Promoting Conference for OSI (Japan)
POSIT	Profiles for Open Systems Internetworking Technology
Positron	Positive electron
POSIX	Portable Operating System Interface for Unix environment
POST	Power-On Self-Test (computers)
Pot	Potentiometer (electronics)
POT	1. Plain Old Telephony (service)
	2. Point Of Termination
	3. Point Of Train (transmission)
POTS	Plain Old Telephone Service
POTS-C	Plain Old Telephone Service-Centralized (ADSL)
POTS-R	Plain Old Telephone Service-Remote (ADSL)
POTV	Plain Old TV (Microsoft)
POV	Peak Operating Voltage (electronics)
Power	Performance Optimization With Enhanced RISK (processor)
Power PC	Performance Optimized With Enhanced RISK Personal Computer
PP	1. Portable Part
	2. Pointer Processor
	3. Polarization-Preserving (optical fiber)
PPA	Production Photographic Area
PPARC	Particle Physics and Astronomy Research Council (UK)
PPBM	Pulse-Polarization Binary Modulation
PPCP	PowerPC Platform
PPD	1. Partial Packet Discard (ATM)
	2. Pay Per Download (software download)
PPDN	Public Packet Data Network
PPDR	Public Protection and Disaster Relief (service)
PPDU	PLCP protocol data unit
PPG	Pulse Pattern Generator
PPGA	Plastic Pin-Grid Array (microchips)

ppi	**p**ixels **p**er **i**nch (displays)
PPI	1. **P**lan **P**osition **I**ndicator (radar)
	2. **P**DH **P**hysical **I**nterface
PPL	**P**ioneer **P**reference **L**icensees (wireless)
ppm	**p**ages **p**er **m**inute (printers)
PPM	1. **P**ulse-**P**osition **M**odulation
	2. **P**ulse-**P**hase **M**odulation
	3. **P**ulse **P**er **M**inute (parameter)
	4. **P**arts **P**er **M**illion (data errors)
	5. **P**eriodic **P**ermanent **M**agnet
	6. **P**seudo-**P**ermanent **M**agnet
	7. **P**eriodic **P**ulse **M**etering
PPN	**P**rocessor **P**ort **N**etwork
PPO	**P**eak **P**ower **O**utput
PPP	1. **P**oint-to-**P**oint **P**rotocol
	2. **P**hased **P**roject **P**lanning
PPP/MP	**P**oint-to-**P**oint **P**rotocol/**M**ultilink **P**rotocol
PPPI	**P**recision **P**lan-**P**osition **I**ndicator
PPPoA	**P**oint-to-**P**oint **P**rotocol **o**ver **A**TM (DSL)
PPPoE	**P**oint-to-**P**oint **P**rotocol **o**ver **E**thernet (DSL)
pps	1. **p**ackets **p**er **s**econd
	2. **p**ulses **p**er **s**econd
PPS	1. **P**ath **P**rotection **S**witch
	2. **P**recise **P**ositioning **S**ervice (GPS)
	3. **P**eripheral **P**ower **S**upply
	4. **P**ermanent **P**resentation **S**tatus
PPSN	**P**ublic **P**acket-**S**witched **N**etwork (frame relay)
PPSR	**P**ath **P**rotection **S**witched **R**ing (MUX)
PPSS	1. **P**ublic **P**acket-**S**witched **S**ervice (frame relay)
	2. **P**ositron **P**ublic **S**afety **S**ystems (U.S.A.)
PPT	**P**recision **P**ad **T**echnology
PPTP	**P**oint-to-**P**oint **T**unneling **P**rotocol (VPNs)
PPV	**P**ay-**P**er-**V**iew (TV broadcasting)
PQA	1. **P**ath **Q**uality **A**nalysis
	2. **P**alm **Q**uery **A**pplication (PDAs)
PQFP	**P**lastic **Q**uad-leaded **F**lat **P**ack (microchips)
PQSCADA	**P**ower **Q**uality **SCADA** (remote control)
PR	1. **P**recipitation **R**adar
	2. **P**ulse **R**ate
	3. **P**seudo-**R**ange (microwave)
	4. **P**erformance **R**ating (processing power)
	5. **P**attern **R**ecognition
PRA	**P**rimary **R**ate **A**ccess (ISDN)
PRACH	**P**hysical **R**andom **A**ccess **Ch**annel

P

PRAM	**P**rogrammable **RAM** (memory)
PRB	1. **P**rimary **R**eference **B**urst
	2. **P**rivate **R**adio **B**ureau
PRBA	**P**ortable **R**echargeable **B**attery **A**ssociation
PRBS	**P**seudo-**R**andom **B**it-**S**equence (data encoding pattern)
PRC	1. **P**rotection, **R**estoration, **C**ombination
	2. **P**rimary **R**eference **C**lock (networking)
PRD	**Pr**eamble **D**etector
Preamp	**Pre-amp**lifier (electronics)
Pre-emph	**Pre-emph**asis (frequency modulation)
Prefs	**pref**erences
PREP	**P**owerPC **R**eference **P**latform (specification)
PREST	Center for **P**olicy **R**esearch in **E**ngineering, **S**cience and **T**echnology. U.K. group
PRF	**P**ulse **R**epetition **F**requency (radar)
PRG	**Pr**eamble **G**enerator
PRI	1. **P**rimary **R**ate **I**nterface (ISDN)
	2. **P**rimary **R**ate **I**SDN (service)
	3. **P**ublic **R**adio **I**nternational
PRI-EOP	**Pr**ocedure **I**nterrupt-**E**nd **O**f **P**age (fax machine)
PRI-MPS	**Pr**ocedure **I**nterrupt-**M**ulti**p**age **S**ignal (fax machine)
PRIDE	**P**olice **R**egionalized **I**nformation and **D**ata **E**ntry (Canada)
PRISM	**P**anchromatic **R**emote-sensing **I**nstrument for **S**tereo **M**apping
PRK	**P**hase-**R**eversal **K**eying (modulation)
PRM	**P**ulse-**R**ate **M**odulation
PRMA	**P**acket **R**eservation **M**ultiple **A**ccess
PRMD	**Pr**ivate **M**anagement **D**omain (X.400)
PRML	**P**artial-**R**esponse **M**aximum-**L**ikelihood (disk storage)
PRN	1. Logical device name for **Pri**nter
	2. **P**seudo-**R**andom **N**oise
PRNG	**P**seudo-**R**andom **N**oise **G**eneration (bluetooth)
PRO	**P**recision **R**ISC **O**rganization
PROFS	**Pr**ofessional **O**ffice **S**ystem (software)
Prolog	**Pro**gramming **log**ic (programming language)
PROM	**P**rogrammable **ROM** (memory)
PROSIGN	**Pro**cedure **Sign**
PROTEL	**Pr**ocedure **O**riented **T**ype **E**nforcing **L**anguage (software)
Protn	**prot**ection
PROWORD	**Pro**cedure **Word**
PRR	**P**ulse **R**epetition **R**ate (parameter)

PRS	1. **P**remium **R**ate **S**ervice
	2. **P**ersonal **R**adio **S**ervice
	3. **P**rimary **R**eference **S**ource (network clock)
	4. **P**attern **R**ecognition **S**ystem
PRSL	**P**rimary **R**ate **S**witch **L**ocator
PRSM	**P**ost **R**elease **S**oftware **M**anager
PRT	**P**lanar **R**esistor **T**echnology
PrtSc	**Pr**int **Sc**reen (PC keyboard)
.ps	Name extension for **P**ost**S**cript printer files (computer)
ps	**p**icosecond (10^{-12} second)
PS	1. **P**ower **S**upply
	2. **P**aging **S**ystem
	3. **P**hase **S**hift
	4. **P**ortable **S**tation
	5. **P**acket **S**witching
PS-ACR	**P**ower **S**um-**A**ttenuation-to-**C**rosstalk **R**atio
PS-NEXT	**P**ower **S**um-**N**ear-**E**nd **CrossT**alk
PS/2	**P**ersonal **S**ystem/**2** (IBM PCs)
PSA	1. **P**oint of **S**ervice **A**ctivation (Inmarsat)
	2. **P**rotected **S**ervice **A**rea (FCC)
	3. **P**ublic **S**ervice **A**greement (Inmarsat)
	4. **P**olysilicon **S**elf-**A**ligned (bipolar transistor)
PSAI	**P**rocessor-to-**S**witch **A**pplications **I**nterface (AT&T)
PSAP	**P**ublic **S**afety **A**nswering **P**oint
PSB	**P**ink-**S**lip **B**lizzard (microwave)
PSC	1. **P**ublic **S**ervices **C**ommittee (IMSO)
	2. **P**acket-**S**witched **C**apable (GMPLS)
	3. **P**ublic **S**ervices **C**ommission (U.S.A.)
	4. **P**ath **S**witching **C**ount
	5. **P**ermanent-**S**plit **C**apacitor (motors)
PSCS	**P**ersonal **S**pace **C**ommunication **S**ervice
PSD	1. **P**ower **S**pectral **D**ensity (carrier frequency)
	2. **P**hase **S**ensitive **D**etector
PSDN	**P**ublic-**S**witched **D**ata **N**etwork
	Packet-**S**witched **D**ata **N**etwork
PSDS	**P**ublic-**S**witched **D**ata **S**ervice (AT&T)
PSDTS	**P**ublic-**S**witched **D**ata **T**ransmission **S**ervice (AT&T)
PSE	1. **P**acket **S**witching **E**xchange
	2. **Plea**se (Morse code transmissions)
psec	**p**ico**sec**ond
PSELFEXT	**P**ower **S**um **E**qual **L**evel **F**ar-**E**nd **CrossT**alk
PSG	**P**rogrammable **S**ound **G**enerator
psi	**p**ounds per **s**quare **i**nch (telephone cables)

P

PSI	1. **P**acket **S**witching **I**nterface
	2. **P**rogram **S**pecific **I**nformation (broadcasting)
PSID	**P**rivate **S**ystem **Id**entifier (signaling)
PSK	**P**hase-**S**hift **K**eying (modulation)
PSL	**P**hysical **S**ignaling **S**ublayer (LANs and MANs)
PSM	1. **P**hase-**S**hift **M**odulation
	2. **P**ulse-**S**pacing **M**odulation
	3. **P**ower **S**upply **M**odule
PSMBS	**P**ublic-**S**afety **M**obile **B**roadband **S**pecifications
Psmc-channel	**P**acket-mode **channel** for **s**ystem **m**anagement and **c**ontrol
PSMM	**P**ilot **S**trength **M**easurement **M**essage
PSN	1. **P**acket-**S**witched **N**etwork
	2. **P**acket **S**witching **N**ode
	3. **P**ublic-**S**witched **N**etwork
	4. **P**rocessor **S**erial **N**umber (Intel Pentium III)
PSNEXT	**P**ower **S**um **NEXT** (cabling)
PSO	1. **P**ublic **S**ervice **O**bligations
	2. **P**orts and **S**hipping **O**rganization (Iran)
PSOP	**P**lastic **S**mall **O**utline **P**ackage (microchips)
PSP	1. **P**CS **S**ervice **P**rovider
	2. **P**ayphone **S**ervice **P**rovider
	3. **P**urchase **S**ervice **P**rovider
	4. **P**rogrammable **S**ignal **P**rocessor
	5. **P**ower **S**upply **P**roduction company (Iran)
PSPDN	**P**acket-**S**witched **P**ublic **D**ata **N**etwork
PSPP	**P**ublic **S**afety **P**artnership **P**roject
PSQM	**P**erceptual **S**peech **Q**uality **M**easure
PSR	1. **P**etabit **S**witch **R**outer
	2. **P**rofessional **S**atellite **R**eceiver
PSRAM	**P**seudo-**S**tatic **RAM** (memory)
PSRCP	**P**ublic **S**afety **R**adio **C**ommunication **P**roject
PSRR	**P**ower **S**upply **R**ejection **R**atio
PSS	1. **P**acket-**S**witched **S**ervice
	2. **P**acket-switched **S**treaming **S**ervice
	3. **P**hysical **S**ignaling **S**ublayer (OSI model)
PSS1	**P**rivate **S**ignaling **S**ystem number **1**
PSTLXN	**P**ublic-**S**witched **Tel**ex **N**etwork
PSTN	**P**ublic-**S**witched **T**elephone **N**etwork
PSU	1. **P**ower **S**upply **U**nit
	2. **P**acket **S**witch **U**nit
PSW	**P**ath **Sw**itch
PSWAC	**P**ublic **S**afety **W**ireless **A**dvisory **C**ommittee
PSWN	**P**ublic **S**afety **W**ireless **N**etwork

PT	1. **P**ayload **T**ype (ATM)
	2. **P**ersonal **T**elecommunications
	3. **P**ath **T**ermination
PTA	**P**rogrammable **T**racking **A**ntenna
PTC	1. **P**ortable **T**eletransaction **C**omputers
	2. **P**ersonal **T**elecommunications **C**enter
	3. **P**acific **T**elecommunications **C**ouncil (U.S.A.)
	4. **P**ositive **T**emperature **C**oefficient (thermistors)
	5. **P**ublic **T**elecoms **C**orporation (Yemen)
PTCE	**P**ermanent **T**est **C**ontrol **E**lement
PTCL	**P**akistan **T**elecoms **C**ompany **L**td
PTE	**P**ath **T**erminating **E**lement (SONET)
PTF	1. **P**atch and **T**est **F**acility
	2. **P**olymer **T**hick-**F**ilm
PTFE	**P**oly**t**etra**f**luoro**e**thylene (cables)
PTI	1. **P**ayload **T**ype **I**ndicator (ATM)
	2. **P**ower **T**errestrial **I**nterface
	3. **P**rimary **T**errestrial **I**nterface
PTK	**P**ars **T**elephone **K**ar company (Iran)
PTM	1. **P**ulse-**T**ime **M**odulation
	2. **P**ath **T**race **M**ismatched (MUX)
	3. **P**acket **T**ransfer **M**ode
PTMPT	**P**oint-**T**o-**M**ulti**p**oint
PTN	**P**ublic **T**elecommunications **N**etwork
PTNX	**P**rivate **T**elecommunications **N**etwork **E**xchange
PTO	1. **P**ublic **T**elephone **O**perators
	2. **P**ublic **T**elecommunications **O**perator
	3. **P**atent and **T**rademark **O**ffice (U.S.A.)
PTP	**P**oint-**T**o-**P**oint (microwave)
PTR	1. **p**oin**t**e**r** (MUX)
	2. **P**roblem **T**racking **R**eport
PTS	1. **P**ersonal **T**elecommunications **S**ystem
	2. **P**rofile **T**est **S**pecification
	3. **P**roceed **T**o **S**elect tone (Inmarsat)
	4. **P**resentation **T**ime **S**tamp (MPEG2 encoder)
PTSE	**P**NNI **T**opology **S**tate **E**lement (ATM)
PTSP	**P**NNI **T**opology **S**tate **P**acket (ATM)
PTT	1. **P**ost, **T**elephone and **T**elegraph administrations
	2. **P**ush-**T**o-**T**alk (wireless systems)
PTTI	**p**recise **t**ime and **t**ime **i**nterval
PTTC	**P**aper **T**ape **T**ransmission **C**ode
PTY	**p**ar**t**y
PU	1. **P**hysical **U**nit (IBM SNA)
	2. **P**ower **U**nit

P

pub	short for **pub**lic (directory)
PUC	1. **P**ublic **U**tilities **C**ommission
	2. **P**ersonal **U**nlocking **C**ode
PUCP	**P**hysical **U**nit **C**ontrol **P**oint (SNA)
PUK	**P**IN **U**nblocking **K**ey (GSM)
PUMA	**P**roduct **U**pgrade **Ma**nager
PUP	**P**ARC **U**niversal **P**acket
PUT	**P**rogrammable **U**nijunction **T**ransistor (semi-conductors)
pV	**p**ico**v**olt (10^{-12} Volt)
PV	**P**hoto**v**oltaic (electronics)
PVA	**P**erigee **V**elocity **A**ugmentation (satellite)
PVC	1. **P**ermanent **V**irtual **C**ircuit (frame relay)
	2. **P**rivate **V**irtual **C**ircuit
	3. **P**ermanent **V**irtual **C**onnection
	4. **P**oly**v**inyl **C**hloride (cables)
	5. **P**remises **V**isit **C**harge
PVCC	**P**ermanent **V**irtual **C**hannel **C**onnection (ATM)
PVD	1. **P**eak **V**oltage **D**etect
	2. **P**hysical **V**apor **D**eposition
PVDF	**P**oly**v**inyl **D**iflouride (cables)
PVDM	**P**acket **V**oice **D**ata **M**odule
PVN	**P**rivate **V**irtual **N**etwork
PVPC	**P**ermanent **V**irtual **P**ath **C**onnection (ATM)
PVR	**P**ersonal **V**ideo **R**ecorder
PVT	**P**erformance **V**erification **T**est
PvtDN	**P**riva**t**e **D**ata **N**etwork
pW	**p**ico**w**att (10^{-12} Watt)
PWB	**P**rinted **W**iring **B**oard
PWD	**P**rint **W**orking **D**irectory
PWL	**P**ass **w**ord **L**isting (file)
PWM	**P**ulse-**W**idth **M**odulation
pWp	**p**ico**w**att, **p**sophometrically weighted
PWR	**p**o**w**e**r** (electronics)
PWS	**P**ower **S**upply
PWT	1. **P**ersonal **W**ireless **T**elecommunications
	2. **P**ersonal **W**ireless **T**elephony
PX	**P**rivate **Ex**change
PXE	**P**reboot **E**xecution **E**nvironment
PXI	**P**CI **Ex**tension for **I**nstrumentation
PXML	**P**rivate **Ex**change **M**aster **L**ist
PXP	**P**acket **Ex**change **P**rotocol
PZT	**P**ie**z**oelectric **T**ransducer (electronics)

Q q

q	1. Symbol for **q**uantum value
	2. Symbol for electric charge in coulomb
Q	1. Symbol for **Q**uality factor (electronics)
	2. **Q**ueue
Q Bit	**Q**ualifier data **Bit** (X.25 packet switching)
Q factor	**Q**uality **factor**
Q.SIG	**Q** Interface **Sig**naling
Q&A	**Q**uestion **and A**nswer (teleconferencing)
Q-band	Radio frequency **band** ranging from 36 to 46 GHz
Q-Tel	**Q**utar **Tel**ecommunications Company (operator)
QA	1. **Q**uality **A**ssurance (hardwares)
	2. **Q**ueued **A**rbitrated (SMDS)
QAM	1. **Q**uadrature **A**mplitude **M**odulation
	2. **Q**uaternary **A**mplitude **M**odulation
	3. **Q**uadrature **A**synchronous **M**ultiplexer (Timeplex)
	4. **Q**ueued **A**ccess **M**ethod
	5. **Q**uad **A**synchronous I/O **M**odule (Intelsat)
QAPSK	**Q**uadrature **A**mplitude **PSK** (modulation)
QAS	**Q**uick **A**rbitrate and **S**elect
QASK	**Q**uadrature **ASK** (modulation)
QBE	**Q**uery **B**y **E**xample (databases)
QBF	"**Q**uick **B**rown **F**ox" (test text message)
QC	1. **Q**uality **C**ontrol (of products)
	2. **Q**uantum **C**ascade (laser)

Q

QCELP	Qualcomm Codebook Excited Linear Prediction (voice coding)
QCIF	Quarter-Common Intermediate Format (video compression)
QCPSK	Quadriphase Coherent PSK (modulation)
QCT	Qualcomm CDMA Technologies
QD	Queuing Delay
QDOS	Quick and Dirty Operating System (Microsoft)
QDR	Quad Data Rate
QDU	Quantizing Distortion Unit (voice quality)
QE	Quantum Efficiency (photomultiplier)
QFA	Quick File Access
QFB	Quad Flat Butt-leaded package (microchips)
QFC	Quantum Flow Control (ATM)
QFD	Quality Function Deployment
QFM	Quadrature Frequency Modulation
QFP	Quad Flat Pack (microchips)
QIC	Quarter Inch Cartridge (tape storage)
QICC	1. Quad Integrated Communications Controller (microchips)
	2. Quad International Communications Corporation
QIFM	Quadrature Intermediate Frequency Mixer
QIP	1. Quadritek Internet Protocol (software)
	2. Quad In-line Package (microchips)
QJDP	QIP Joint Developer Program
QL	Query Language (programming)
QLLC	Qualified Logical Link Control (X.25)
QM	1. Quadrature Modulation
	2. Qualification Model
	3. Quantum Mechanics
	4. Queue Management
QML	Qualified Manufacturers List (of products)
QMS	Queue Management System
QO	Quartz Oscillator
QO-CDMA	Quasi-Orthogonal CDMA (access)
QoR	Query-on-Release (databases)
QoS	Quality of Service
QoSR	Quality of Service Routing (IETF)
QPAM	Quadrature Phase-Amplitude Modulation
QPL	1. Qualcomm Pure-Voice Library
	2. Qualified Products List
QPM	Quasi-Phase-Matched
QPR	Quadrature Partial Response
QPRS	Quadrature Partial Response Shift

QPSK	Quadrature **PSK** (modulation)
QPSX	Queued Packet Synchronous Exchange (LANs)
QR	1. Queuing Requirements
	2. Quasi-Random
QRA	Quick Response Alarm
QRNS	Quadratic Residue Number System
QRSS	1. Quasi-Random Sequence Signal
	2. Quasi-Random Signal Source
QS	Quasi-Synchronous
QSAM	Quadrature Sideband Amplitude Modulation
QSDG	Quality of Service Development Group
QSIG	Q-Interface Signaling (protocol)
QSL	Morse Question code for 'Do you Acknowledge Receipt of a good Signal Level of my station?' (Ham Radio)
QSLing	The act of receiving QSL card of a short wave radio station by a listener (Ham Radio)
QSLs	QSL cards of a short wave radio station (Ham Radio)
QSM	1. Quad Synchronous Module
	2. Queued Serial Module
QSOP	Quad Small-Outline Package (microchips)
QSS	Quasi-Stellar Radio Source (Intelsat)
.qt	Name extension for Quick Time files (computer)
QT	Qualification Test
QTAG	Quality Test Action Group
QTAM	Queued Telecommunications Access Method (IBM)
QTC	Quick Time Conference (Apple Macintosh)
QTCP	Quad Tape Carrier Package (microchips)
QTSS	Quick Time Streaming Server
QTVR	Quick Time Virtual Reality
QuAD	Quorum Associate Distributor
Quad IW	**Quad Inside-Wire**
QUALDIR	**Qual**ification **Dir**ective (wireless)
QUICC	**Qu**ad Integrated Communications Controller
QUIL	**Qu**ad In-Line (microchips)
QUIP	**Qu**ad In-line Package (microchips)
QVGA	Quad Video Graphics Array (monitors standard)
QW	Quantum Well
QWERTY	**Q,W,E,R,T,Y** (left side, top row of letter keys on the keyboard)
QWIP	Quantum Well Infrared Photodetector
QWR	Quarter-Wave Rule
QZSS	Quasi-Zenith Satellite System (Japan)

Q

R r

r	Symbol for **r**evolution
r/m	**r**evolutions **p**er **m**inute
r/s	**r**evolutions **p**er **s**econd
R	1. Symbol for **R**esistor or **R**esistance (electronics)
	2. Symbol for **R**oentgen (unit of X-rays exposure dose)
R&D	**R**esearch **and D**evelopment
R&E	1. **R**esearch **and E**ducation
	2. **R**esearch **and E**ngineering
R&O	**R**eport **and O**rder
R&QA	**R**eliability **and Q**uality **A**ssurance (Intelsat)
R&S	**R**esearch **and S**tatistics
R&TTE	**R**adio **and T**elecommunications **T**erminal **E**quipment directive
R-BGAN	**R**egional **BGAN** terminal (Inmarsat)
R-channel	**R**andom-access **channel**
R/T	1. **R**eceive **and T**ransmit
	2. **R**eal-**T**ime
R/W	**R**ead **and W**rite (memories)
RA	1. **R**eal **A**udio
	2. **R**eturn **A**uthorization
	3. **R**ate **A**rea
	4. **R**emote **A**ccess
	5. **R**outing **A**rea

R

RA#	Return Authorization **number**
RA-EN	**R**adio **A**mateur **E**mergency **N**etwork
RA/TDMA	**R**andom-**A**ccess mode **TDMA** (access)
RAAN	**R**ight **A**scension of the **A**scending **N**ode (satcom)
RAB	**R**adio **A**dvertising **B**ureau
RABAL	**R**adiosonde **Ball**oon (meteorology)
RABC	**R**adio **A**dvisory **B**oard of **C**anada
RAC	1. **R**adio **A**mateurs of **C**anada
	2. **R**emote **A**ccess **C**oncentrator
	3. **R**epeat **A**utomated **C**ommand
RACE	1. **R**esearch in **A**dvanced **C**ommunications in **E**urope
	2. **R**andom-**A**ccess **C**omputer **E**quipment
RACES	**R**adio **A**mateur **C**ivil **E**mergency **S**ervice
RACF	1. **R**esource **A**ccess **C**ontrol **F**acility
	2. **R**adio **A**ccess **C**ontrol **F**unction (PCS)
RACH	**R**andom-**A**ccess **Ch**annel (GSM)
RACON	**Ra**dar transponder bea**con**
RACS	1. **R**emote **A**ccess **C**ontrol **S**ervices
	2. **R**emote **A**ccess **C**alibration **S**ervices/System
	3. **R**emote **A**ccess **C**omputing **S**ervices/System
rad	**rad**ian (unit of angle)
RAD	1. **R**apid **A**pplication **D**evelopment (SQL databases)
	2. **R**adiation **A**bsorbed **D**ose (medicine)
	3. **Rad**iance
	4. **R**andom-**A**ccess **D**evice
	5. **R**ecorded **A**nswering **D**evice
	6. **R**emote **A**ntenna **D**river
RADA	**R**andom-**A**ccess **D**iscrete **A**ddress (Intelsat)
RADAN	**Ra**dar **D**oppler **A**utomatic **N**avigation system
radar	**ra**dio **d**etecting **a**nd **r**anging
RADARSAT	**Radar Sat**ellite (Canada)
RADB	**R**outing **A**rbiter **D**atabase
RADHAZ	**Rad**iation **Haz**ard (electromagnetism)
RADIAC	**Ra**dioactivity **D**etection, **I**dentification,
	And **C**omputation
RADINT	**Rad**ar **Int**elligence
RADIR	**R**andom-**A**ccess **D**ocument **I**ndexing and **R**etrieval
RADIUS	**R**emote **A**uthentication **D**ial-**I**n **U**ser **S**ervice (ISPs)
RADL	1. **Rad**io **L**aboratory
	2. **R**eticular **A**gent **D**efinition **L**anguage (programming)
RADNOS	**Rad**io fadeout (**no** signal) caused by **S**olar explosions
radome	**ra**dar **dome**
RADP	**R**emote **A**ccess **D**ata **P**rocessing
RADSL	**R**ate-**A**daptive **DSL** (access)

RADTR	**rad**iator
RAE	**R**afsanjan **A**sre **E**lectronic company (Iran)
RAF	**R**ate **A**djustment **F**actor
RAG	**R**equest Handling and **A**ssignment **G**eneration (Intelsat)
RAI	**R**emote **A**larm **I**ndication
RAID	**R**edundant **A**rray of **I**ndependent **D**isks (data storage)
RAIM	**R**eceiver **A**utonomous **I**ntegrity **M**onitoring (GPS)
RAIN	**R**edundant **A**rray of **I**ndependent **N**etworks
RAIS	**R**edundant **A**rray of **I**ndependent **S**ystems
RAM	1. **R**andom-**A**ccess **M**emory (data storage)
	2. **R**adar-**A**bsorbing **M**aterial (microwave)
	3. **R**emote **A**ccess **M**ultiplexer
	4. **R**eliability, **A**ccessibility, and **M**aintainability
RAMAC	**R**andom **Ac**cess (IBM)
RAMARK	**R**adar **Mark**er
RAMDAC	**R**andom-**A**ccess **M**emory **D**igital-to-**A**nalog **C**onverter
RAMS	**R**andom-**A**ccess **M**easuring **S**ystem
RAN	1. **R**egional **A**rea **N**etwork
	2. **R**ecorded trunk **An**nouncement
	3. **R**eturn **A**uthorization **N**umber (defective hardware)
RAND	1. **Rand**om number (GSM)
	2. **R**ural **A**rea **N**etwork **D**esign
	3. **R**easonable **A**nd **N**on-**D**iscriminatory (licensing)
RANP	**R**egional **A**ir **N**avigation **P**lan (aviation)
RAO	1. **R**adio **A**stronomy **O**bservatory
	2. **R**evenue **A**ccounting **O**ffice
RAP	1. **R**emote **A**ccess **P**oint
	2. **R**oute **A**ccess **P**rotocol
	3. **R**adar-**A**bsorbing **P**aint (microwave)
RAPCON	**R**adar **Ap**proach **Con**trol
RAPID	**R**eserved **A**lternate **P**ath with **I**mmediate **D**iversion (frame relay)
RARC	**R**egional **A**dministrative **R**adio **C**onference (Intelsat)
RARE	**R**éseaux **A**ssociés pour la **R**echerche **E**uropéenne (France)
RARP	**R**everse **A**ddress **R**esolution **P**rotocol (networking)
RAS	1. **R**adio **A**stronomy **S**tation
	2. **R**emote **A**ccess **S**erver
	3. **R**emote **A**ccess **S**ervice
	4. **R**andom-**A**ccess **S**torage
	5. **R**egistration, **A**dmission, and **S**tatus (protocol)
	6. **R**ow **A**ddress **S**trobe (logic signal)
	7. **R**ussian **A**cademy of **S**ciences

RASC	1. **R**adio **A**mateur **S**atellite **C**orporation of North America
	2. **R**oyal **A**stronomy **S**ociety of **C**anada
raser	**r**adio **a**mplification by **s**timulated **e**mission of **r**adiation
RASSP	**R**apid prototyping of **A**pplication **S**pecific **S**ignal **P**rocessors
RAT	**R**emote **A**ccess **T**rojan (network computer)
RATCC	**R**adar **A**ir **T**raffic **C**ontrol **C**enter
RATP	**R**eliable **A**synchronous **T**ransfer **P**rotocol
RATS	**R**adio **A**mateur **T**elecommunications **S**ociety
RATT	**Ra**dio **T**eletypewriter
RAVE	**R**eal-time **A**udio/**V**isual **E**nvironment
RAW	**R**ead **A**fter **W**rite (memories)
rawin	1. **r**adar **win**d
	2. **r**adio **win**d
RAWOL	**Ra**dar **W**ith**o**ut **L**ine-of-sight
RAX	**R**ural **A**utomatic **E**xchange
RAYDAC	**Ray**theon **D**igital **A**utomatic **C**omputer
RAYNet	**R**adio **A**mateur **E**mergency **Net**work
RB	1. **R**eference **B**urst
	2. **R**adar **B**eacon
	3. **R**everse **B**attery
RB1	**R**eference **B**urst **1**
RB2	**R**eference **B**urst **2**
RBB	**R**esidential **B**road**b**and (service)
RBBS	**R**emote **B**ulletin **B**oard **S**ystem
RBC	**R**adiation **B**oundary **C**onditions
RBDS	**R**adio **B**roadcast **D**ata **S**ystem
RBER	**R**esidual **B**it-**E**rror **R**ate (transmission)
RBOC	**R**egional **B**ell **O**perating **C**ompany
RBR	**R**adar **B**ind **R**ange
RBS	1. **R**obbed-**B**it **S**ignaling
	2. **R**adio **B**ase **S**tation
	3. **R**adar **B**ind **S**peed
RBT	**R**eceive **B**urst **T**iming
RBV	**R**eturn **B**eam **V**idicon camera (remote sensing)
RBW	1. **R**everse **B**and **W**orking
	2. **R**esolution **B**and **w**idth (oscilloscopes)
RBWG	**R**esidential **B**roadband **W**orking **G**roup (ATM)
RbXO	**R**ubidium-**C**rystal **O**scillator
RC	1. **R**emote **C**ontrol
	2. **R**adio **C**ontrol
	3. **R**esource **C**ontroller
	4. **R**estricted **C**hannel

5. **R**eference **C**lock
6. **R**eflection **C**oefficient (fiber optics)
7. **R**ate **C**enter
8. **R**esistance–**C**apacitance (electronic circuits)

RCA   1. **R**egional **C**alling **A**rea (of a telephone company)
2. **R**emote **C**ontrol **A**ccess
3. **R**adio **C**orporation of **A**merica cables (company)

RCAT   **R**adio **C**ommunications **A**nalysis **T**est
RCC   1. **R**escue **C**oordination **C**enter (COMSAR)
2. **R**ectangular **C**hip **C**arrier
3. **R**adio **C**ommon **C**arrier (cellular networks)
4. **R**egional **C**ommonwealth in the field of **C**ommunications
5. **R**adiocommunications **C**onsultative **C**ouncil (Australia)
6. **R**educed **C**omplexity **C**omputing

RCDD   **R**egistered **C**ommunications **D**istribution **D**esigner
RCEE   **R**esource **C**ontrol **E**xecution **E**nvironment (Telcordia Technologies)
RCF   1. **R**adio **C**ontrol **F**unction (PCS)
2. **R**adio **C**hrétiennes en **F**rance (broadcaster)
3. **R**emote **C**all **F**orwarding (service)
4. **R**emote **C**ontrol **F**acility

RCG   **R**everberation-**C**ontrolled **G**ain (circuit)
RCI   **R**emote **C**ard **I**dentifier
RCIT   **R**esearch **C**enter **I**nformation **T**echnology
RCL   **R**estrictive **C**abling **L**icense (Australia)
1. **rec**a**l**l (cellular phone)
2. **R**untime **C**ontrol **L**ibrary

RCM   1. **R**emote **C**arrier **M**odule
2. **R**elease **C**omplete **M**essage
3. **R**adar-**C**ounter**m**easure (electronics)

RCP   1. **R**emote **C**opy **P**rotocol
2. **R**emote **C**opy **P**rogram (Berkeley UNIX)
3. **R**emote **C**ontrol **P**anel
4. **R**adar **C**ommunication **P**rocessor
5. **R**eal-time **C**ontrol **P**rogram
6. **R**adio **C**ontrol **P**oint
7. **R**ight-**C**ircular **P**olarization (antennas)

RCS   1. **R**evision **C**ontrol **S**ystem
2. **R**emote **C**ontrol **S**ystem
3. **R**adar **C**ross-**S**ection
4. **R**adio **C**orporation of **S**ingapore (broadcasting)
5. **R**adio **C**ommunications **S**ervice

R

RCT	1. Remote Craft Terminal
	2. Reduced Carrier Transmission
RCTL	Resistor–Capacitor–Transistor Logic (digital electronics)
RCU	Remote Concentration Unit
RCV	receive
RCVO	Receive Only (satcom)
RCVR	receiver
rd	rad (unit of absorbed radiation exposure dose)
Rd-channel	Random-access channel (data communication)
RD	1. Routing Domain (ATM)
	2. Routing Diagram
	3. Routing Directory
	4. Directional Radiobeacon
	5. Received Data (logic signal)
	6. Reflection Density (fiber optics)
RDB	1. Receive Data Buffer
	2. Remote Database
RDBMS	Relational Database Management System
RDBS	Routing Database System
RDC	1. Remote Digital Concentrator
	2. Redirect Confirm packet (CDPD)
RDCCH	Reverse Digital Control Channel (cellular networks)
RDF	1. Radio Direction Finder/Finding (equipment)
	2. Rate Decrease Factor (ATM)
RDI	1. Remote Defect Indication (ATM)
	2. Restricted Digital Information
	3. Restricted Digital Information (ISDN)
RDIST	Remote file Distribution Program
RDM	Receive Driver Module
RDMA	Remote Direct Memory Access
RDMS	Relational Database Management System (ICT)
RDN	Radio Data Network
RDO	Remote Data Objects (Visual Basic)
RDP	1. Radar Data Processing
	2. Reliable Data Protocol
RDPC	Remote Data Port Card
RDQ	Redirect Query packet (CDPD)
RDR	Redirect Request packet (CDPD)
RDRAM	Rambus Dynamic RAM (memory)
RDS	1. Radio Data System
	2. Radio Determination Service
RDSI	The Spanish equivalent for ISDN
RDSS	Radio Determination Satellite Service (ITU-T)

RDT	1. **R**ecall **D**ial **T**one
	2. **R**emote **D**igital **T**erminal
	3. **R**equest **D**ata **T**ransfer (control character)
	4. **R**omanian **D**omestic **T**elephony
RE	**R**adio **E**xchange
REA	**R**ural **E**lectrification **A**dministration
REAC	**R**eeves **E**lectronic **A**nalog **C**omputer
READ	**R**elative **E**lement **A**ddress **D**esignate
Rec	**rec**ommendation (ITU)
RECAPSS	**Re**mote **Ca**ble **P**air **S**witching **S**ystem
ReCCIT	**Re**search **C**enter for **C**ommunications and **I**nformation **T**echnology
RECO	**R**esources (people), **E**quipment, **C**ircuits, and **O**ther (IBM)
RECON	**Recon**naissance
RECT	**rect**ifier (electronics)
RED	1. **R**andom **E**arly **D**etection (QoS)
	2. **R**eflection **E**lectron **D**iffraction (crystals)
REED	**R**estricted **E**dge-**E**mitting **D**iode
REG	1. **R**ange **E**xtender with **G**rain (transmission)
	2. **Reg**enerator
REGEDIT	**Reg**istry **Edit**or
REGNOT	**Reg**istration **Not**ification (wireless)
REJ	**Rej**ect (link protocol command)
REL	**rel**ease (message)
RELP	**R**esidually **E**xcited **L**inear **P**rediction (voice coding)
RELURL	**Rel**ative **URL**
REM	1. **R**emote **E**quipment **M**odule
	2. **R**ing **E**rror **M**onitor (LANs)
	3. **R**oentgen **E**quivalent **M**an
Remod	**remod**ulation
REMOS	**Re**sources **M**anagement **O**nline **S**ystem
REN	1. **R**inger **E**quivalence **N**umber (telephone set)
	2. **Ren**ame the current file (MS DOS Command)
REO	**R**emovable **E**rasable **O**ptical
Rep	**rep**eater
REP	1. **R**oentgen **E**quivalent **P**hysical
	2. **R**ight-**E**lliptical **P**olarization (antennas)
REPACCS	**Re**mote **C**able-**P**air **C**ross-**C**onnect **S**ystem
REPROM	**Re**programmable **ROM** (memory)
REQ	**req**uest
RES	1. **R**emote **E**arth **S**tation
	2. **R**egional **E**arth **S**tation
	3. **R**esidential **E**nhanced **S**ervice

R

	4. **res**erved (ATM)
	5. **R**adio **E**quipment and **S**ystem
RET	**ret**urn
RETMA	**R**adio-**E**lectronics-**T**elevision **M**anufacturers Association
REV	**rev**erse
REW	**rew**ind
REX	**R**outine **Ex**ercise
REXX	**Re**structured **Ex**tended **Ex**ecutor (programming language)
RF	1. **R**adio **F**requency (10 kHz to 3 MHz range)
	2. **R**eference **F**requency
	3. **R**ange **F**inder
	4. **R**ating **F**actor
	5. **R**aster **F**ile
RFA	1. **R**adio **F**requency **A**llotment
	2. **R**equest **F**or **A**ction
	3. **R**eceive **F**rame **A**cquisition
	4. **R**emote **F**ile **A**ccess
RFAC	**R**estricted **F**orced **A**uthorization **C**ode
RFB	**R**adio **F**requency **B**oard
RFBP	**R**equest **F**or **B**usiness **P**lan (contracts)
RFC	1. **R**equest **F**or **C**omments (Internet)
	2. **R**equired **F**or **C**ompliance
	3. **R**adio **F**requency **C**hoke (electronics)
RFCA	**R**adio **F**requency **C**hannel **A**llotment
RFCOH	**R**adio **F**rame **C**omplementary **O**ver**h**ead bit
RFD	**R**equest **F**or **D**iscussion (Internet)
RFE	**R**adio **F**ree **E**urope (broadcaster)
RFF	1. **R**adio **F**requency **F**ingerprinting
	2. **R**aster **F**ile **F**ormat
RFHMA	**R**andom **F**requency **H**opping **M**ultiple **A**ccess
RFI	1. **R**equest **F**or **I**nformation (contracts)
	2. **R**adio **F**requency **I**nterference
	3. **R**adio **F**rance **I**nternationale (broadcaster)
	4. **R**emote **F**ailure **I**ndication
RFICs	**R**adio **F**requency **I**ntegrated **C**ircuit**s**
RFID	**R**adio **F**requency **Id**entification
RFMD	**RF** **M**icro **D**evices company (China)
RFMU	**R**adio **F**requency **M**onitoring **U**nit
RFO	**R**eseaux **F**rance **O**utre-mer (broadcaster)
RFOI	**R**adio **F**iber **O**ptic **I**ntegration
RFP	1. **R**equest **F**or **P**roposal (contracts)
	2. **R**adio **F**ixed **P**art (GSM)

RFQ	Request For Quotation (contracts)
RFS	1. Receive Frame Synchronization
	2. Request For Service
	3. Ready For Service
	4. Remote File Sharing (networking)
	5. Radio Frequency System
	6. Radio Frequency Simulator
	7. Radio Frequency Shift (broadcasting)
	8. Range Finding System
	9. Real File Store
RFT	1. Radio Frequency Transceiver
	2. Receive Frame Timing
RFT DCA	Revisable-Form-Text Document Content Architecture
RFTS	Remote File Transfer System
	Radio Frequency Test Set
RFU	Radio Frequency Unit
RG	1. Radio Guide
	2. Ring Generator
RGB	Red–Green–Blue (TV and CRT monitors)
RGO	Royal Greenwich Observatory
RGP	Raster Graphics Processor
RGS	Route Guidance System
RH	1. Request Header
	2. Response Header
RHC	Regional Holding Company
RHCP	Right-Hand Circular Polarization (microwave)
RHEED	Reflected High-Energy Electron Diffraction
RHEP	Right-Hand Elliptical Polarization
RHET	Resonant-tunneling Hot-Electron Transistor (semi-conductors)
Rhm	Roentgen-per-hour-at-one-meter
RHP	Reconfigurable Hardware Products
RHR	1. Radar Horizon Range
	2. Radio Horizon Range
RI	1. Ring-Indicator (RS232)
	2. Radio Interference
	3. routing indicator
RIA	Routing Indicator Allocation
RIAA	Recording Industry Association of America
RIACS	Research Institute for Advanced Computer Science
RIAS	Research Institute for Advanced Studies
RIB	1. Routing Information Base (BGP)
	2. RenderMAN Interface Bytestream
RIBE	Reactive Ion Beam Etching

R

RIC	1. **R**egional **I**nformation **C**enter
	2. **R**afsanjan **I**ndustrial **C**omplex (Iran)
RIE	**R**eactive **I**on **E**tching
RIF	1. **R**outing **I**nformation **F**ield (LANs)
	2. **R**ate **I**ncrease **F**actor (ATM)
RIFF	**R**aster **I**mage **F**ile **F**ormat
RII	**R**outing **I**nformation **I**ndicator
RILD	**R**emote **I**SDN **L**ine **D**rawer
RIM	1. **R**emote **I**ntegrated **M**ultiplexer
	2. **R**esearch **I**n **M**otion
RIME	**R**elay**N**et **I**nternational **M**essage **E**xchange
RIMM	**R**ambus **I**n-line **M**emory **M**odule (Macintosh, PCs.)
RIMPATT	**R**ead **IMPATT** (diode)
RIMS	**R**anging and **I**ntegrity **M**onitoring **S**tation (navigation)
RIN	**R**elative **I**ntensity **N**oise
RIO	**R**eusable **I**nformation **O**bject
RIP	1. **R**outing **I**nformation **P**rotocol (networking)
	2. **R**emote **I**maging **P**rotocol
	3. **R**aster **I**mage **P**rocessor
	4. **R**efractive **I**ndex **P**rofile (waveguides)
RIPE	**R**eseaux **IP** **E**uropeens (a group)
RIPL	**R**emote **I**ntelligent **P**arallel **L**oad (method)
RIPscript	**R**emote **I**maging **P**rotocol **script** (language programming)
RIPSO	**R**evised **I**nterconnection **P**rotocol **S**ecurity **O**ption
RIR	**R**egional **I**nternet **R**egistries
RIS	**R**etroreflector **I**n **S**pace (remote sensing)
RISC	1. **R**educed-**I**nstruction-**S**et **C**omputing (processor chips)
	2. **R**educed **I**nstruction **S**et **C**hip
RISLU	**R**emote **I**ntegrated **S**ervices **L**ine **U**nit (Lucent Technologies)
RISR	**R**emote **I**nterrupt **S**ervice **R**outine
RiSU	**R**emote **i**ndoor **S**ervice **U**nit (cable telephony)
RIT	**R**ate of **I**nformation **T**ransfer
RITL	**R**adio **I**n **T**he **L**oop (cable telephony)
RIU	**R**ing **I**nterface **U**nit
RJ	**R**egistered **J**ack
RJE	**R**emote **J**ob **E**ntry (terminal)
RJEP	**R**emote **J**ob **E**ntry **P**rotocol
RKA	**R**ossiskaja **K**osmitaska **A**gentura (Russian Space Agency)
RL	1. **R**eturn **L**oss
	2. **R**adio **L**ocator

3. **R**otating **L**inear (antenna pattern)

4. **R**esistor-**C**oil (circuit)

RLAN   Radio **LAN**

RLB   **R**emote **L**oop-**B**ack

RLC   1. **R**elease **C**omplete message

2. **R**esistance–**I**nductance–**C**apacitance (electronic circuits)

3. **R**un **L**ength **C**oding

4. **R**emote **L**ine **C**oncentrator

RLCE   **R**emote **L**ine **C**oncentrate **E**quipment

RLCM   **R**emote **L**ine **C**oncentrating **M**odule

RLCS   **R**emote **L**ive **C**all **S**creening (Panasonic phone)

RLE   1. **R**un **L**ength **E**ncoding (data compression)

2. **R**esearch **L**aboratory of **E**lectronics

RLES   **R**egional **L**and **E**arth **S**tation (Inmarsat)

RLG   **R**ing **L**aser **G**yroscope

RLIN   **R**esearch **L**ibraries **I**nformation **N**etwork

RLL   1. **R**un **L**ength **L**imited (data encoding)

2. **R**adio **L**ocal **L**oop

RLLS   **R**adio **L**ocation **L**and **S**tation

RLMS   **R**adio **L**ocation **M**obile **S**tation

RLN   **R**emote **LAN** **N**ode

RLOGIN   **R**emote **Log-In** application

RLP   1. **R**adio **L**ink **P**rotocol (TIA)

2. **R**esource **L**ocation **P**rotocol

RLQ   **R**ate-**L**imiting **Q**ueues

RLR   **R**eceive **L**oudness **R**ating

RLS   1. **r**ele**a**se button (telephone set)

2. **R**emote **L**ive **S**creening (telephone feature)

RLSD   **R**eceived **L**ine **S**ignal **D**etect

RLT   **R**elease **L**ink **T**runk

RM   1. **R**esource **M**anagement (ATM)

2. **R**eference **M**odel (OSI model)

RMA   1. **R**adio **M**anufacturers **A**ssociation

2. **R**andom **M**ultiple **A**ccess

3. **R**emote **M**emory **A**ccess

4. **R**eturned **M**erchandise **A**uthorization code

5. **R**eturned **M**erchandise **A**uthorization **N**umber

RM-Cell   **R**esource **M**anagement **C**ell (ATM)

RMAS   **R**emote **M**emory **A**dministration **S**ystem

RMATS   **R**emote **M**aintenance **A**nd **T**est **S**ystem

RMCP   **R**emote **M**ail **C**hecking **P**rotocol

RME   **R**emote **M**ediation **E**quipment

RMF   **R**emote **M**anagement **F**acility

R

RMI	1. **R**adio **M**agnetic **I**nterference
	2. **R**esource **M**anager **I**nterface
RMII	**R**educed **M**edia **I**ndependent **I**nterface
RMM	**R**eal-**M**ode **M**apper (Windows)
RMON	**R**emote Network **Mon**itoring (networking)
RMOS	**R**efractory **M**etal-**O**xide **S**emiconductor
RMP	**R**oving **M**onitor **P**ort
rms	**r**oot-**m**ean-**s**quare (output power parameter)
RMS	**R**ecord **M**anagement **S**ystem
RMS-D1	**R**emote **M**easurement **S**ystem **D**igital **1**
RMTP	**R**eliable **M**ulticast **T**ransport **P**rotocol
RMU	**R**emote **M**ask **U**nit
RMW	**R**ead-**M**odify-**W**rite (software)
RNA	1. **R**ing **N**o **A**nswer (modems)
	2. **R**emote **N**etwork **A**ccess
RNC	**R**adio **N**etwork **C**ontroller (cellular networks)
RNCC	**R**egional **N**etwork **C**ontrol **C**enter
RND	**R**adio **N**etwork **D**esign
RNE	**R**ailway **N**etwork **E**xchange
RNG	**R**andom **N**umber **G**enerator
RNLS	**R**adio **N**avigation **L**and **S**tation
RNMS	**R**adio **N**avigation **M**obile **S**tation
RNO	**R**adio **N**etwork **O**ptimization
RNP	**R**adio **N**etwork **P**lanning
RNR	**R**eceiver **N**ot **R**eady (link protocol command)
RNS	**R**adio **N**avigation **S**atellite (ITU)
RNSS	**R**adio **N**avigation **S**atellite **S**ervice (ITU)
RNX	**R**estricted **N**umeric **E**xchange
RO	1. **R**outing **O**rganization (Inmarsat)
	2. **R**eceive-**O**nly (earth station)
	3. **R**ead-**O**nly (ATM)
	4. **R**emote **O**peration
	5. **R**ecovery **O**peration
	6. **R**ing-**O**ff
R/O	**R**adio **O**fficer (shipping)
ROA	**R**ecognized **O**perating **A**gency (ITU)
ROADS	**R**obust **O**pen **A**rchitecture **D**istributed **S**witching
ROBO	**R**emote **O**ffice/**B**ranch **O**ffice market
ROC	1. **R**ing **O**perations **C**enter
	2. **R**ate **O**f **C**onvergence
	3. **R**ecord **O**f **C**omments
	4. **R**egional **O**perating **C**ompany (Bell)
ROD	**R**ewritable **O**ptical **D**isk
RODB	**R**untime **O**bject **D**ata **B**ase

ROE	**R**eturn **O**n **E**nquiry
ROH	1. **R**eceiver **O**ff **H**ook (telephone set)
	2. **R**inger **O**ff **H**ook
ROHS	**R**estriction **O**f **H**azardous **S**ubstances (directive)
ROI	**R**eturn **O**n **I**nvestment
RoIP	**R**adio **o**ver **IP** (service)
ROLC	**R**outing **O**ver **L**arge **C**louds
ROM	1. **R**ead-**O**nly **M**emory (data storage)
	2. **R**ough **O**rder of **M**agnitude
ROM BIOS	**R**ead-**O**nly **M**emory **B**asic **I**nput/**O**utput **S**ystem
RON	**R**outing **O**rganization **N**umber
ROP	**R**aster **Op**eration (graphics)
RORA	**R**egion **O**riented **R**esource **A**llocation
ROS	1. **R**ead-**O**nly **S**torage (data storage)
	2. **R**emote **O**perations **S**ervice
ROSAT	**Ro**entgen **Sat**ellite
ROSE	**R**emote **O**perations **S**ervice **E**lement (protocol)
ROTA	**rota**ting schedule (call centers)
ROTHR	**R**elocatable **O**ver-**T**he-**H**orizon **R**adar
ROTL	**R**emote **O**ffice **T**est **L**ine
ROTS	1. **R**otary **O**ut-**T**runk **S**witch
	2. **R**ugged **O**ff-**T**he-**S**helf
ROV	**R**emotely **O**perated **V**ehicle
ROW	**R**emote **O**rder-**W**ire
RP	1. **R**egistration **P**oint
	2. **R**adio **P**art (GSM)
	3. **R**ange **P**rocessor
	4. **R**apid **P**rototyping
RPAD	**R**emote **P**layout **A**nd **D**istribution
RPC	**R**emote **P**rocedure **C**all (programming)
RPDU	**R**esponse **P**rotocol **D**ata **U**nit
RPE	**R**adio **P**aging **E**quipment
RPE-LTP	**R**egular **P**ulse **E**xcitation—**L**ong-**T**erm **P**rediction (GSM)
RPF	**R**everse **P**ath **F**orwarding (Internet)
RPG	**R**eport **P**rogram **G**enerator (programming language)
RPI	**R**adio **P**hysical **I**nterface
RPL	1. **R**emote **P**rocedure **L**oad
	2. **R**emote **P**rogram **L**oad
	3. **R**epair **P**arts **L**ist
rpm	**r**evolutions **p**er **m**inute (hard drives)
RPM	1. **R**emote **P**acket **M**odule
	2. **R**ed-hat **P**acket **M**anager (files)
RPN	1. **R**everse **P**olish **N**otation (as in HP calculators)

R

	2. Residual Phase Noise
RPOA	Recognized Private Operating Agency (common carrier)
RPOL	Role Playing Online
RPOP	Remote Post Office Protocol
RPQ	Request for Price Quotation (contracts)
RPR	Resilient Packet Ring
RPRINTER	Remote Printer (Novell NetWare)
rps	revolutions per second
RPS	1. Repetitive Pattern Suppression (data compression)
	2. Ring Protection Switching (SONET)
	3. Radar Processing System
	4. Radio Protection Switch
RPT	repeat
RPV	Remotely Piloted Vehicle
RQC	Repair and Quick Clean (in industry)
RQS	Rate Quote System
RR	1. Radio Regulations (ITU)
	2. Radio Resource
	3. Receiver Ready (link protocol command)
	4. Return Rate
	5. Rural Route
	6. Round Robin (Internet)
RRC	1. Radar Resolution Cell
	2. Raised-Root Cosine (mathematics)
RRD	Remote Reconciliation Descriptor
RRE	1. Receive Reference Equivalent
	2. Radar Range Equation
RREI	Radio-Relay Equipment Interface
RRI	Request Randomization Index
RRM	Radio Resource Management (UMTS)
RRME	Radio Resource Management Entity (AT&T)
RRN	Routing Recording Number (AT&T)
RROCP	Restricted Radio Operator's Certificate of Proficiency (shipping)
RROP	Restricted Radiotelephone Operator Permit (FCC)
RRPI	Radio-Relay Physical Interface (MUX)
RRR	Radio-Relay Regenerator (MUX)
RRRP	Radio-Relay Reference Point (MUX)
RRS	1. Remote Radar Station
	2. Radio-Relay System
RRSF	RACF Remote Sharing Facility (PCS)
RRSP	Resource Reservation Setup Protocol
RRST	Radio Regenerator Section Termination

RRT	**R**adio-**R**elay **T**erminal
RS	1. **R**emote **S**ite
	2. **R**educed **S**lope (optical fiber)
	3. **R**emote **S**tation
	4. **R**egenerator **S**ection (MUX)
	5. **R**ecommended **S**tandard
	6. **R**ead **S**olomon code (error correction)
	7. **R**ecord **S**eparator (character control code)
	8. **R**adio **S**atellite
	9. **R**emote **S**ingle-layer (ATM)
RS-1	**R**ussian amateur **S**atellite **1**
RSA	1. **R**eliable **S**ervice **A**rea (cellular networks)
	2. **R**ural **S**ervice **A**rea
	3. **R**ural **S**tatistical **A**rea (FCC)
	4. **R**emote **S**torage **A**rea
	5. **R**epair **S**ervice **A**nswering (AINs)
	6. **R**ivest-**S**hamir-**A**dleman (encryption algorithm)
RSAC	**R**ecreational **S**oftware **A**dvisory **C**ouncil
RSB	**R**epair **S**ervice **B**ureau
RSC	1. **R**emote **S**witching **C**enter
	2. **R**epair **S**ervice **C**enter
	3. **R**adio **S**tation **C**entral
RSCC	**R**ussian **S**atellite **C**ommunication **C**ompany
RSCH	**R**adio **S**ervice **Ch**annel Unit
RSD	**R**emote **S**pecific **D**escriptor
RSE	1. **R**emote **S**ubscriber **E**quipment
	2. **R**emote **S**ingle-layer **E**mbedded test method (ATM)
RSF	**R**ostered **S**taff **F**actor (call centers)
RSFG	**R**oute **S**erver **F**unctional **G**roup (ATM)
RSGB	**R**adio **S**ociety of **G**reat **B**ritain
RSH	**R**emote **Sh**ell (UNIX)
RSID	**R**esidential **S**ystem **Id**entifier (cellular networks)
RSL	1. **R**equest and **S**tatus **L**ink (PBXs)
	2. **R**eceived **S**ignal **L**evel (radio receiver)
	3. **R**adio **S**ignaling **L**ink (GSM)
RSM	1. **R**emote **S**witching **M**odule (AT&T)
	2. **R**adio **S**ubsystem **M**anagement (wireless)
Rsmc-channel	**R**andom-access **channel** for **s**ystem **m**anagement and control
RSN	1. **R**adio **S**tation **N**odal
	2. **R**obust **S**ecurity **N**etwork (802.11i)
RSOH	**R**egenerator **S**ection **O**verhead (MUX)
RSPI	**R**adio **S**DH **P**hysical **I**nterface (MUX)

R

RSPX	**R**emote **S**equenced **P**acket E**x**change (Novell NetWare)
RSRB	**R**emote **S**ource **R**oute **B**ridging
rss	**r**oot **s**um **s**quare
RSS	**R**emote **S**witching **S**ystem
RSSI	1. **R**eceived **S**ignal **S**trength **I**ndicator
	2. **R**eceived **S**ignal **S**trength **I**ntensity
RST	1. **R**adio **S**tation **T**erminal
	2. **R**egenerator **S**ection **T**ermination (MUX)
	3. **R**eadability, **S**trength, and **T**one (Ham Radio)
	4. **R**unning **S**tatus **T**able (broadcasting)
RST-RR	**R**egenerator **S**ection **T**ermination for STM-1 **R**adio-**R**elay (MUX)
RSU	1. **R**emote **S**ubscriber **U**nit
	2. **R**emote **S**ervice **U**nit (cable telephony)
	3. **R**emote **S**witching **U**nit
RSUP	**R**eliable **S**AP **U**pdate **P**rotocol (Cisco)
RSVP	**R**esource Re**s**er**v**ation **P**rotocol
RSW	1. **R**otary **Sw**itch
	2. **R**elative **S**pectral **W**idth
RSZI	**R**egional **S**ubscription **Z**one **I**dentity (GSM)
RT	1. **R**adio **T**elephony
	2. **R**eal-**T**ime
	3. **R**ise-**T**ime (pulses)
	4. **R**untime
	5. **R**emote **T**erminal or **T**ermination (DLC)
	6. **R**ecorder **T**one
	7. **R**outing **T**able
RT-VBR	**R**eal-**T**ime **V**ariable **B**it **R**ate (ATM)
RTA	**R**emote **T**runk **A**rrangement
RTANI	**R**eal-**T**ime **ANI**
RTB	**R**egional **T**est **B**ed
RTC	1. **R**eceive **T**iming **C**ontroller
	2. **R**untime **C**ontrol (SCSA)
	3. **R**untime **C**ode
	4. **R**emote **T**unable **C**ombiner
	5. **R**eal-**T**ime **C**lock
RTCA	1. **R**equirements and **T**echnical **C**haracteristics for Aviation
	2. **R**adio **T**echnical **C**ommission of **A**eronautics
RTCP	**R**eal-**T**ime **C**onferencing **P**rotocol
RTD	1. **R**ound-**T**rip **D**elay (ATM)
	2. **R**esonant-**T**unneling **D**iode
	3. **R**eal-**T**ime **D**isplay
	4. **R**esistance-**T**emperature **D**etector (probe)

RTDF	Radio Telephone Distress Frequency
RTDM	1. Real-Time Data Migration
	2. Rough Terrain Diffusion Model
RTDMS	Real-Time Database Management System
RTDP	Radar Target Data Processor
RTDS	Real-Time Data System
RTE	1. Registered Terminal Equipment
	2. Remote Terminal Emulation
RTF	1. Rich Text Format
	2. Radio Terminal Function
RTFD	Recommended Technical Framework Document
RTFS	Remote Throttled First Service (ATM)
RTG	Radio Telegraphy
RTI	1. Radiation Transfer Index (transmission)
	2. Real-Time Interrupt (electronics)
RTIP	Real-Time IP
RTK	Real-Time Kinematic Surveying (GSM)
RTL	1. Resistor–Transistor Logic (digital electronics)
	2. Relative Transmission Level
	3. Radio Television Luxemburg (broadcaster)
	4. Runtime License
RTM	1. Ready To Manufacture
	2. Runtime Monitor
RTMA	Radio–Television Manufacturers Association
RTML	1. Real-Time Markup Language (programming)
	2. Rich Text Markup Language (programming)
	3. Robotic Telescope Markup Language (programming)
	4. Remote Telescope Markup Language (programming)
RTMOS	Room Temperature MOS (semiconductors)
RTMP	Routing Table Maintenance Protocol (networking)
RTN	Routing Transport Number (AT&T)
RTNDA	Radio–Television News Directors Association
RTNDF	Radio–Television News Directors Foundation
RTNR	Ring Tone No Reply
RTO-IS	Ready To Order—In Service
RTOS	Real-Time Operating System
RTP	1. Real-time Transport Protocol (IETF)
	2. Rapid Transport Protocol (IETF)
	3. Routing Table Protocol (Banyan Systems)
	4. Release To Pivot
	5. Radio E Televisao de Portugal (broadcaster)
RTS	1. Request-To-Send (RS232 signal)
	2. Real-Time System
	3. Reliable Transfer Service (OSI model)

R

RTSE	Reliable Transfer Service Element (OSI model)
RTSP	Real-Time Streaming Protocol (IETF)
RTSU	Remote Terminal Sub-Unit
RTT	1. Round-Trip Time (ATM)
	2. Radio and Telecommunications Terminal
	3. Radio Transmission Technologies
	4. Remote Traffic Terminal
RTTU	Remote Trunk Test Unit
RTTY	Radio Teletype
RTU	1. Remote Terminal Unit
	2. Remote Termination Unit
	3. Remote Telemetry Unit
	4. Right To Use
RTV	Real-Time Video
RTX	Request to **Transmit**
RU	1. Request Unit
	2. Response Unit
	3. Rack Unit
RUA	Remote User Agent (X.400)
RUBS	Resource Usage Billing System (Inmarsat)
RUC	Resource Utilization Charge (Inmarsat)
RUR	Repair, Update, and Refurbish (in industry)
RURL	Relative Uniform Resource Locator
RUS	Rural Utilities Service
RVA	1. Recorded Voice Announcement
	2. Reactive Volt–Ampere (electrical engineering)
RVC	Reverse Voice Channel
RVDT	Rotary Variable Differential Transducer
RVE	Reference Vector Equalization
RVP	Remote Voice Port
RVPC	Remote Voice Port Card
RVPP	Revenue Volume Pricing Plan (finance)
RW	1. Read and Write (data storage)
	2. Re-Writable (compact disks)
	3. Remote Workstation
RWhois	Referral Whois (service)
RWI	1. Radio and Wire Integration
	2. Radio-Wire Interface
RWR	Radar Warning Receiver
RWT	Recorder Warning Tone
RWW	Read-While-Write (data storage)
RX	**Receiver** or **Reception**
RXD	**Receive**-Data (RS-232 signal)
RXLEV	**Receiving Lev**el on the air-interface (GSM)

RXQUAL	**Receiving Qual**ity on the air-interface (GSM)
RZ	**R**eturn-to-**Z**ero level (data encoding)
RZ-AMI	**R**eturn-to-**Z**ero level—**A**lternate **M**ark **I**nversion (data encoding)
RZL	**R**eturn-to-**Z**ero **L**evel (data encoding)

R

S s

s	1. Symbol for **s**econd (of time)
	2. Symbol for **s**ynchronous
S	1. Fixed-**S**atellite service (ITU-T)
	2. Symbol for **S**trength of a signal
	3. Symbol for **S**ource (FETs)
	4. Symbol for **S**iemens (unit of conductance)
	5. Symbol for **S**urface area
S port	**S**lave **port** (FDDI Architecture)
S SEED	**S**ymmetric **S**elf **E**lectro-optic **E**ffect **D**evice (switching)
S&F	**S**tore-**and-F**orward (messaging technique)
S&H	**S**ample-**and-H**old (PCM)
S&R	**S**tore-**and-R**etrieve (messaging technique)
S+N+D	**S**ignal **plus** **N**oise **plus** **D**istortion
S+N/N	**S**ignal **plus** **N**oise-**to-N**oise (ratio)
S-ALOHA	**S**lotted **ALOHA** (protocol)
S-band	Radio frequency **band** ranging from 1700 to 2360 MHz
S-BCCH	**S**ystem **B**roadcast **C**ontrol **Ch**annel
S-CDMA	**S**ynchronous **CDMA** (access)
S-DAB	**S**atellite **D**igital **A**udio **B**roadcasting
S-HTTP	**S**ecure **H**ypertext **T**ransfer **P**rotocol
S-ISUP	**S**atellite **ISD**N **U**ser **P**art (SS7)
S-PCN	**S**atellite **P**ersonal **C**ommunications **N**etwork
S-Video	**S**uper **Video**

S

S/C	Signal-to-Cross talk (ratio)
S/D	1. Synchro-to-Digital (electronics)
	2. Signal-to-Distortion (ratio)
S/DMS	SONET Digital Multiplex System
S/EOS	Standard Earth Observation Satellite
S/H	Sample-and-Hold (ADC)
S/I	Signal-to-Interference (ratio)
S/MAIL	Secure MAIL
S/MIME	Secure Multipurpose Internet Mail Extension (e-mail attachments)
S/N	Signal-to-Noise power (ratio)
S/P	Serial-to-Parallel
S/P DIF	Sony and Philips Digital Interface Format
S/Q	Signal Quality Value (Intelsat)
S/R	Symbol Rate (satellite TV broadcasting)
S/S	Stop and Start (control character)
S/W	Software
S/WAN	Secure Wide Area Network
SA	1. Service Agent
	2. Source Address (LANs)
	3. Section Adaptation Function (MUX)
	4. Space Application and meteorology service (ITU-T)
	5. Selective Availability (GPS)
	6. Security Associations
SA-RR	Section Adaptation for STM-1 Radio-Relay (MUX)
SAA	1. Supplemental Alert Adapter (AT&T)
	2. Systems Application Architecture (IBM standard)
	3. Standard Association of Australia
SAAL	Signaling ATM Adaptation Layer
SABER	Situation Awareness Beacon Reply
SABM	Set Asynchronous Balanced Mode command (X.25)
SABME	Set Asynchronous Balanced Mode Extended (X.25)
SAC	1. Satellite Access Controller
	2. Single Attached Concentrator (FDDI or CDDI)
	3. Single Attachment Concentrator (FDDI)
	4. Service Access Code (INs)
	5. Special Area Code
	6. STN-to-ATM Converter (MUX)
	7. Shelf Alarm Card
	8. Special Access Code (Inmarsat)
	9. Subscriber Acquisition Cost
SACCH	Slow Associated Control Channel (GSM)
SACD	Supper-Audio Compact Disk
SAD	Serial Analog Delay Line

SADL	**S**ynchronous **A**uto **D**ial **L**anguage (Racal Vadic)
SAE	**S**ociety of **A**utomotive **E**ngineering
SAF-TE	**S**CSI **A**ccessed **F**ault—**T**olerant **E**nclosures
SAFDT	**S**tore **A**nd **F**orward **D**ata **T**ransmission
SAFE	**S**ecurity **A**nd **F**reedom through **E**ncryption (Act)
SAFEAct	**S**ecurity **A**nd **F**reedom through **E**ncryption **Act**
SAFENet	**S**urvivable **A**daptable **F**iber **O**ptic **E**mbedded **Net**work (U.S. Navy)
SAG	**S**elf-**A**ligned **G**ate technology
SAGE	**S**emi-**A**utomatic **G**round **E**nvironment (air defense)
SAI	1. **S**erving **A**rea **I**nterface
	2. **S**tatus Computer **A**pplication **I**nterface (PBX switch)
	3. **S**tandards **A**ustralia **I**nternational Ltd.
SAIC	**S**cience **A**pplications **I**nternational **C**orporation (U.S.A.)
SAID	**S**peech **A**ctivated **I**ntelligent **D**ialing
SAIL	**S**tanford **A**rtificial **I**ntelligence **L**aboratory
SAIS	**S**ection **A**larm **I**ndication **S**ignal (MUX)
SAM	1. **S**elf-**A**dministered **M**aintenance (technology)
	2. **S**erving **A**rea **M**ultiplex
	3. **S**ervice **A**ctivation **M**anual (Inmarsat)
	4. **S**witched **A**ccess with **M**ultiplexer
	5. **S**erial **A**ccess **M**emory
	6. **S**ubscriber **A**ccess **M**anagement
SAMA	**S**pread **ALOHA** **M**ultiple **A**ccess
SamaCom	**Sama Com**munications Company (U.A.E.)
SAN	1. **S**torage **A**rea **N**etwork (Cisco)
	2. **S**atellite **A**ccess **N**ode (Inmarsat)
SANZ	**S**tandard **A**ssociation of **N**ew **Z**ealand
SAP	1. **S**ervice **A**ccess **P**oint (Internet)
	2. **S**atellite **A**ccess **P**rovider (Satcom)
	3. **S**ervice **A**dvertizing **P**rotocol (networking)
	4. **S**ession **A**nnouncement **P**rotocol
	5. **S**ecure **A**udio **P**ath (Microsoft)
	6. **S**erver **A**pplication **P**rogramming
	7. **S**oftware **A**cquisition **P**laning
SAPI	1. **S**ervice **A**ccess **P**oint **I**dentifier (ISDN)
	2. **S**peech **A**pplication **P**rogramming **I**nterface
SAPS	**S**ervice **A**ccess **P**rocessing **S**ystem
SAR	1. **S**earch **A**nd **R**escue (IMO)
	2. **S**earch **A**nd **R**eplace (softwares)
	3. **S**egmentation **A**nd **R**eassembly (ATM)
	4. **S**ynthetic-**A**perture **R**adar (remote sensing)
	5. **S**pecific **A**bsorption **R**ate (radio wave exposure)

S

	6. **S**urface-**A**rray **R**ecording (optical discs)
SARA	**S**oftware **A**rchitecture for **R**euse in **A**ccess
SARAH	**S**earch **A**nd **R**escue **A**nd **H**oming
SARBE	**S**earch **A**nd **R**escue **Be**acon
SAREX	**S**huttle **A**mateur **R**adio **Ex**periment
SARFT	**S**tate **A**dministration of **R**adio, **F**ilm, and **T**elevision (China)
SARM	**S**et **A**synchronous **R**esponse **M**ode (X.25)
SARP	**S**earch **A**nd **R**escue **P**rocessor
SARPs	**S**tandards **A**nd **R**ecommended **P**ractices and **P**rocedure**s** (ICAO)
SARSAT	**S**earch **a**nd **R**escue **S**atellite-**A**ided **T**racking
SART	**S**earch **A**nd **R**escue **T**ransponder
SARTS	**S**witched **A**ccess **R**emote **T**est **S**ystem
SAS	1. **S**atellite **A**ccess **S**tation (Inmarsat)
	2. **S**urvivable **A**daptive **S**ystem
	3. **S**ubscriber **A**lerting **S**ignal
	4. **S**ingle **A**ttached **S**tation (FDDI)
	5. **S**imple **A**ttachment **S**cheme
	6. **S**ilicon **A**symmetrical **S**witch
	7. **S**everely errored frame/**A**larm indication **S**ignal
	8. **S**witched **A**ccess **S**ervice
SASE	**S**pecific **A**pplication **S**ervice **E**lement
SASG	**S**pecial **A**utonomous **S**tudy **G**roup (ITU-T)
SASI	**S**hugart **A**ssociates **S**ystem **I**nterface
SASL	**S**imple **A**uthentication and **S**ecurity **L**ayer (Internet)
SASMO	**S**yrian **A**rab **S**tandardization and **M**etrology **O**rganization
SASO	**S**audi **A**rabian **S**tandards **O**rganization
SASS	**S**easat-**A** **S**atellite **S**catterometer (remote sensing)
SAT	1. **sat**ellite
	2. **S**ite **A**cceptance **T**ests
	3. **S**tand-**A**lone **T**erminal
	4. **S**ubscriber **A**ccess **T**erminal (SMDS)
	5. **S**upervisory **A**udio **T**one (cellular networks)
SATA	**S**erial **ATA** (specification)
SATAN	**S**ecurity **A**dministrator **T**ool for **A**nalyzing **N**etworks (networking)
SATCOM	**Sat**ellite **Com**munications
SatMex	**Sat**elites **Mex**icanos
SatNav	**Sat**ellite **Nav**igation
satphone	**sat**ellite tele**phone**
SatVOD	**Sat**ellite **V**ideo-**O**n-**D**emand (service)
SAU	**S**ignaling **A**ccess **U**nit

SAW	Surface Acoustic Wave (filter)
SB	1. Synchronous Burst (GSM)
	2. Secondary Body (orbit)
	3. Signal Battery lead
SB-ADPCM	Sub-Band Adaptive Differential PCM (modulation)
SBAN	Services-oriented Building Area Network
SBAS	Satellite-Based Augmentation System (navigation)
SBC	1. Sub-Band Coding
	2. Subsequent Bill Company
	3. Single-Board Computer
SBCA	Satellite Broadcasting and Communications Association (U.S.A.)
SBCF	Single-Bit Cipher Feedback
SBCS	Satellite-Based Cellular System
SBD	Smart Battery Data
SBE	Society of Broadcast Engineers
SBIC	Shared Bus Interface Controller
SBM	1. Shared Buffer Memory (Alcatel)
	2. Subnet Bandwidth Manager (signaling scheme)
SBO	Service-Based Operator (VPNs)
SBR	1. Spaceborne Radar
	2. Subsystem Backup Routing
SBRC	Subsystem Backup Routing Control
SBS	1. Satellite Base Station (Inmarsat)
	2. Smart Battery System
	3. Shift Bit Select
SBTS	Sistema Brasilero de Telecomunicaceos por Satellite (Brazil)
SBUR	Software Blocking/Unblocking Reception
SBus	Sun Microsystems data Bus
SBUS	Software Blocking/Unblocking Sending
SBUV	Solar Backscatter Ultra-Violet Experiment (remote sensing)
SC	1. Service Channel
	2. Sub-Committee (ISO)
	3. Suppressed Carrier (modulation)
	4. scrambler
	5. System Controller
	6. Secondary Center (switching)
SC242	Slot Connector 242
SC330	Slot Connector 330
SCA	1. Selective Call Acceptance
	2. Signal Communications Axis (military)
	3. Supplemental Communications Authority

S

4. **S**ubsidiary **C**ommunications **A**uthorization (FCC regulation)

5. **S**ub-**C**arrier **A**dapter

6. **S**pecial **C**ustomer **A**rrangement (billing)

SCADA   **S**upervisory **C**ontrol **A**nd **D**ata **A**cquisition (remote control)

SCAI   **S**witch to **C**omputer **A**pplications **I**nterface (protocol)

SCAM   **SCSI** **C**onfigured **Auto**matically (scuzzy card devices)

SCAN   **S**witched **C**ircuit **A**utomatic **N**etwork

SCAP   **S**ilicon **C**apacitive **A**bsolute **P**ressure (sensor)

SCAPI   **SCSA** **A**pplication **P**rogramming **I**nterface

SCbus   **SCSA** **bus** (computer architecture)

SCC
1. **S**atellite **C**ontrol **C**enter (Inmarsat)
2. **S**atellite **C**ommunications **C**ontrol System
3. **S**erial **C**ommunications **C**ontroller
4. **S**uper**c**omputing **C**enter
5. **S**pecialized **C**ommon **C**arrier (AT&T)
6. **S**ignal **C**hannel **C**odec
7. **S**ignal **C**ommunications **C**enter
8. **S**mart-**C**ard **C**hip
9. **S**pace **C**ommunications **C**orporation (Japan)
10. **S**audi **C**ommunications **C**ommission (Saudi Arabia)
11. **S**ystem **C**ontrol **C**omputer (CATV)
12. **S**tandards **C**ouncil of **C**anada

SCCE   **S**ervice **C**ircuits **C**ontrol **E**lement

SCCP   **S**imple **C**lient **C**ontrol **P**rotocol

SCCS
1. **S**ecure **C**ode **C**ontrol **S**ystem
2. **S**witching **C**enter **C**ontrol **S**ystem

SCD
1. **S**ignaling **C**hannel **D**escriptor
2. **S**ource-**C**ontrolled **D**rawing (semiconductors)

SCDMA   **S**ynchronous **CDMA** (access)

SCDPI   **SCSA** **D**evice **P**rogramming **I**nterface (computer architecture)

SCE
1. **S**ervice **C**reation **E**nvironment (INs)
2. **S**torage **C**ontrol **E**lement

SCEF   **S**ervice **C**reation **E**nvironment **F**unction (INs)

Scelbi   **S**cientific, **e**lectronic, and **bi**ological (early computers)

SCEP   **S**ervice **C**reation **E**nvironment **P**oint (INs)

SCF
1. **S**hared **C**hannel **F**eedback (wireless)
2. **S**ervice **C**ontrol **F**acility

SCFL   **S**ource-**C**oupled **FET** **L**ogic (digital electronics)

SCH   **S**ynchronization **Ch**annel (GSM)

SCI
1. **S**calable **C**oherent **I**nterface (transmission)
2. **S**erial **C**ommunication **I**nterface

SCI-PHY	**S**aturn-**C**ompliant **I**nterface—**Phy**sical (protocol)
SCIA	**S**mart-**C**ard **I**ndustry **A**ssociation
SCICT	**S**upreme **C**ouncil of **ICT** (Iran)
SCIM	**S**ingle-**C**hip **I**ntegration **M**odule
SCL	1. **S**upervisory **C**ontrol **L**anguage (programming)
	2. **S**pace-**C**harge **L**ayer
	3. **S**erial **C**lock
SCM	1. **S**ubscriber **C**arrier **M**odule
	2. **S**ervice **C**ircuit **M**odule
	3. **S**ingle-**C**hip **M**odule
	4. **S**tation **C**lass **M**ark (cellular networks)
SCMS	1. **S**mart-**C**ard **M**anufacturing **S**ystem
	2. **S**erial **C**opyright **M**anagement **S**ystem (TV broadcasting)
SCOI	**S**tanding **C**ommunications **O**perating **I**nstruction
SCONUL	**S**ociety of **Co**llege, **N**ational and **U**niversity **L**ibraries (U.K. ICT)
scope	1. oscillo**scope**
	2. radar**scope**
SCORM	**S**hareable **C**ontent **O**bject **R**eference **M**odel
SCOT	**S**ealed **C**hips **O**n **T**ape
SCOTS	1. **S**urveillance and **C**ontrol of **T**ransmission **S**ystems
	2. **S**witched **C**ircuit-**O**riented and **T**racking **S**ystem (MCI)
SCP	1. **S**ystem **C**ontrol **P**rocessor
	2. **S**atellite **C**ommunications **P**rocessor (Northern Telecom)
	3. **S**ession **C**ontrol **P**rotocol
	4. **S**ervice **C**ontrol **P**oint (INs)
	5. **S**witching **C**ontrol **P**oint (GSM)
	6. **S**ignal **C**ontrol **P**oint (AINs)
SCPC	**S**ingle **C**hannel **P**er **C**arrier
SCPC-FM	**SCPC F**requency **M**odulation
SCPC-PSK	**SCPC P**hase **S**hift **K**eying (modulation)
SCPI	**S**tandard **C**ommands for **P**rogrammable **I**nstruments
SCR	1. **S**ilicon-**C**ontrolled **R**ectifier (semiconductors)
	2. **S**ustainable **C**ell **R**ate (ATM)
	3. **S**ystem **C**lock **R**eference
	4. **SCSA C**all **R**outer
SCR NRZ	**Scr**ambled **N**on-**R**eturn-to-**Z**ero (data encoding)
SCRL	**S**ignal **C**orps **R**adio **L**aboratories
SCS	1. **S**ilicon-**C**ontrolled **S**witch (semiconductors)
	2. **S**tructured **C**abling **S**ystem
	3. **S**cattering **C**ross **S**ection

S

	4. **S**plit **C**harging **S**ervice
	5. **S**yrian **C**omputer **S**ociety
SCSA	**S**ignal **C**omputing **S**ystem **A**rchitecture
SCSI	**S**mall **C**omputer **S**ystems **I**nterface (scuzzy card)
SCSI3	**SCSI** version **3** (scuzzy card)
SCSR	**S**ingle-**C**hannel **S**ignaling **R**ate
SCSU	**S**upervisory, **C**ontrol & **S**witching **U**nit
SCT	1. **S**ystem **C**ontrol **T**erminal
	2. **S**ecretaria de **C**omunicaciones y **T**ransportes (Mexico)
SCTE	1. **S**ociety of **C**able **T**elevision **E**ngineers, Inc. (standards)
	2. **S**erial **C**lock **T**ransmit **E**xternal (modems)
ScTP	**S**creened **T**wisted **P**air (cabling)
SCTP	1. **S**imple **C**omputer **T**elephony **P**rotocol (Internet)
	2. **S**tream **C**ontrol **T**ransmission **P**rotocol
SCU	1. **S**ystem **C**ontrol **S**ignal **U**nit
	2. **S**ervice **C**hannel **U**nit
	3. **S**ignaling and **C**ontrol **U**nit
	4. **S**ignaling **C**hannel **U**nit
	5. **S**elector **C**ontrol **U**nit
SCVF	**S**ingle-**C**hannel **V**oice **F**requency
SCWD	**S**pontaneous **C**all **W**aiting **D**isplay
SCXI	**S**ignal **C**onditioning **E**xtension for **I**nstrumentation
SD	1. **S**pace-**D**iversity (MUX)
	2. **S**ignal **D**egrade (MUX)
	3. **S**tarting **D**elimiter (LANs)
SDAP	**S**ervice **D**iscovery **A**pplication **P**rofile (Bluetooth)
SDARS	**S**atellite **D**igital **A**udio **R**eceiver **S**ervice
SDBGA	**S**uper-**D**issipation **B**all-**G**rid **A**rray (microchips)
SDC	1. **S**ynchro-to-**D**igital **C**onverter
	2. **S**ecurity **D**atabase **C**omputer
SDCA	**S**hort **D**istance **C**harging **A**rea
SDCC	**S**ection **D**ata **C**ommunication **C**hannel (MUX)
SDCCH	**S**tand-alone **D**edicated **C**ontrol **Ch**annel (GSM)
SDCD	**S**econdary **D**ata **C**arrier **D**etect
SDCM	**S**erial **D**ata **C**onverter **M**odules
SDE	1. **S**ynchronization **D**istribution **E**xpander
	2. **S**ubmission and **D**elivery **E**ntity (X.400)
	3. **S**ecure **D**ata **E**xchange (IEEE)
SDF	1. **S**pecialized **D**atabase **F**unctions
	2. **S**tandard **D**elay **F**ormat
	3. **S**tandard **D**elay **F**ile
	4. **S**ub-**D**istribution **F**rame (trunk cable and LAN)

	5. **S**ervice **D**ata **F**unction
	6. **S**ervice **D**efinition **F**unction
SDH	**S**ynchronous **D**igital **H**ierarchy (MUX)
SDI	1. **S**erial **D**ata **I**nterface
	2. **SD**H physical **I**nterface
	3. **S**ynchronization **D**istribution **I**nterface
SDIO	**S**ecure **D**igital **I**nput/**O**utput
SDIP	**S**hrink **D**ual **I**nline **P**ackage (microchips)
SDK	**S**ystem **D**eveloper's **K**it (ICT)
SDL	**S**pecification and **D**escription **L**anguage (ITU-T)
SDLA	**S**uccessive **D**etection **L**og **A**mplifier
SDLC	1. **S**ynchronous **D**ata **L**ink **C**ontrol (protocol)
	2. **S**oftware **D**escription **L**anguage **C**
SDLLC	**S**ynchronous **D**ata **L**ogical **L**ink **C**ontrol (Cisco)
SDLT	**S**uper **DLT** (tape data format)
SDLVA	**S**uccessive **D**etection **L**og **V**ideo **A**mplifier (microwave)
SDM	1. **S**ystem **D**efinition **M**anual (Inmarsat)
	2. **S**ub-rate **D**ata **M**ultiplexing
	3. **S**pace **D**ivision **M**ultiplexing
	4. **S**uperNode **D**ata **M**anager
SDMA	1. **S**pace **D**ivision **M**ultiple **A**ccess
	2. **S**tation **D**etail **M**essage **A**ccounting
SDMF	**S**ingle **D**ata **M**essage **F**ormat
SDMI	**S**ecure **D**igital **M**usic **I**nitiative
SDN	1. **S**oftware **D**efined **N**etwork
	2. **S**econdary **D**istribution **N**etwork
SDNP	**S**ustainable **D**evelopment **N**etworking **P**rogram (Bangladesh)
SDNS	**S**ecure **D**ata **N**etwork **S**ystem
SDOs	**S**tandard **D**evelopment **O**rganization**s**
SDP	1. **S**ession **D**escription **P**rotocol (IETF)
	2. **S**ervice **D**ata **P**oint (INs)
	3. **S**ervice **D**elivery **P**oint
	4. **S**ervice **D**iscovery **P**rotocol (INs and Bluetooth)
	5. **S**imple **D**iscovery **P**rotocol (Martin Hamilton)
	6. **SAR** **D**ata **P**rovider (COMSAR)
	7. **S**everely **D**isturbed **P**eriod (MUX)
SDPC	**S**ynchronous **D**ata **P**ort **C**luster
SDR	1. **S**oftware **D**efined **R**adio (FCC)
	2. **S**ignal-to-**D**istortion **R**atio
	3. **S**ession **D**etail **R**ecord
	4. **S**pecial **D**rawing **R**ight (billing)
SDRAM	**S**ynchronized **D**ynamic **RAM** (memory)
SDRP	**S**ource **D**emand **R**outing **P**rotocol

S

SDS	1. **S**hort **D**ata **S**ervice (TETRA)
	2. **S**DH **D**ata **S**ub-network
SDSAF	**S**witched **D**igital **S**ervices **A**pplications **F**orum
SDSC	**S**tandards and **D**ata **S**ervices **C**ommittee
SDSL	1. **S**ymmetrical **DSL** (access)
	2. **S**ingle-line **DSL** (access)
SDT	1. **S**tructured **D**ata **T**ransfer
	2. **S**ervice **D**escription **T**able (broadcasting)
SDTI	**S**erial **D**ata **T**ransport **I**nterface (data standard)
SDTP	**S**erial **D**ata **T**ransport **P**rotocol
SDTV	**S**tandard-definition **D**igital **Tele**vision (broadcasting)
SDU	1. **S**ervice **D**ata **U**nit (OSI model)
	2. **S**atellite **D**ata **U**nit (Inmarsat)
´ **SDUS**	**S**mall-scale **D**ata **U**tilization **S**tation (remote sensing)
SDV	**S**witched **D**igital **V**ideo (network architecture)
SDVB	**S**witched **D**igital **V**ideo **B**roadcasting
SDVN	**S**witched **D**igital **V**ideo **N**etwork
SDX	**S**torage **D**ata **Acceleration** (technology)
SDXC	**S**ynchronous **D**igital **Cross** **C**onnect (MUX)
SE	1. **S**witching **E**lement (ATM)
	2. **S**ystem **E**ngineering
	3. **S**oftware **E**ngineering
.sea	Name extension for **s**elf-**e**xtracting Macintosh **a**rchive files
SEA–ME–WE	**S**outh **E**ast **A**sia–**M**iddle **E**ast–**W**est **E**urope (cable consortium)
SEA	**S**elf-**E**xtracting **A**pplication
SEAC	**S**tandard **E**astern **A**utomatic **C**omputer
SEAL	**S**imple and **E**fficient **A**daptation **L**ayer (ATM)
SEC	1. **S**DH **E**quipment **C**lock
	2. **S**ynchronous **E**quipment **C**lock
	3. **S**ecurity **E**ncoding **C**omputer
	4. **S**ervice **E**ntrance **C**able
SECABS	**S**mall **E**xchange **C**arrier **A**ccess **B**illing **S**pecifications
SECAM	**Se**quential **C**olor **A**nd **M**emory (French TV broadcast format)
SECC	**S**ingle **E**dge **C**ontact **C**artridge (processor cartridge)
SECO	**Se**quential **C**ontrol
SECOR	**Se**quential **C**ollation **O**f **R**ange
SECORD	**Se**cure voice **CORD** board (no more exists)
SECS	**S**emiconductor **E**quipment and Material International's **C**ommunications **S**tandard
SECTEL	**Se**cure **Tel**ephone
SED	**S**moke-**E**mitting **D**iode

SEDFB	Surface-Emitting Distributed Feedback (laser)
SEED	Self-Electro-optic Effect Device (switches)
SEF	1. Support Entity Function
	2. Severely Errored Framing (SONET)
	3. Secure Explicit Forwarding (transmission)
SEFS	Severely Errored Framing Seconds
SEI	1. Software Engineering Institute
	2. Secondary Electron Imaging
	3. Shiraz Electronics Industries
SEL	1. selector (ATM)
	2. Surface-Emitting Laser
Selcall	Selective calling
SELED	Surface-Emitting Light-Emitting Diode
Selsyn	Self-synchronous
SELV	Safety Extra Low Voltage
SEM	1. Space Environment Monitor (remote sensing)
	2. Scanning Electron Microscope
	3. Synchronous Equipment Management (MUX)
	4. Standard Electronic Module
SEM-E	Standard Electronic Module E-type (avionics)
SEMF	Synchronous Equipment Management Function (MUX)
SEMI	Semiconductor Equipments and Materials International
SEP	Symbol Error Probability
SEPP	1. Secure Electronic Payment Protocol (bank card transactions)
	2. Software Engineering for Parallel Processing (project)
SEPT	Signaling Endpoint Translator (SS7)
SEQUEL	Structured English Query Language (programming)
SER	Satellite Equipment Room
SERN	Software Engineering Research Network
SERVORD	Service Order
SES	1. Ship Earth Station (Inmarsat)
	2. Satellite Earth Station
	3. Source End Station (ATM)
	4. Severely Errored Seconds (MUX)
SESR	Severely Errored Seconds Ratio (MUX)
SET	1. Secure Electronic Transaction (protocol)
	2. Secure Encryption Technology
	3. Single Electron Transistor
SETA	Southeastern Telecommunications Association
SETAMS	System Engineering, Technical assistance And Management Services

SETI	Search for Extraterrestrial Intelligence
SETS	Synchronous Equipment Timing Source
SEU	Single Event Upset
SEVAS	Secure Voice Access System
SF	1. Single-Frequency (signaling)
	2. Secondary Frequency
	3. Shared Frequency (ITU-T)
	4. Super Frame
	5. Signal Failure (MUX)
SFBI	Shared Frame-Buffer Interface
SFC	1. Switch Fabric Controller
	2. Shannon–Farro Coding (data compression)
SFD	1. Saturation Flux Density
	2. Start Frame Delimiter
SFDR	Spurious Free Dynamic Range (receiver)
SFG	Simulated Facility Group
SFH	Slow Frequency Hopping (GSM)
SFINX	Service for French Internet Exchange
SFM	Synthesizer Filter Module
SFN	Single-Frequency Network (broadcasting)
SFQL	Structured Full-text Query Language (programming)
SFR	Société Française du Radiotéléphone (French GSM operator)
SFROM	Smart Flash ROM (memory)
SFS	Standard Frequency and Time signal Satellite (ITU)
SFT	1. Single-Frequency Tone
	2. System Fault Tolerance
SFTA	Scalable Fault Tolerance Architecture
SFTP	Simple File Transfer Protocol
SFTS	Standard Frequency and Time Signal
SFU	Store-and-Forward Unit (Inmarsat)
SG	1. Super Group (MUX)
	2. Study Group (ITU-T)
	3. Signaling Gateway
	4. Signal Ground
SG2	Study Group 2 (ITU-T)
SGA	Solder-Grid Array (microchips)
SGCC	Shahid Ghandi Communication Cable Company (Iran)
SGCP	Simple Gateway Control Protocol (Telcordia Technologies)
SGDF	Super-Group Distribution Frame
SGI	Silicon Graphics Incorporated (company)
SGLS	Space Ground Link System

.sgm	The MS-DOS/Windows3.x name extension for **SGML** files
.sgml	Name extension for **SGML** files (computer)
SGML	**S**tandard **G**eneralized **M**arkup **L**anguage (programming)
SGMP	**S**imple **G**ateway **M**onitoring **P**rotocol
SGRAM	**S**ynchronous **G**raphics **RAM** (memory)
SGSN	**S**erving **GPRS S**upport **N**ode (cellular networks)
SGW	**S**ervice **G**ate**w**ay
SHA	1. **S**ecure **H**ash **A**lgorithm (digital signature)
	2. **S**idereal **H**our **A**ngle (earth rotation)
	3. **S**ample-and-**H**old **A**mplifier (ADC)
SHARP	**S**elf-**H**ealing **A**lternate **R**oute **P**rotection (SONET)
SHDSL	**S**ingle-pair **H**igh-bit-rate **DSL** (access)
SHF	**S**uper-**H**igh **F**requency (3–30 GHz range)
SHG	**S**econd **H**armonic **G**enerator
SHIELD	**S**ilicon **H**ybrids with **I**nfrared **E**xtrinsic **L**ong-wave-length **D**etector
SHIRAN	**S**-band **H**igh-accuracy **Ran**ging
SHOALS	**S**canning by **H**ydrographic **O**perational **A**irborne **L**idar **S**urvey
SHODOP	**Sho**rt-range **Dop**pler
shoran	**sho**rt-**r**ange **a**id to **n**avigation
SHOT	**S**ociety for the **H**istory **O**f **T**echnology
SHP	**S**ignaling **H**andoff **P**oint (SS7 networks)
SHPD	**S**ample and **H**old **P**hase **D**etector
SHPO	**S**tate **H**istoric **P**reservation **O**fficer
SHR	**S**elf-**H**ealing **R**ing
SHT	1. **S**hort **H**old **T**ime
	2. **S**uppressed **H**igh-level differential **T**ransfer (circuit)
SHTML	**S**erver-parsed **HTML** (programming)
SHVIA	**S**atellite **H**ome **V**iewer **I**mprovement **A**ct
Si	**Si**licon (semiconductors)
Si-APD	**Si**licon **A**valanche **P**hoto **D**iode (semiconductors)
Si–Ge	**Si**licon–**Ge**rmanium (semiconductors)
Si₃N₄	**Si**licon **N**itride (semiconductors)
SI	1. **S**hift-**I**n (character control code)
	2. **S**ervice **I**ndicator (SS7)
	3. **S**ystemé **I**nternational D'unites (French)
	4. **I**nternational **S**ystem of Units
	5. **S**emi-**I**nsulator
SIA	1. **S**atellite **I**ndustry **A**ssociation (standards organization)
	2. **S**ecurities **I**ndustry **A**ssociation
	3. **S**emiconductor **I**ndustry **A**ssociation (U.S.A.)

S

SIBB	Service Independent Building Blocks (Telcordia Technologies)
SIBH	Semi-Insulated Buried Heterostructure (laser)
SiC	Silicon Carbide (semiconductors)
SIC	1. System Interface Cluster
	2. Station Identification Code
	3. Selectively Implanted Collector
	4. Service Initiation Charge
	5. Service Information Compiler (TV broadcast)
	6. Standard Industrial Classification (numbering code)
SICL	Standard Instrument Control Library
SID	1. System Identifier
	2. Security ID
	3. Silence Descriptor (GSM)
	4. Society for Information Display
SIDN	Security Industry Digital Network
SIDR	SES ID Record
SIF	1. Standard Image Format (video images)
	2. SONET Interoperability Forum (U.S. standards organization)
	3. Signaling Information Fields (SS7)
.sig	Name extension for the e-mail signature files (computer)
SIG	1. signal (transmissions)
	2. Special Interest Group (SDR Forum)
	3. SMDS Interest Group
SIGCOMM	Special Interest Group on data Communications
SiGe	Silicon–Germanium (semiconductors)
SIGGRAPH	Special Interest Group on Computer Graphics
SIGINT	Signal Intelligence
SII	System Integration Interface (NTT)
SIIA	Software & Information Industry Association
SILC	Selective Incoming Load Control
SILO	Signal Intercept from Low Orbit
SILS	Standard for Interoperable LAN Security (IEEE)
SIM	1. Subscriber Identity Module (GSM)
	2. Single Interface Module (NEC)
	3. Society for Information Management
SIMD	Single-Instruction, Multiple-Data (processor instruction)
SIMM	Single In-line Memory Module (Macintosh PCs)
SiN	Silicon Nitride (semiconductors)
SINAD	Signal-to-Noise-And-Distortion (ratio)
SINCGARS	Single-Channel Ground and Airborne Radio Systems

SingTel	**Singapore Tel**ecommunications Ltd. (company)
SinoSat	**Sino Sat**ellite Communications (company)
SINPO	**S**ignal strength, **I**nterference, **N**oise, **P**ropagation, and **O**verall quality (signal quality rating system for radio signals)
SINS	**S**hip's **I**nertial **N**avigation **S**ystem
SiO	1. **Si**licon **O**xide (semiconductors)
	2. **S**erial **I**nput and **O**utput (port)
	3. **S**cientific or **I**ndustrial **O**rganization (ITU)
	4. **S**ervice **I**nformation **O**ctet (SS7)
SIP	1. **S**DMS **I**nterface **P**rotocol
	2. **S**ession **I**nitiation **P**rotocol
	3. **S**ingle **I**nline **P**ackage (microchips)
	4. **S**MDS **I**nterface **P**rotocol
SIPO	**S**erial-**I**n **P**arallel-**O**ut (port)
SIPP	1. **S**imple **I**nternet **P**rotocol **P**lus
	2. **S**ingle **I**nline **P**inned **P**ackage (microchips)
SIR	1. **S**ustained **I**nformation **R**ate (SMDS)
	2. **S**ubstrate-**I**ncident **R**ecording
	3. **S**erial **I**nfra**r**ed data transmission system (Hewlett-Packard)
	4. **S**peaker **I**ndependent **R**ecognition
	5. **S**ignal-to-**I**nterference **R**atio
	6. **S**urface **I**nsulation **R**esistance
	7. **S**tepped **I**mpedance **R**esonator
SIS	1. **S**ound-**I**n-**S**ync (transmission)
	2. **S**tandardiseringen **I S**verige (Sweden)
SISO	**S**oft **I**nput–**S**oft **O**utput technique (FEC)
.sit	Name extension for a Macintosh file compressed with **S**tuff**IT**
SIT	1. **S**imple **I**nternet **T**ransition (protocol)
	2. **S**atellite **I**nteractive **T**erminal
	3. **S**pecial **I**nformation **T**one
SITA	**S**ociété **I**nternationale de **T**élécommunication **A**eronautique
SIVR	**S**peaker **I**ndependent **V**oice **R**ecognition
SK Telecom	**S**outh **K**orea **Telecom** (company)
SKAM	**S**kills, **K**nowledge, **A**ccess, and **M**otive
SKED	**Sched**ule (Morse code transmissions)
SKIP	**S**imple **K**ey management for **I**nternet **P**rotocol
SKU	**S**tock-**K**eeping **U**nit (E-Commerce)
SKW	**S**atellite-**K**eeping **W**indow (satcom)
SL	1. **s**ate**l**lite
	2. **S**ynchronous **L**ine

S

	3. stripline
	4. Session Layer
SLA	Service Level Agreement (contracts)
SLAM	Scanning Laser Acoustic Microscope
SLAR	Side-Looking Airborne Radar
SLARP	Serial Line Address Resolution Protocol (Cisco)
SLC	1. Subscriber Loop Carrier
	2. Subscriber Line Charge
	3. Signaling Link Code (SS7)
	4. Straight-Line Capacitance (electronics)
	5. System Life Cycle
	6. Simple Line Code (transmission)
	7. Single Layer Ceramic
	8. Surface Laminar Circuit
SLCC	1. Subscriber Line Carrier Circuit
	2. Subscriber Loop Carrier Circuit
SLD	1. Super Luminescent Diode
	2. Secure Level Debugger
SLDRAM	Sync Link Dynamic RAM (memory)
SLE	Satellite Link Emulator
SLED	1. Single Large Expensive Disk
	2. Single Large Expensive Drive
	3. Superluminescent LED
SLEE	Service Logic Execution Environment (AINs)
SLF	1. Super Low Frequency
	2. Straight-Line Frequency (electronics)
SLI	Service Logic Interpreter (INs)
SLIC	1. Subscriber Line Interface Circuit (VoIP)
	2. Subscriber Line Interface Card
	3. Standard Linear Integrated Circuit (electronics)
Slicc	Slightly larger than IC carrier (microchips)
SLICE	Simulation Language for Integrated Circuit Emphasis
SLIP	Serial Line Internet Protocol
SLIP/PPP	Serial Line Internet Protocol to Point-to-Point Protocol
SLM	1. Service Level Management
	2. System Load Module
	3. Signal Label Mismatched (SONET)
	4. Spatial Light Modulator
	5. Single Longitudinal Mode (laser)
SLMQW	Strained Layer Multiple Quantum Well
SLOs	Service Level Obligations (contracts)
SLP	1. Service Location Protocol (IETF)
	2. Service Logic Program
SLR	1. Send Loudness Rating

	2. **S**ource **L**ocal **R**eference (SS7)
	3. **S**ervice **L**evel **R**eporter
SLS	1. **S**ignaling **L**ink **S**election code (SS7)
	2. **S**oftlanding **L**inux **S**ystem
SLSA	**S**ingle-**L**ine **S**witching **A**pparatus
SLSI	**S**uper **L**arge-**S**cale **I**ntegration (microchips)
SLT	1. **S**ingle-**L**ine **T**elephone
	2. **S**ubscriber **L**ine **T**erminal
	3. **S**ubscriber **L**ine **T**est
SLTE	**S**ubscriber **L**ine **T**erminal **E**quipment
SLW	**S**traight-**L**ine **W**avelength (electronics)
SM	1. **S**witch **M**odule
	2. **S**ervice **M**odule (ADSL)
	3. **S**ingle **M**ode (optical fiber)
	4. **S**ynchronous **M**ultiplexer (MUX)
	5. **S**pectrum **M**anagement (ITU-T)
	6. **S**ession **M**anagement
SMA	1. **S**pectrum **M**anagement **A**gency
	2. **S**emiconductor **M**anufacturers **A**ssociation (U.K.)
	3. **S**urface-**M**ounted **A**ssembly
SMAE	**S**ystem **M**anagement **A**pplication **E**ntity
SMAP	**S**ervice **M**anagement **A**ccess **P**oint (INs)
SMART	1. **S**elf-**M**onitoring **A**nalysis and **R**eporting **T**echnology
	2. **S**tress-**M**arginality and **A**ccelerated-**R**eliability **T**esting
	3. **S**tandard **M**odule **A**vionics **R**epair and **T**est (software)
SMAS	**S**witched **M**aintenance **A**ccess **S**ystem
SMASE	**S**ystem **M**anagement **A**pplication **S**ervice **E**ntity
SMATV	**S**atellite **M**aster-**A**ntenna **T**ele**v**ision
SMB	1. **S**erver **M**essage **B**lock (networking protocol)
	2. **S**hared-**M**emory **B**uffer
SMBus	**S**ystem **M**anagement **Bus** (Intel standard)
SMC	1. **S**ystem **M**anagement **C**enter
	2. **S**tandard **M**anagement **C**ommittee
	3. **S**urface-**M**ount **C**omponents (electronics)
	4. **S**AR **M**ission **C**oordinator (COMSAR)
	5. **S**ubscriber **M**anagement **C**enter
SMCC	**S**un **M**icrosystems **C**omputer **C**ompany
SMD	1. **S**urface-**M**ount **D**evice (electronics)
	2. **S**older **M**ask **D**efined
	3. **S**torage **M**odule **D**rive
SMDA	**S**tation **M**essage **D**etail **A**ccounting
SMDI	1. **S**tation **M**essage **D**esk **I**nterface (data link)
	2. **S**implified **M**essage **D**esk **I**nterface (data link)

S

SMDR	Station Message Detail Recording
SMDS	Switched Multimegabit Data Services (MANs)
SMDSU	Switched Multimegabit Data Service Unit
SME	1. Signaling Management Equipment
	2. Security Management Entity
	3. Small-to-Medium Enterprise
SMEMA	Surface-Mount Equipment Manufacturers Association
SMF	1. Service Management Function (TMN)
	2. Single-Mode Fiber
	3. Sub-Multiframe (MUX)
SMFA	Specific Management Function Area
SMG	1. Super Master Group (MUX)
	2. Special Mobile Group (former name of GSM)
SMI	1. Structure of Management Information
	2. Small- and Medium-sized Industries
	3. System Management Interrupts (power)
	4. Sample Matrix Inversion
SMII	Serial Media Independent Interface
SMIL	Synchronized Multimedia Integrated Language (programming)
SMIP	Strategic Management Information Plan
SMIS	Society for Management Information Systems
SMM	System Management Mode (power)
SMMR	Scanning Multichannel Microwave Radiometer
SMMS	Spatial Metadata Management System
SMN	Synchronous Management Network
SMO	Stabilized Master Oscillator
SMOBC	Solder Mask Over Bare Copper
SMOP	Small Matter Of Programming
SMP	1. Service Management Point (INs)
	2. Symmetric Multiprocessing (ICT)
	3. SCSA Message Protocol
	4. Simple Management Protocol
	5. Sub-Miniature Push-on (connectors)
	6. Surface-Mount Package (microchips)
SMPGA	Surface-Mount Pin-Grid Array (microchips)
SMPI	SCSA Message Protocol Interface
SMPP	Short Message Peer-to-Peer (networking protocol)
SMPS	Switched-Mode Power Supply
SMPTE	Society of Motion Picture and Television Engineers
SMPTE	Society of Motion Picture and Television Engineers (U.S.A.)
SMR	Specialized Mobile Radio
SMRP	Simple Multicast Routing Protocol

SMRT	1. Single Message-unit Rate Timing
	2. Surface-Mount and Reflow Technology
SMS	1. Short Message Service (GSM)
	2. Satellite Multiservice
	3. Subscribers Management System
	4. Service Management System (INs)
	5. SDH Management Sub-network
SMS-MO	SMS Mobile Originated (GSM)
SMS-MT	SMS Mobile Terminated (GSM)
SMS-PP	SMS Point-to-Point (GSM)
SMSA	Standard Metropolitan Statistical Area (FCC)
SMSC	Short Message Service Center (GSM)
SMSCB	SMS Cell Broadcast (GSM)
SMSCH	Short Message Service Channel (GSM)
SMSK	Serial MSK (modulation)
SMSR	Side-Mode Suppression Ratio (fiber optics)
SMT	1. Surface-Mount Technology (electronics)
	2. Simultaneous Multitasking
	3. Station Management (FDDI Architecture)
SMTA	1. Single-line Multiextension Telephone Apparatus
	2. Surface-Mount Technology Association
SMTP	Simple Mail Transfer Protocol (Internet)
SMU	1. Subscriber Carrier Module-100 URBAN
	2. Surface-Measure Unit
SMU-R	Subscriber Carrier Module-100 URBAN—Remote
SMUG	Spokane Microcomputer User Group
SN	1. Switching Node
	2. Services Node (AINs)
	3. Subscriber Number
	4. Sequence Number (ATM)
SNA	1. Systems Network Architecture (IBM)
	2. Scalar Network Analyzer
	3. SDH Network Aspects (MUX)
SNAC	1. Single Network Access Code (Inmarsat)
	2. SMS/800 Number Administration Committee
SNACP	SNA Control Protocol
SNADS	SNA Distribution Services (IBM)
SNAP	1. Sub-Network Access Protocol (IEEE)
	2. Systems for Nuclear Auxiliary Power
SNC	Sub-Network Connection (MUX)
SNC/I	Sub-Network Connection with Inherent Monitoring (MUX)
SNC/N	Sub-Network Connection with Non-intrusive Monitoring (MUX)

S

SNCP	1. **S**atellite **N**etwork **C**ontrol **P**rocessor
	2. **S**ub-**N**etwork **C**onnection **P**rotection (MUX)
.snd	Name extension for **S**un a**nd** Next Machine audio files
SND	**s**e**nd** (cellular language)
SNDCF	**S**ub-**N**etwork **D**ependent **C**onvergence **F**unction
SNDCP	**S**ub-**N**etwork **D**ependent **C**onvergence **P**rotocol
SNEK	**S**atellite **N**etwor**k** node computer
SNES	**S**pecial **N**etwork **E**xchange **S**mall
SNet	**S**ONET **Net**work (Northern Telecom)
SNET	**S**outhern **N**ew **E**ngland **T**elephone Corporation
SNG	**S**atellite **N**ews **G**athering system
SNI	1. **S**ystem **N**etwork **I**nterconnection (IBM)
	2. **S**ubscriber **N**etwork **I**nterface (SMDS)
´ **SNIR**	**S**ignal-to-**N**oise plus **I**nterference **R**atio (signal quality)
SNM	**S**ub-**N**etwork **M**anagement
SNMP	**S**imple **N**etwork **M**anagement **P**rotocol (networking)
SNOBOL	**S**tring-**O**riented Sym**bo**lic **L**anguage (programming)
SNOC	**S**atellite **N**etwork **O**perations **C**enter
SNPA	**S**ub-**N**etwork **P**oint of **A**ttachment
SNPP	**S**imple **N**etwork **P**aging **P**rotocol (IETF)
SNR	1. **S**ignal-to-**N**oise **R**atio (signal quality)
	2. **S**ub-**N**etwork **R**outer
	3. **S**aved **N**umber **R**edial
SNRM	**S**et **N**ormal **R**esponse **M**ode (command)
SNVT	**S**tandard **N**etwork **V**ariable **T**ype
SO	1. **S**hift **O**ut (character control code)
	2. **S**erving **O**ffice
	3. **S**erial **O**utput
	4. **S**mall-**O**utline
SO-DIMM	**S**mall **O**utline-**D**ual **I**n-line **M**emory **M**odule (Maintosh PCs)
SOA	**S**emiconductor **O**ptical **A**mplifier
SOAP	**S**imple **O**bject **A**ccess **P**rotocol
SOB	**S**mall-**O**utline **B**utt-leaded package (microchips)
SOC	1. **S**ystem **O**perator **C**onsole
	2. **S**tatement **O**f **C**ompliance
	3. **S**tate **O**f **C**harge
	4. **S**ystem-**O**n-a-**C**hip (microchips)
SoD	**S**ervice-**o**n-**D**emand
SOD	**S**ource-**O**ver-**D**rain (FETs)
SOF	**S**tart **O**f **F**rame (TDMA)
SOFAR	**So**und **F**ixing **A**nd **R**anging
SOG	**S**ource-**O**ver-**G**ate (FETs)
SOH	1. **S**tart **O**f **H**eader (control character)

	2. **S**ection **O**ver**h**ead (MUX)
SOHO	1. **S**mall **O**ffice, **H**ome **O**ffice (computer market)
	2. **So**lar and **H**eliospheric **O**bservatory
SOI	1. **S**ignal **O**perating **I**nstruction
	2. **S**ilicon-**O**n-**I**nsulator (semiconductors)
	3. **S**econd-**O**rder **I**ntercept
SOIC	**S**mall-**O**utline **I**ntegrated **C**ircuit package (microchips)
SOJ	**S**mall-**O**utline ICs with **J**-leads (microchips)
SOLA	**S**hort, **O**pen, **L**oad, and **A**ir-capacitor
SOLAS	**S**afety **O**f **L**ife **A**t **S**ea (IMO convention)
Solion	**Sol**ution **ION** (electrochemical device)
SOLR	**S**hort-**O**pen-**L**ine-**R**eciprocal (calibration)
SOLT	**S**hort-**O**pen-**L**oad-**T**hrough (calibration)
SOLV	**Sol**enoid **V**alue (electronics)
SOM	1. **S**tart **O**f **M**essage
	2. **S**ystem **O**bject **M**odel (IBM architecture)
SOMO	**S**mall **O**ffice, **M**edium **O**ffice (computer market)
SON	**S**ervice **O**rder **N**umber (local exchange carrier)
sonar	**so**und **n**avigation **a**nd **r**anging
SONET	**S**ynchronous **O**ptical **Net**work (standard)
SONIA	**S**un Microsystems, **O**racle Corp., **N**etscape Corp., **I**BM, **A**pple, Inc.
SOP	1. **S**tandard **O**perating **P**rocedure
	2. **S**mall-**O**utline **P**ackage (microchips)
SOPC	**S**ystem **O**n a **P**rogrammable **C**hip (microchips)
SOR	**S**tart **O**f **R**ecord
SORA	**S**atellite-**O**riented **R**esource **A**llocation
SORF	1. **S**tart **O**f **R**eceive **F**rame
	2. **S**mall-**O**utline **R**adio **F**requency (microchips)
SORMF	**S**tart **O**f **R**eceive **M**ultiframe
SOS	1. **S**ave **O**ur **S**ouls or Ship (international distress signal)
	2. **S**ilicon-**O**n-**S**apphire (semiconductors)
SOSCARD	**S**ecure **O**perating **S**ystem **CARD**
SOSIC	**S**ilicon-**O**n-**S**apphire **I**ntegrated **C**ircuit (semiconductors)
SOSTEL	**So**lid-**St**ate **E**lectronic **L**ogic (digital electronics)
SOSUS	**So**und **Su**rveillance **S**ystem
SOT	**S**mall-**O**utline **T**ransistor (semiconductors)
SOTF	**S**tart **O**f **T**ransmit **F**rame
SOTMF	**S**tart **O**f **T**ransmit **M**ultiframe
SOVA	**S**oft-**O**utput **V**iterbi **A**lgorithms
SP	1. **S**ervice **P**rovider
	2. **S**upport **P**rocessor
	3. **S**tream **P**rotocol

S

	4. **S**ignal **P**rocessor
	5. **S**pan **P**rocessor
	6. **S**ignaling **P**oint
	7. **S**ignal **P**resent
	8. **S**witch **P**ort (Ericson)
SP3	**Third-order S**uppression
SP3T	Single-**P**ole **3** **T**hrow (switches)
SPA	1. **S**pecial **P**hotographic **A**rea
	2. **S**oftware **P**ublishers **A**ssociation (U.S. standards organization)
	3. **S**hared **P**rinter **A**ccess (ISDN)
Spacecom	**S**pace **Com**munications Corporation
SPACH	**S**MS messaging, **P**aging, and **A**ccess response **Ch**annel
SPADATS	**Spa**ce **D**etection **A**nd **T**racking **S**ystem
SPADE	**S**CPC **P**CM Multiple **A**ccess **D**emand—assignment **E**quipment
SPAG	**S**tandard **P**romotion and **A**pplication **G**roup (Europe)
SPAM	**Sp**ecialized **A**utomated **M**ail (e-mail)
SPAN	**S**witched **P**ort **An**alyzer (Cisco)
SPAP	**S**ecure **P**assword **A**uthentication **P**rotocol
SPARC	**S**calable **P**rocessor **Arc**hitecture (Sun Microsystems)
SPARS	**S**ociety of **P**rofessional **A**udio **R**ecording **S**tudios
SPATA	**Sp**eech **A**nd d**ata**
SPB	1. **S**atellite **P**rocessor **B**oard
	2. **S**ynchronous **P**ort **B**lock (MUX)
SPBX	**S**atellite **P**rivate **B**ranch E**x**change
SPC	1. **S**tored **P**rogram **C**ontrol (telephone system)
	2. **S**emi-**P**ermanent **C**onnection
	3. **S**ignaling **P**oint **C**ode (SS7)
	4. **S**tatistical **P**rocess **C**ontrol
SPCAS	**SPC A**llocation **S**ervice
SPCM	**S**ubscriber **PCM** (modulation)
SPCP	**Sp**ectrum **C**ellular error-correction **P**rotocol
SPCS	**S**tored **P**rogram **C**ontrolled **S**witch
SPD	1. **S**ecurity **P**roduction **D**evelopment
	2. **S**erial **P**resence **D**etect (SDRAM memories)
	3. **sp**rea**d**er
SPDIP	**S**hrink **P**lastic **D**ual **I**nline **P**ackage (microchips)
SPDM	**S**calable **P**olynomial **D**elay **M**odel
SPDT	Single-**P**ole **D**ouble-**T**hrow (switches)
SPDU	**S**ession **P**rotocol **D**ata **U**nit
SPE	1. **S**emi-**P**ublic **E**nvironment
	2. **S**witch **P**rocessing **E**lement
	3. **S**ignal **P**rocessing **E**lement

	4. **S**ynchronous **P**ayload **E**nvelope (MUX)
spec	**spec**ification
SPEC	1. **S**peech **P**redictive **E**ncoded **C**ommunication
	2. **S**tandard **P**erformance **E**volution **C**orporation
SPECT	**S**ingle-**P**hoton **E**mission-**C**omputed **T**omography (medicine)
SPF	1. **S**hortest **P**ath **F**irst (routing algorithm)
	2. **S**tateful **P**acket **F**iltering
SPG	**S**ignal **P**rocessing **G**ain
SPHIGS	**S**imple **PHIGS**
SPI	1. **S**ervice **P**rovider **I**nterface
	2. **S**erial **P**eripheral **I**nterface (Motorola)
	3. **S**erial-to-**P**arallel **I**nterface (electronics)
	4. **S**ecurity **P**arameters **I**ndex
	5. **S**ynchronous **P**ublic **I**nput
	6. **S**ynchronous **P**ixel **I**nterface
	7. **S**DH **P**hysical **I**nterface (MUX)
	8. **S**tateful **P**acket **I**nspection (firewall monitoring system)
Spice	**S**imulation **p**rogram with **i**ntegrated **c**ircuits **e**mphasis
SPID	1. **S**ervice **P**rofile **Id**entifier (ISDN)
	2. **S**ervice **P**rotocol **Id**entifier (ISDN)
SPIE	1. **S**ociety of **P**hotometric **I**ndustry **E**ngineers
	2. **S**ociety of **P**hoto-optical **I**nstrumentation **E**ngineering
SPIN	**S**ervice **P**rovider **I**dentification **N**umber
SPINA	**S**ubscriber **P**ersonal **I**dentification **N**umber **A**ccess
SPINI	**S**ubscriber **P**ersonal **I**dentification **N**umber **I**ntercept
SPIRIT	1. **S**pecial **P**urpose **I**ntegrated **R**emote **I**ntelligence **T**erminal
	2. **S**ecure **P**lanning and **I**ntegrated **R**esources for **I**nformation **T**echnology (PHS of Japan)
SPIROU	**S**ignalization **P**our l' **I**nterconnexion des **R**eseaux **Ou**verts
SPISN	**S**ecurity **P**rotocols to explore security **I**ssues in **S**ensor **N**etworks
SPITE	**S**witching **P**rocessing **I**nterface **T**elephony **E**vent
SPKR	**sp**ea**k**e**r**
SPL	**S**ound **P**ressure **L**evel (sound power)
SPLD	**S**imple **P**rogrammable **L**ogic **D**evice (digital electronics)
SPLIC	**S**pecial **P**urpose **L**inear **I**ntegrated **C**ircuit (electronics)
SPM	1. **S**elf-**P**hase **M**odulation
	2. **S**atellite **P**ropulsion **M**odule
	3. **S**pectrum **P**eripheral **M**odule

S

	4. Service Provider Messages
	5. Subscriber Private Meter
	6. Scanning Probe Microscope
SPMT	Single-Pole Multiple-Throw (switches)
SPN	1. Service Provider Networks
	2. Subscriber Premises Network
SPNA	Secure Public Networks Act (U.S. Senate)
SPNE	Signal Processing Network Equipment
SPNI	Service Provider Network Identifier
SPOC	1. SAR Point Of Contact (COMSAR)
	2. Single-Point-Of-Contact
SPOI	Signaling Point Of Interface (SS7)
SPOOL	Simultaneous Peripheral Operations Online
SPOT	1. Sequential Proxy Optimization Technique
	2. Secondary-location Point of Termination
	3. Smart Personal Object Technology (Microsoft)
SPP	1. Service Provision Point (IN)
	2. Scalable Parallel Processing (multiprocessing architecture)
	3. Sequenced Packet Protocol
	4. Signal Processing Platform
	5. Serial Port Profile (Bluetooth)
SPRE	1. Special Prefix code
	2. Satellite Position Reporting Equipment
SPS	1. Standard Positioning Service (GPS)
	2. Signaling Protocol and Switching
	3. Sound-Processing Software
	4. Standby Power System
	5. Samples Per Second
SPST	Single-Pole Single-Throw (switches)
SPT	Stationary Plasma Thruster
SPTS	Single-Program Transport Stream (MPEG-2)
SPU	Satellite Position Uncertainty
SPUD	Small Planned Unit Development
SPUDT	Single-Phased Uni-Directional Transducer
SPX	Sequenced Packet Exchange (protocol)
Sq	Square (SI Units)
SQAM	1. Staggered Quadrature Amplitude Modulation
	2. Sound Quality Assessment Material
SQC	Statistical Quality Control (products)
SQCIF	Sub-Quarter Common Intermediate Format (video compression)
SQE	Signal Quality Error (Ethernet)
SQFP	Shrink Quad Flat Pack (microchips)

SQL	**S**tructured **Q**uery **L**anguage (databases)
SQPSK	**S**taggered **Q**uadrature **PSK** (modulation)
SQUID	**S**uperconducting **Qu**antum **I**nterference **D**evice
sr	**s**te**r**adian
SR	1. **S**et and/or **R**eset (flip-flops)
	2. **S**canning **R**adiometer
	3. **S**pectral **R**esponsivity
	4. **S**hift **R**egister (digital electronics)
	5. **S**econdary **R**adar
	6. **S**peech **R**ecognition
	7. **S**ource **R**outing
	8. **S**elective **R**epeat
SRA	**S**ystem **R**eliability **A**rchitecture
SRAM	**S**tatic **RAM** (memory)
SRAPI	**S**peech **R**ecognition **A**pplication **P**rogramming **I**nterface
SRB	1. **S**econdary **R**eference **B**urst
	2. **S**ource **R**oute **B**ridging (LANs)
	3. **S**ynchronization **R**eference **B**urst
SRC	1. **S**trategic **R**eview **C**ommittee (ETSI)
	2. **S**tupid **R**ich **C**ustomer (telecom products)
	3. **S**emiconductor **R**esearch **C**orporation
SRD	**S**uper-**R**adiant **D**iode
SRD Industries	**S**aamaaneh **R**ahe **D**our **Industries** (Iran)
SRDC	**S**ub-**R**ate **D**igital **C**ross connect
SRDM	**S**ub-**R**ate **D**ata **M**ultiplexer
SRE	**S**end **R**eference **E**quivalent
SREJ	**S**elective **Rej**ect
SRES	**S**igned **Res**ponse (GSM)
SRF	1. **S**pecial **R**esource **F**unction (AINs)
	2. **S**pecifically **R**outed **F**rame
SRGF	**S**peech **R**ecognition **G**rammar **F**ormat
SRGP	**S**imple **R**aster **G**raphics **P**ackage (software)
SRI	**S**o**rr**y (Morse code transmission)
SRL	1. **S**tructured **R**eturn **L**oss (cable)
	2. **S**hift **R**egister **L**atch (microchips)
SRM	1. **S**ub-**R**ate **M**ultiplexing
	2. **S**ervice **R**esource **M**anagement
	3. **S**elective **R**inging **M**odule
	4. **S**ignaling **R**oute **M**anagement
SRMS	**S**ervice **R**equest **M**anagement **S**ystem
SRO	**S**tandards **R**elated **O**rganization
SRP	1. **S**patial **R**euse **P**rotocol (fiber optics)
	2. **S**ource **R**outing **P**rotocol

S

	3. **S**uggested **R**etail **P**rice
SRR	**S**earch and **R**escue **R**egion (COMSAR)
SRS	1. **S**pace **R**esearch **S**ervice
	2. **S**tatistical **R**epository **S**ystem
	3. **S**hared **R**egistry **S**ystem
	4. **S**timulate **R**oman **S**cattering (optical fiber)
SRSC	**S**ilkroad **R**efractive **S**ynchronization **C**ommunication
SRT	1. **S**tation **R**inging **T**ransfer
	2. **S**ource **R**oute **T**ransparent bridging (Ethernet, Token Ring)
SRTS	1. **S**ynchronous **R**esidual **T**ime **S**tamp (clock recovery)
	2. **S**econdary **R**equest **T**o **S**end
SRTT	**S**aco **R**iver **T**elegraph & **T**elephone Company
SRU	**S**earch and **R**escue **U**nit (COMSAR)
SS	1. **S**upplementary **S**ervices (GSM)
	2. **S**pread **S**pectrum
	3. **S**atellite **S**witching
	4. **S**election **S**ignal
	5. **S**ubscriber **S**tation
SS-CDMA	**S**pread-**S**pectrum **CDMA** (access)
SS-LORAN	**S**kywave-**S**ynchronized **LORAN**
SS-TDMA	1. **S**pread-**S**pectrum **TDMA** (access)
	2. **S**atellite-**S**witched **TDMA** (access)
SS1	**S**ignaling **S**ystem Number **1**
SS2	**S**ignaling **S**ystem Number **2**
SS7	**S**ignaling **S**ystem Number **7**
SSA	**S**erial **S**torage **A**rchitecture (IBM)
SSAC13	**S**ignaling **S**ystem **A**lternating **C**urrent No. **13**
SSAC15	**S**ignaling **S**ystem **A**lternating **C**urrent No. **15**
SSAP	**S**ession **S**ervice **A**ccess **P**oint
SSAS	**S**hip **S**ecurity **A**larm **S**ystem (IMO)
SSAT	**S**ystem **S**oftware **A**cceptance **T**est
SSB	1. **S**ingle**S**ide**b**and (modulation)
	2. **S**uperframe **S**hort **B**urst
SSB-SC	**S**ingle-**S**ide**b**and **S**uppressed **C**arrier (modulation)
SSB/AM	**S**ingle-**S**ide**b**and operation **with** **A**mplitude Modulation
SSBW	**S**urface-**S**kimming **B**ulk **W**ave
SSCC	**S**elective **S**equence **C**ontrol **C**omputer (IBM)
SSCF	**S**ervice **S**pecific **C**oordination **F**unction (ISDN)
SSCOP	**S**ervice **S**pecific **C**onnection-**O**riented **P**rotocol (ATM)
SSCP	1. **S**ervice **S**witching and **C**ontrol **P**oint (INs)
	2. **S**ystem **S**ervices **C**ontrol **P**oint (networking)
SSCS	**S**ervice **S**pecific **C**onvergence **S**ublayer (ATM)
SSCSG	**S**pread-**S**pectrum **C**ode **S**equence **G**enerator

SSD	1. **S**hared **S**ecret **D**ata
	2. **S**ervice **S**election **D**ashboard (Cisco)
	3. **S**imultaneous **S**ignal **D**etection
	4. **S**olid-**S**tate **D**isk
	5. **S**olid-**S**tate **D**rive
	6. **S**top–**S**tart **D**istortion
SSDA	**S**equential **S**imilarity **D**etection **A**lgorithm (remote sensing)
SSDH	**S**atellite **SDH**
SSDR	**S**top–**S**tart **D**istortion **R**atio
SSE	**S**treaming **S**IMD **E**xtensions (Intel processors)
SSET	**S**witching **S**ystem **E**xchange **T**ermination
SSF	1. **S**ervice **S**witching **F**unction (INs)
	2. **S**ub-**S**ervice **F**ield (SS7)
SSFDC	**S**olid-**S**tate **F**loppy **D**isk **C**ard
SSG	**S**mall-**S**ignal **G**ain
SSH	1. **S**ecure **S**hell
	2. **S**ecurity **S**erver **H**ost
	3. **S**imultaneous **S**ample-and-**H**old (data acquisition)
SSI	1. **S**mall-**S**cale **I**ntegration (microchips)
	2. **S**erver **S**ide **I**ncludes (code)
SSID	**S**hips' **S**tation **Id**entification number
SSL	**S**ecure **S**ockets **L**ayer (Internet)
SSM	1. **S**econd **S**urface **M**irror
	2. **S**ynchronous **S**tatus **M**essage
SSMA	**S**pread-**S**pectrum **M**ultiple **A**ccess
SSMF	**S**tandard **S**ingle-**M**ode **F**iber
SSML	**S**peech **S**ynthesis **M**arkup **L**anguage (programming)
SSMM	**S**olid-**S**tate **M**ass **M**emory
SSMP	**S**imple **S**creen **M**anagement **P**rotocol
SSMS	**S**pace **S**pectrum **M**onitoring **S**ystem
SSN	1. **S**ub**s**ystem **N**umber (SS7)
	2. **S**tation **S**erial **N**umber
SSO	1. **S**ingle **S**ign-**O**n (networking)
	2. **S**un **S**ynchronous **O**rbit (Sun Microsystems)
SSOG	**S**atellite **S**ystem **O**perations **G**uide (Intelsat)
SSOP	**S**hrink **S**mall-**O**utline **P**ackage (microchips)
SSP	1. **S**ervice **S**witching **P**oint (INs)
	2. **S**ignal **S**witching **P**oint (AINs)
	3. **S**torage **S**ervice **P**rovider
	4. **S**ervice **S**ignaling **P**oint
SSPA	**S**olid-**S**tate **P**ower **A**mplifier
SSPE	**S**ingle **S**olid-**P**hase **E**pitaxy
SSPI	**S**ociety of **S**atellite **P**rofessionals **I**nternational (U.S.A.)

S

SSR	1. **S**ES **S**tatus **R**ecord
	2. **S**olid-**S**tate **R**elay (electronics)
	3. **S**econdary **S**urveillance **R**adar
	4. **S**urface **S**earch **R**adar
	5. **S**calable **S**ampling **R**ate
SSRM	**S**pace **S**egment **R**esources **M**anagement
SSRP	**S**imple **S**erver **R**edundancy **P**rotocol
SSRT	**S**ub-**S**econd **R**esponse **T**ime
SSS	**S**witching **S**ubsystems
SST	1. **S**pread-**S**pectrum **T**echnology
	2. **S**olid-**S**tate **T**echnology
	3. **S**imultaneous **S**elf-**T**est
	4. **S**tandard **S**tar **T**racker (remote sensing)
SSTL	1. **S**tub-**S**eries-**T**erminated **L**ogic (digital electronics)
	2. **S**urrey **S**atellite **T**echnologies **L**td (U. K.)
SSTO	**S**ingle **S**tage **T**o **O**rbit
SSTP	**S**witched **S**ervices **T**ransport **P**rotocol
SSTV	**S**low-**S**can **T**ele**v**ision
SSU	1. **S**ubsequent **S**ignal **U**nit
	2. **S**ynchronization **S**ource **U**nit
	3. **S**tratospheric **S**ounding **U**nit
	4. **S**ession **S**upport **U**tility
SSUC	**S**pace **S**egment **U**tilization **C**harge
SSUPS	**S**olid-**S**tate **UPS** (power)
SSUS	**S**olid **S**panning **U**pper **S**tage
SSV	**S**ub-**S**uper**v**isory equipment
ST	1. **S**ignaling **T**erminal
	2. **S**ystem **T**est
	3. **st**art (signal)
	4. **S**tuffing **T**able (broadcasting)
	5. **S**cheduled **T**ransfer
	6. **S**traight **T**ip (connectors)
ST2+	**St**ream protocol version **2+**
ST-Bus	**S**erial **T**elecom **Bus**
STA	1. **S**panning **T**ree **A**lgorithm (IEEE)
	2. **S**cience and **T**echnology **A**gency
	3. **S**pecial **T**emporary **A**uthority
STALO	**Sta**blized **L**ocal **O**scillator (electronics)
STAR	1. **S**peech **T**echnology **A**nd **R**esearch Laboratory
	2. **S**pecial **T**elecommunications **A**ction for **R**egional Development
STAR TAP	**S**cience, **T**echnology, **A**nd **R**esearch **T**ransit **A**ccess **P**oint
STATMUX	**Stat**istical **Mu**ltiple**x**er

STB	Set-Top Box (integrated TV receiver decoder)
STC	1. Sub-Technical Committee (ETSI)
	2. Society of Telecommunications Consultants
	3. Saudi Telecoms Company (Saudi Arabia)
	4. System Time Clock
	5. Sensitivity Time Control
	6. Supplemental Type Certification (aviation)
STD	1. Subscriber Trunk Dialing (telephone exchange)
	2. Synchronous Time Division
	3. standard
	4. State Transition Diagram
STDM	1. Statistical Time-Division Multiplexing (transmission)
	2. Synchronous Time-Division Multiplexing (transmission)
STE	1. Spanning Tree Explorer
	2. Special Test Equipment
	3. Section Terminating Equipment (SONET)
	4. Syrian Telecommunications Establishment (Syria)
STECS	Synchronous Terminal Equipment Control System
STEP	Short-Term Error Performance
STEREO	Solar Terrestrial Relations Observatory
STFS	Standard Time and Frequency Signal (service)
STG	Scale To Gray (graphics)
STI	1. Secondary Terrestrial Interface
	2. Speech Transmission Index
STICI	Self-Teaching Imperative Communicating Interface (E-learning)
STID	Service Termination Identifier (ISDN)
STK	SIM Toolkit (GSM)
STL	1. Standard Template Library (C++ programming)
	2. Studio-to-Transmitter Link (broadcast)
	3. Schottky Transistor Logic (digital electronics)
	4. Standard Telegraph Level
STM	1. Synchronous Transport Module (MUX)
	2. Service Traffic Management
STM-0	Synchronous Transport Module level-0 (51.48 Mb/s MUX)
STM-1	Synchronous Transport Module level-1 (155.52 Mb/s MUX)
STM-4	Synchronous Transport Module level-4 (622.080 Mb/s MUX)
STM-16	Synchronous Transport Module level-16 (2488.32 Mb/s MUX)

S

STM-N	Synchronous Transport Module level-**N**
STM-NC	Synchronous Transport Module level-**N** Concatenated
STM-RR	Synchronous Transport Module for **R**adio-**R**elay (MUX)
STMR	Sidetone Masked loudness Rating
STN	Super-Twisted Nematic display (LCD)
STOP	Satellite Tracking of Polluters
STP	1. Shielded Twisted Pair (cable)
	2. Signaling Transfer Point (SS7)
	3. Spanning Tree Protocol (Ethernet switching)
STP-A	Shielded Twisted Pair-A
STPC	Southwest Technical Products Corporation
STR	Symbol Timing Recognition
STRADIS	Structured Analysis, Design, and implementation of Information Systems
STRICOM	Simulation, Training, and Instrumentation Command (U.S. Army)
STRIFE	Stress and life (testing)
STS	1. Satellite Transportation System
	2. Space Transportation System
	3. Synchronous Transport Signal (SONET)
	4. Shared Tenant Services
STS-1	Synchronous Transport Signal 1 (SONET)
STS-3	Synchronous Transport Signal 3 (SONET)
STS-n	Synchronous Transport Signal n (SONET)
STSC	1. Software Technology Support Center
	2. Station-To-Station Connectivity
	3. Scandinavian Telecommunications Satellite Comittee
STSK	Scandinavian Telecommunications Satellite Committee
STSP	Solar Terrestrial Science Programme (remote sensing)
STT	1. Secure Transaction Technology (ICT)
	2. Set-Top Terminal
STTL	Schottky Transistor-to-Transistor Logic (digital electronics)
STU	1. Secure Telephone Unit
	2. Set-Top Unit (TV reception)
STU-III	Secure Telephone Unit 3rd generation
STUN	Serial Tunnel
STV	1. Subscriber TV
	2. Subscription TV
STW	Surface Transverse Wave (microwave)
STX	1. Start of Text (control character)
	2. Start Transmission

SU	1. **S**ignaling **U**nit (RF modem)
	2. **S**ubscriber **U**nit
	3. **S**ervice **U**ser
SUB	**sub**stitute (character control code)
subNMS	**sub**-**N**etwork **M**anagement **S**ystem
SUERM	**S**ignal **U**nit **E**rror **R**ate **M**onitor (SS7)
SUI	**S**peech **U**ser **I**nterface (multimedia)
SUIRG	**S**atellite **U**sers **I**nterference **R**eduction **G**roup
SUMAC	**Su**per-HIPPI **M**edia **A**ccess **C**ontroller
SunOS	**Sun** Microsystems' variety of **O**perating **S**ystem
SUPARCO	**S**pace & **Up**per **A**tmosphere **R**esearch **Co**mmission (Pakistan)
SUPV	**Sup**er**v**ision
SUS	1. **S**tation **U**nit **S**haring
	2. **S**ilicon **U**nilateral **S**witch
SUT	1. **S**ystem **U**nder **T**est (ATM)
	2. **S**ervice **U**ser **T**able
SV	**S**uper**v**isory Unit
SVC	1. **S**witched **V**irtual **C**ircuit (ATM and frame relay)
	2. **S**witched **V**irtual **C**onnection
	3. **S**mart **V**irtual **C**ircuit
	4. **S**oft **V**irtual **C**ircuit
SVCC	**S**witched **V**irtual **C**hannel **C**onnection
SVCD	**S**uper **V**ideo **C**ompact **D**isk
SVD	**S**imultaneous **V**oice and **D**ata
SVGA	**S**uper **V**ideo **G**raphics **A**dapter (800×600 pixels monitor)
SVHS	**S**uper **VHS** (video format)
SVHS-C	**S**uper **VHS**—**C**ompact (video format)
SVID	**S**ystem **V** **I**nterface **D**efinition
SVIIX	**S**ilicon **V**alley **I**nternational **I**nternet **Ex**change
SVN	**S**ubscriber **V**erification **N**umber (ISDN)
SVOD	**S**tandard **V**ideo-**O**n-**D**emand
SVP	**S**urge **V**oltage **P**rotector
SVPC	**S**witched **V**irtual **P**ath **C**onnection
SVS	1. **S**hort **V**oice **S**ervice (GSM)
	2. **S**witched **V**oice **S**ervice
SVU	**S**iran **V**irtual **U**niversity (Syria)
Sw	**sw**itch
SW	1. **S**hort **W**ave (radio broadcast)
	2. **S**tanding **W**ave
	3. **S**tation **W**ire
	4. **sw**itch
SWAC	**S**tandard **W**estern **A**utomatic **C**omputer

S

SWACT	Switch of Activity
SWAG	Scientific Wild Ass Guess
SWAN	Satellite WAN (networking)
SWAP	1. Simple Workflow Access Protocol
	2. Shared Wireless Access Protocol
SWATS	Standard Wireless AT command Set
SWC	Service Wire Center
SWDL	Software Download
SWEDAC	Swedish Board for Technical Accreditation
SWHK	Switch Hook
SWIFT	Society for Worldwide Interbank Financial Telecommunications
SwiftBB	Swift64 Broadband (Inmarsat)
SWIGs	Special Working Interest Groups
SWIP	Shared Who is Project (ARIN)
SWIR	Short Wavelength Infrared Radiometer (remote sensing)
SWL	Short Wave Listener (amateur radio)
SWLing	Short Wave Listening (amateur radio)
SWLs	Short Wave Listeners (amateur radio)
SWM	Software Module
SWR	Standing-Wave Ratio (transmission lines parameter)
SWS	Structured Wiring System
SWVR	Standing-Wave Voltage Ratio (transmission lines)
SXGA	Super-extended Graphics Adapter (1600×1200 pixel monitors)
SXGA+	Super-extended Graphics Adapter Plus (1800×1440 monitors)
SXS	1. Step-by-Step Switching (telephony)
	2. Synchronous Cross-connect System
SYLK file	Symbolic Link file (spreadsheet data)
symlink	symbolic link
SYN	Synchronous idle (character control code)
Sync	1. Synchronization
	2. Synchronous Control Character
synchro	synchronous
synth	synthesizer
.sys	Name extension for system configuration files (computer)
sysop	systems operator (BBS)
SYSREQ	System Request
systray	system tray (Windows taskbar)
SYU	Synchronization signal Unit

T t

t	**t**ime (mathematics)
T	1. Symbol for prefix **t**era-, denoting 2^{40} or 10^{12}
	2. Symbol for **T**esla (unit of magnetic induction)
	3. Symbol for **Tr**unk level
	4. Symbol for **T**ransformer (circuit diagrams)
T Carrier	Trunk **Carrier**
T Connector	**T**-shaped **Connector**
TPAD	**T**erminal **P**acket **A**ssembler/**D**isassembler
T&A	**T**est **and** **A**ccept
T&C	**T**elemetry **and** **C**ommand (satcom)
T&E	**T**echnology **and** **E**valuation Lab
T&EC	**T**ests **and** **E**valuation **C**enter (Intelsat)
T&L	**T**eaching **and** **L**earning (e-learning)
T&M	**T**est **and** **M**easurement
T&T	**T**hrane **and** **T**hrane (company)
T-Berd	Another name for **T-1** Carrier
T-BGA	**T**ape **B**all **G**rid **A**rray (microchips)
T-Bird	Another name for **T-1** Carrier
T-Carrier	Trunk **Carrier** (T-1 and T-2)
T-CCS	**T**ransparent **C**ommon **C**hannel **S**ignaling
T-channel	**T**DMA **channel**
T-Load	**T**echnology **Load**

T

T-1	The North American **T**runk Carrier level-**1** transmission line operating at 1.544 Mbps
T-1 CAS	**T-1 C**hannel-**A**ssociated **S**ignaling
T-1C	**T**runk level-**1 C**ombined
T-2	The North American standard for Digital Signal level-**2** transmission line operating at 6.312 Mbps
T-3	The North American standard for Digital Signal level-**3** transmission line operating at 44.736 Mbps
T-4	The North American standard for Digital Signal level-**4** transmission line operating at 274.176 Mbps
T/H	**T**rack **and H**old
T/R	**T**ransmit **and R**eceive
T/V	**T**hermal **V**acuum
T²L	**T**ransistor–**T**ransistor **L**ogic (digital electronics)
T3POS	**T**ransaction **P**rocessing **P**rotocol for **P**oint **O**f **S**ale
TA	1. **T**erminal **A**dapter (ISDN)
	2. **T**erminal **A**cquisition
	3. **T**iming **A**dvance (GSM)
	4. **T**echnical **A**ssembly (ETSI)
	5. **T**echnical **A**dvisory (Telcordia Technologies)
	6. **T**elecommunication **A**dministration
	7. **T**ransistor **A**mplifier
TAAPM	**T**wo-**A**xis **A**ntenna **P**ositioning **M**echanism (Intelsat)
TAB	1. **T**one **A**bove **B**and
	2. **T**ape-**A**utomated **B**onding
	3. **T**echnical **A**pplication **B**ulletin
TABS	**T**elemetry **A**synchronous **B**lock **S**erial protocol (AT&T)
TAC	1. **T**est **A**ccess **C**ontroller
	2. **T**erminal **A**ccess **C**ontroller
	3. **T**ype **A**pproval **C**ode (GSM mobile station)
TACACS	**T**erminal **A**ccess **C**ontroller **A**ccess **C**ontrol **S**ystem (IETF)
tacan	**tac**tical **a**ir **n**avigation (military)
TACC	**T**racking **A**nd **C**ontrol **C**enter
TACOS	**T**ransmission **A**ccurate-purchase **C**ontrol **S**ystem
TACS	**T**otal **A**ccess **C**ommunications **S**ystem (cell phone standard)
TACSAT	**Tac**tical **Sat**ellite (U.S.A.)
TACSIM	**Tac**tical **Sim**ulation (military)
TACT	**T**rend **A**nalysis for **C**ircuit **T**roubles
TAD	**T**elephone **A**nswering **D**evice
TADIL	**Ta**ctical **D**ata **I**nformation **L**ink (military)
TADIXS	**Ta**ctical **D**ata **I**nformation **E**xchange **S**ystem (military)
TADM	**T**raffic **A**nalysis **D**ata **M**anagement

TADS	1. Teletypewriter Automatic Dispatch System
	2. Test And Display System
TADSS	Tactical Automatic Digital Switching System (messaging)
TAE	Trans-Asia Europe (Cable Consortium)
TAF	1. Targeted Accessibility Fund
	2. Terminal Adaptation Function (GSM)
TAFAS	Trunk Answer From Any Station
TAG	Technical Advisory Group (GSM)
TAI	International Atomic Time
TAL	Telephone Acoustic Loss
TALI	Tekelec Adaptation Layer Interface (Tekelec)
TAM	1. Telephone Answering Machine
	2. Telephone Account Management
	3. Telephony Access Module
	4. Telecommunication Access Method
	5. Total Assignment Message (Intelsat)
TANC	1. Thrust Actuated Nutation Damping Control (Intelsat)
	2. Thruster Active Nutation Control (Intelsat)
TANDM	Traffic Analysis Data Management
TANE	Telephone Association of New England
TAO	1. Telephony Application Object
	2. Track At Once (CD data storage)
TAP	1. Transit Access Point
	2. Test Access Part
	3. Transferred Account Procedure (GSM)
	4. Telocator Alphanumeric Protocol (messaging)
	5. Terrain Analysis Package (software)
	6. Two-step Approval Process (ETSI)
TAPAC	Terminal Attachment Program Advisory Committee
TAPI	Telephony Application Program Interface
TAPS	TETRA Advanced Packet Service
.tar	Name extension for the **TAR** program files (computer)
TAR	1. Total Accounting Rate
	2. Tape Archive
TARD	Towed Active Radar Device
TARE	Telegraph Automatic Relay Equipment
TARGA	Truevision Advanced Raster Graphics Adapter
TARM	Telephone Answering and Recording Machine
TARP	Target ID Address Resolution Protocol (SONET)
TARS	Turn Around Ranging Station
TAS	1. Telecommunication Authority of Singapore
	2. Telephone Answering Service
	3. Terminal Acquisition Support

T

TASC	Telecommunications **A**larm, **S**urveillance, and **C**ontrol system
TASI	Time **A**ssignment **S**peech **I**nterpolation (telephone technique)
TASO	Television **A**dvisory **S**tandards **O**rganization
TAT	Trans-**A**tlantic **T**elephone (cable)
TAV	Transverse **A**coustic-wave **V**oltage
TAXI	Transparent **A**synchronous **Transmitter and Receiver** Interface
Tb	Terabit (2^{40} or 1024 times gigabit)
TB	1. Tera**b**yte
	2. Traffic **B**urst
	3. Transparent **B**ridging (ATM)
TB/s	Tera**b**yte **per s**econd
TB1	Traffic Burst **1**
TBn	Traffic Burst **n**
TBA	1. **T**ransmit **B**urst **A**cquisition
	2. **T**o **B**e **A**nnounced later on
TBB	1. **T**ransnational **B**roadband **B**ackbone
	2. **T**elecommunications **B**onding **B**ackbone
TBBIBC	**TBB** Interconnecting **B**onding **C**onductor
TBC	**T**o **B**e **C**onfirmed later on
TBD	1. **T**o **B**e **D**efined later on
	2. **T**o **B**e **D**ecided later on
TBE	Transient **B**uffer **E**xposure (number of cells)
TBM	Transport **B**andwidth **M**anager (Northern Telecom)
TBOP	Transparent **B**it-**O**riented **P**rotocol
TBOS	Telemetry **B**yte-**O**riented **S**erial (protocol)
TBR	1. Technical **B**asis for **R**egulation (ETSI)
	2. Timed **Br**eak
TBS	Transmit **B**urst **S**ynchronization
TC	1. Transmission **C**ontrol code
	2. Transmission **C**onvergence (ATM)
	3. Tandem **C**onnection
	4. trans**c**oder
	5. Terrestrial **C**hannel
	6. Toll **C**enter
	7. Temperature **C**oefficient (electronics)
	8. tele**c**ommand
	9. Time **C**ommitted (frame relay)
	10. Timing **C**lock
	11. Telecommunications **C**loset
	12. Technical **C**ommittee (ETSI)
	13. Transmission **C**oefficient (parameter)

TC SMG	**T**echnical **C**ommittee, **S**pecial **M**obile **G**roup (ETSI)
TC&F	**T**elecom **C**ertification **& F**iling, Inc.
TC&R	**T**elemetry, **C**ommand, **and R**anging
TCA	1. **Tele**communications **A**ssociation (U.S.A.)
	2. **T**arget **C**hannel **A**dapter (fiber optics)
	3. **T**hreshold **C**rossing **A**lert
TCAM	**Tele**communications **A**ccess **M**ethod (IBM)
TCAP	**T**ransaction **C**apabilities **A**pplication **P**art (SS7)
TCAS	1. **T**-**C**arrier **A**dministration **S**ystem
	2. **T**raffic **A**lert and **C**ollision and **A**voidance **S**ystem
TCB	**T**rusted **C**omputer **B**ase
TCC	1. **T**runk **C**ontrol **C**ircuit
	2. **T**emperature **C**oefficient of **C**apacitance (electronics)
	3. **T**elephone **C**ountry **C**ode
TCCF	**T**actical **C**ommunications **C**ontrol **F**acility
TCCGB	**T**elephone **C**ard **C**lub of **G**reat **B**ritain
TCCM	**T**otal **C**ountry **C**onnectivity **M**easure
TCE	1. **T**homson **C**onsumer **E**lectronics (company)
	2. **T**echnical **C**omputing **E**nvironment
	3. **T**hermal **C**oefficient of **E**xpansion
TCF	1. **T**raining **C**heck **F**rame
	2. **T**echnical **C**ontrol **F**acility
TCG	**T**eleport **C**ommunications **G**roup (SONET)
TCH	**T**raffic **Ch**annel (GSM)
TCI	1. **Tele**communications, **I**nc. (company)
	2. **T**elecommunications **C**ompany of **I**ran (operator)
TCIC	1. **T**elecommunication **& C**omputer **I**ndustries **C**onsortium (Iran)
	2. **T**runk **C**ircuit **I**dentification **C**ode (SS7)
TCIF	**Tele**communications **I**ndustry **F**orum (standards organization)
TCL	**1. T**ool **C**ommand **L**anguage (programming)
	2. **T**erminal **C**oupling **L**oss
TCLw	**W**eighted **T**erminal **C**oupling **L**oss
TCL/TK	**T**ool **C**ommand **L**anguage **T**ool**k**it
TCLE	**T**emperature **C**oefficient of **L**inear **E**xpansion
TCM	1. **T**rellis-**C**oded **M**odulation
	2. **T**ime-**C**ompression **M**ultiplexing (transmission)
	3. **T**raveling **C**lass **M**ark
	4. **Tele**communications **M**anager
TCM-4D	**T**rellis-**C**oded **M**odulation **4**-**D**imensional
TCN	**T**able of **C**ollective **N**umbers
TCNS	**Th**omas **C**onrad **N**etworking **S**ystem

T

TCO	1. Tandem Connection Overhead (SONET)
	2. Total Cost of Ownership (Network Computer)
TCP	1. Transmission Control Protocol (ARPAnet)
	2. Termination Connection Point
	3. Tape Carrier Package (microchips)
TCP/IP	Transmission Control Protocol/Internet Protocol (DARPA)
TCPA	Telephone Consumer Protection Act
TCR	1. Transaction Confirmation Report
	2. Telemetry, Command, and Ranging (Intelsat)
	3. Temperature Coefficient of Resistance (electronics)
	4. Tagged Cell Rate (ATM)
TCS	1. Transmission Convergence Sublayer (ATM)
	2. Tactical Communications System
	3. Traffic Control System
TCSF	Telecommunications and Customer Service Forum (Canada)
TCTS	Trans-Canada Telephone System
TCU	1. Timing Control Unit
	2. Teletypewriter Control Unit
	3. Telematics Control Unit
	4. Table of Closed User group
TCUA	Terminal Control Unit for A-interface
TCV	Table of Address Conversion
TCVXO	Temperature-Compensated Voltage-controlled **Crystal** Oscillator
TCW	Taping Command Word
TCXO	Temperature-Compensated **Crystal** Oscillator (electronics)
TD	1. Time Division
	2. Time Delay
	3. Tunnel Diode (electronics)
TD-CDMA	Time Division CDMA (access)
TD-SCDMA	Time Division Synchronous CDMA (access)
TDAC	Time Domain Alias Cancellation
TDAS	Traffic Data Administration System
TDC	1. Time Division Controller
	2. Table of Dial Codes
TDCS	Theater Deployable Communications System
TDD	1. Telecommunications Device for the Deaf
	2. Time Division Demultiplexed (cellular networks)
	3. Time Division Duplexing (transmission)
TDDRA	Telephone Disclosure and Dispute Resolution Act
TDDS	Time Division Duplex System

TDE	Time Delay Equalizer
TDEL	Technical Development and Evaluation Laboratory
TDEV	Time Deviation
TDF	1. Trunk Distributing Frame
	2. Transborder Data Flow
	3. Telediffusion de France
TDHS	Time Domain Harmonic Scaling
TDI	1. Trade Data Interchange
	2. Transmit Digital Intertie
TDM	1. Time Division Multiplexing (transmission)
	2. Trunk Data Module
TDM0	**Primary** NCS **TDM** carrier (transmission)
TDM1	**Secondary** NCD **TDM** carrier (transmission)
TDMA	Time Division Multiple Access
TDMA/RA	**TDMA with** Random Assignment (access)
TDMA/SS	**TDMA with** Satellite Switching (access)
TDMA-3	A **TDMA** system supporting **3** calls in a single carrier (access)
TDMA-6	A **TDMA** system supporting **6** calls in a single carrier (access)
TDMQPSK	Time Division Multiplex Quadrature **PSK** (modulation)
TDMS	1. Time Division Multiplex System
	2. Technical Document Management System
	3. Transmission Distortion Measuring Set
TDOA	Time Difference Of Arrival (radio direction finding)
TDP	1. Telocator Data Protocol
	2. Tag Distribution Protocol
	3. Transmit Digital Processing (circuit)
TDQM	Total Data Quality Management
TDR	1. Time-Domain Reflectometer (test equipment)
	2. Temperature-Dependent Resistor (electronics)
TDRS	Tracking and Data Relay Satellite (remote sensing)
TDRSS	Tracking and Data Relay Satellite System (NASA)
TDS	1. Test-case Definition System
	2. Time Division Switching
	3. Terrestrial Data Service
	4. Terrestrial Digital Service
	5. Transmission Data Service
TDSAI	Transit Delay Selection And Indication
TDT	1. Total Delay Time
	2. Time and Data Table (broadcasting)
TDU	1. Tape Drive Unit
	2. Topology Database Update
TDWR	Terminal Doppler Weather Radar

T

TE	1. Terminal Equipment (ISDN)
	2. Terminal Endpoint
	3. Transit Exchange
	4. Trunk Exchange
	5. Telecom Egypt (operator)
	6. Transverse Electric waveform (waveguides)
TE1	Terminal Equipment type 1 (ISDN)
TE2	Terminal Equipment type 2 (ISDN)
TEA	Transferred Electron Amplifier (electronics)
TEA laser	Transversely-Excited Atmospheric-pressure **laser**
TEC	1. Total ionospheric Electron Content
	2. Thermo-Electric Cooler
	3. Thermal Electric Cooler
	4. Telecommunications and Electronic Consortium
TECF	Traffic Editor Control File
TED	1. Threshold Extension Demodulator
	2. Transferred Electron Device
	3. Transmission Electron-Diffraction contrast microscopy
	4. Traffic Engineering Database (fiber optics)
	5. Trunk Encryption Device
TEDIS	Trade Electronic Data Interchange System (ICT)
TEF	Telecommunications Entrance Facility
TEFG	Terminal Endpoint Functional Group
TEGFET	Two-dimensional Electron Gas FET (semiconductors)
TEHO	Tail End Hop-Off (traffic engineering)
TEI	Terminal Endpoint Identifier (ISDN)
TEK	Traffic Encryption Key
TEL	telephone
TELCOs	Telephone Companies
TELEC	Telecom Engineering Center (company)
telecast	television broadcast (transmission)
Telecom	telecommunications
Telecoms	telecommunications
TELESA	Telecom Services Association (Japan)
TELESAT	Canadian Domestic Satellite Telecommunications System
teleran	television radar air navigation
Telex	Teletypewriter exchange (service)
TELNet	Teletypewriter Network protocol (DARPA)
TELRIC	Total Element Long-Run Incremental Cost
TELSET	Telephone Set
TEM	1. Transverse Electromagnetic waveform (waveguides)
	2. Transmission Electron Microscope

TEMCO	**T**elecommunications **E**quipment **M**anufacturing **Co**mpany (Iran)
tempco	**temp**erature **co**efficient
TEN	1. **T**ransformer **E**xciting **N**etwork
	2. **T**rans-**E**uropean Telecommunications **N**etwork
TENet	**T**exas **E**ducational **Net**work (e-learning)
TEO	1. **T**ransferred **E**lectron **O**scillator
	2. **T**elephone **E**quipment **O**rder
TERENA	**T**rans-**E**uropean **R**esearch and **E**ducation **N**etworking **A**ssociation (e-learning)
TERM	1. **T**elegraph **E**rror **R**ate **M**easuring (equipment)
	2. **Ter**minal **M**ode
TERRA	**T**andem Accelerator for **E**nvironmental and **R**adar and **R**adiolocation **A**nalysis (remote sensing)
TES	**T**est **E**xecution **S**ystem
TESC	**Te**chnology **S**ub-**C**ommittee
TESP	**T**elecommunications **E**lectric **S**ervice **P**riority
TET	**T**elegraph **E**nd **T**erminal (equipment)
TETRA	1. **T**errestrial **T**runked **Ra**dio (new name)
	2. **T**rans-**E**uropean **T**runked **Ra**dio system (old name)
TeV	**T**era**e**lectron**v**olt (10^{12} eV)
TEW	**T**rapped **E**lectromagnetic **W**ave
Texel	**Tex**ture **el**ement (textured 3D objects)
TF	1. **T**ime signals and **F**requency standards emission
	2. **T**ransfer **F**unction (mathematics)
TFD	**T**hin-**F**ilm **D**etector (semiconductors)
TFEL	**T**hin-**F**ilm **E**lectro-**l**uminescent (semiconductors)
TFF	**T**hin-**F**ilm **F**ilter
TFJ	**T**élévision **F**rançaise **J**uvie (broadcaster)
TFL	**T**hin-**F**ilm **L**aser
TFLOPS	**T**era (one trillion) **Fl**oating-point **Op**eration**s** (computer speed)
TFM	**T**amed **F**requency **M**odulation
TFOM	1. **T**hin-**F**ilm **O**ptical **M**odulator
	2. **T**hin-**F**ilm **O**ptical **M**ultiplexer
TFOS	**T**hin-**F**ilm **O**ptical **S**witch
TFOW	**T**hin-**F**ilm **O**ptical **W**aveguide
TFT	**T**hin-**F**ilm **T**ransistor (semiconductors)
TFTP	**T**rivial **F**ile **T**ransfer **P**rotocol
TFTS	**T**errestrial **F**light **T**elephone **S**ystem
TFU	**T**ime and **F**requency **U**nit
TG	1. **T**ask **G**roup (IEEE)
	2. **T**runk **G**roup (switching)
	3. **T**erminator **G**roup

T

.tga	Name extension for **Tar**ga format files (graphics)
TGA	Short form for **Tar**ga (raster graphics file format)
TGB	1. Trunk **G**roup **B**usy (switching)
	2. Telecommunications **G**rounding **B**usbar
TGC	Transmission **G**roup **C**ontrol (IBM SNA)
TGE	Terminator **G**roup—type **E**
TGF	1. Terminator **G**roup—type **F**
	2. Through **G**roup **F**ilter
TGM	Trunk **G**roup **M**ultiplexer
TGS	Test-case **G**eneration **S**ystem
TGW	Trunk **G**roup **W**arning
TH	Transmission **H**eader (SNA)
THAAD	**T**heater **H**igh-**A**ltitude **A**rea **D**efense
THB	**T**emperature **H**umidity **B**ias
THD	**T**otal **H**armonic **D**istortion (parameter)
THD+N	**T**otal **H**armonic **D**istortion **plus N**oise
THEOS	**Th**ai **E**arth **O**bservation **S**ystem (Thailand)
thermistor	**therm**al re**sistor** (electronics)
THF	**T**remendously **H**igh **F**requency (300–3000 GHz range)
THI	**T**elephone **H**andset **I**ntegrator
thicknet	**thick** Ether**net** coaxial cable
thinnet	**thin** Ether**net** coaxial cable
THL	**T**rans-**H**ybrid **L**oss
THz	**T**era**hertz** (10^{12} Hz)
TI	1. **T**errestrial **I**nterface
	2. **T**ransaction **I**dentifier (GSM)
TIA	1. **T**elecommunications **I**ndustry **A**ssociation (U.S.A.)
	2. **T**ime-**I**nterval **A**nalyzer
	3. **I**nternational **A**tomic **T**ime
TIB	**T**one-**I**n-**B**and (modulation)
TIBS	**T**actical **I**nformation **B**roadcast **S**ervice
TIC	1. **T**erminal **I**nternational **C**enter
	2. **T**oken-**R**ing **I**nterface **C**oupler (IBM)
	3. **T**elecommunications **I**nfrastructure **C**ompany (Iran)
TIC Trunk	**T**andem **I**nter**L**ATA **C**onnecting **Trunk**
TICL	**T**emperature-**I**nduced **C**able **L**oss
TID	1. **T**erminal **Id**entification (ISDN)
	2. **T**arget **Id**entifier
TIE	1. **T**elephone **I**nterconnect **E**quipment
	2. **T**errestrial **I**nterface **E**quipment
	3. **T**rusted **I**nformation **E**nvironment (encryption)
	4. **T**ime-**I**nterval **E**rror

TIES	1. Time Independent Escape Sequence (modems)
	2. Telecom Information Equipment and Services
.tif	Name extension for **TIF** image files (graphics)
TIF	1. Tagged Image File (image format)
	2. Terminal Interface node
TIFF	Tagged Image File Format (graphics)
TIFF-F	Tagged Image File Format—Fax (graphics)
TIGA	Texas Instruments Graphics Architecture
TIIAP	Telecommunications and Information Infrastructure Assistance Program
TIM	1. Terrestrial Interface Module
	2. Teletyper Input Method
	3. Telecom Italia Mobile
	4. Trace Identifier Mismatch
	5. Télecom Italia Mobile (Italian GSM operator)
TIMA	The Interactive Media Alliance
TIMS	Transmission Impairment Measurement Set
TiN	Titanium Nitride (semiconductors)
TINA	Telecommunications Information Networking Architecture (TMN)
TINA-C	Telecommunications Information Networking Architecture Consortium (TMN)
TINS	Thermal Imaging Navigation Set
TIP	1. Transaction Internet Protocol
	2. TETRA Interoperability Profile
	3. Terminal Interface Processor
TIPHON	Telecommunications and IP Harmonization Over Networks (ETSI)
TIPI	Telephone Industry Price Index
TIQ	Telrate International Quotations
TIR	Total Indicated Reading
TIRKS	Trunks Integrated Records Keeping System (Telcordia Technologies)
TIROS	Television Infrared Observation Satellite (U.S.A.)
TIS	Technical Information Sheet
TISOC	Telecom Industry Services Operating Center (Bell Atlantic)
TISSS	Tester Independent Software Support System (DoD program)
TIU	Terrestrial Interface Unit
TJF	Test Jack Frame (PBX)
TL	1. Tie Line (power system)
	2. Transmission Level
	3. Transport Layer

TL1	Transaction Language 1
TLB	1. Terminal Loop-Back
	2. Test Loop-Back
	3. Translation-Look-aside Buffer
TLD	Top-Level Domain (Internet)
TLDN	Temporary Local Directory Number (wireless networks)
TLEC	Terminating Local Exchange Carrier
TLF	Trunk Link Frame (crossbar exchanges)
TLI	Transport Layer Interface
TLM	1. telemetry
	2. Transmission Line Matrix
	3. Triple-Layer Metal
TLMA	Telematic Agent (X.400)
TLP	Transmission Level Point
TLS	1. Transparent LAN Service
	2. Transport Layer Security (protocol)
	3. Telemetry Land Station
TLSU	Token-Ring LAN Service Unit (ATM)
TLV	Type-Length-Value (encoding rules)
TLX	telex
TM	1. Tracking Module
	2. Traffic Management (network events)
	3. Trouble Management (ATM)
	4. Transverse Magnetic waveform (waveguides)
	5. Terminating Multiplexer (SONET)
	6. Terminal Mobility
	7. Thematic Mapper (remote sensing)
	8. telemetry
	9. Transverse Magnetic (wave)
TM/TC	Telemetry and Telecommand
TMA	1. Telecommunication Managers Association
	2. Tower Mounted Amplifier
TMC	1. Timeslot Management Channel (cordless messaging)
	2. Tracking Mode Coupler (antenna, earth station)
	3. Traffic Management Center
	4. Transfer Mode Converter
TMD	Tactical MAP Display
TME	Telocator Message Entry (protocol)
TMF	Telemanagement Forum (U.S. standards organization)
TMGB	Telecommunications Main Grounding Busbar
TMGT	telemanagement
TMI	TRMM Microwave Imager (remote sensing)
TMIB	Traffic Message Interchange Bus

TML	**T**elephony **M**arkup **L**anguage (programming)
TMN	1. **T**elecommunications **M**anagement **N**etwork
	2. **T**elecomunicaçoes **M**oveis **N**acionais (Portugal GSM operator)
	3. **T**raffic **M**anagement **N**etwork
TMP	1. **T**est **M**anagement **P**rotocol (ATM)
	2. **T**heoretical **M**id-**P**oint (private line circuit)
TMR	**T**runked **M**obile **R**adio
TMRC	**T**echnology **M**arketing and **R**esearch **C**ouncil
TMRS	**T**raffic **M**easurement and **R**ecording **S**ystem
TMS	1. **T**ime **M**ultiplexed **S**witch (AT&T)
	2. **T**imes **M**icrowave **S**ystems (company)
	3. **T**OPS **M**essage **S**witch
	4. **T**rouble **M**anagement **S**ystem
	5. **T**eleport **M**onitoring **S**ystem (Intelsat)
	6. **T**ransmission **M**easuring **S**et
TMSI	**T**emporary **M**obile **S**ubscriber **I**dentity (GSM)
TMSL	**T**est and **M**easurement **S**ystems **L**anguage
TMU	**T**erminal **M**ake-**U**p
TMUX	trans**mu**ltiple**x**er
TN	1. **T**elephone **N**umber
	2. **T**ime slot **N**umber (GSM)
	3. **T**erminal **N**umber (Nortel)
	4. **T**ransit **N**etwork
	5. **T**wister **N**ematic (display technology)
TNA	**T**elematics **N**etwork for **A**dministrations
TNC	**T**readed **N**aval **C**onnector
TNDS	**T**otal **N**etwork **D**ata **S**ystem
TNID	**T**errestrial **N**etwork **Id**entification code
TNL	**T**erminal **N**et **L**oss
TNLCD	**T**wisted-**N**ematic **L**iquid-**C**rystal **D**isplay
TNPP	**T**elocator **N**etwork **P**aging **P**rotocol
TNR	**T**ransmission **N**ot **R**eady
TNS	**T**ransit **N**etwork **S**election (ATM)
TNSP	**T**ransmission **N**etwork **S**ervice **P**rovider
TNT	**T**ransparent **N**etworking **T**ransport (service)
TNX	1. **T**etra **N**ode E**x**change
	2. **T**ha**n**ks (Morse code transmissions)
TO	1. **T**elecommunication **O**perator
	2. **T**ransmit **O**nly
	3. **T**andem **O**ffice
	4. **T**oll **O**ffice
	5. **T**ime **O**ut (protocols)
TOA/NPI	**T**ype **O**f **A**ddress/**N**umbering **P**lan **I**dentifier

T

TOC	Technical Operating Center
TOCC	Technical and Operational Control Center (Intelsat)
TOD	Time Of Day
TOF	1. Time Out Factor (ATM)
	2. Top-Of-File
TOGAF	The Open Group Architectural Framework
TOH	Transport Overhead (SONET)
TΩ	Teraohm (electronics)
TOI	Third-Order Intercept
TOK	Test Okay
TOM	1. Thin-film Optical Modulator
	2. Thin-film Optical Multiplexer
	3. Telecom Operations Map
TOMS	Total Ozone Mapping Spectrometer
TOP	1. Technical and Office Protocol (Ethernet)
	2. Thin Outline Package
TOPS	1. Traffic Operator Position System (Northern Telecom)
	2. Transfer Orbit and Payload-testing Support
ToS	Type of Service (DARPA)
TOS	1. Transfer Orbit Stage (satcom)
	2. Thin-film Optical Switch
TOSR	Telecommunications Office of Shovak Republic
TOVS	TIROS Operational Vertical Sounder (remote sensing)
TOW	Thin-film Optical Waveguide
TP	1. Twisted Pair
	2. Transport Protocol (ISO)
	3. Tunneling Protocol (VPNs)
	4. Transition Point
	5. transponder (satcom)
	6. Test Point
	7. Toll Plant
	8. Transaction Processing
TP monitor	Transaction Processing **monitor**
TP-DDI	Twisted-Pair Distributed Data Interface
TP-MIC	Twisted-Pair Media Interface Connector
TP-PMD	Twisted-Pair Physical Media Dependent technology
TP0	Transmission Protocol 0 (OSI model)
TP4	1. Transmission Protocol 4 (OSI model)
	2. Transport Protocol class 4 (ISO)
TPA	Telephone Pioneers of America
TPAD	Terminal Packet Assembler/Disassembler
TPC	1. Turbo Port Card
	2. TOPS Position Controller

	3. Transaction Processing Council
	4. Transmit Power Control
TPC-D	Transaction Processing Council Benchmark-**D**
TPCC	Third-Party Call Control (ATM)
TPDDI	Twisted-Pair Distributed Data Interface
TPDU	Transport Protocol Data Unit (OSI model)
TPEX	Twisted-Pair Ethernet **Transceiver**
TPF	Twists Per Foot (wires and cables)
TPG	Test Pattern Generator (TV broadcast)
TPI	1. Tracks Per Inch (data recording parameter)
	2. Telephone Plant Index
TPM	1. Traffic Processor Module
	2. Total Productive Maintenance
	3. Teleprocessing Monitor
	4. Terminating Point Masterfile
	5. Transactions Per Minute (ICT)
TPMR	Trunked Professional Mobile Radio
TPN	Totally Private Network
TPOA	Telecommunications Private Operating Agency
TPON	Telephony over Passive Optical Network
TPOS	Training, Planning & Operational Support (NCS)
TPQFP	Test-Pad Quad Flat Pack (microchips)
TPR	transponder
TPRC	Telecommunications Policy Research Conference
TPS	1. Telephone Preference Service
	2. Télévision Par Satellite (France)
TPU	Time Processor Unit
TPV	Third-Party Verification (FCC)
TPWG	Technology Policy Working Group
TQ	Toll Quality
TQC	Total Quality Control (of products)
TQFP	Thin Quad Flat Pack (microchips)
TQM	Total Quality of Management (ICT)
TR	1. Technical Requirement
	2. Token Ring (LANs)
	3. Trouble Report
	4. Technical Reference (Telcordia Technologies)
	5. transistor (circuit diagrams)
	6. Transfer Request
	7. Transmit–Receive
	8. Transient Response
TR tube	Transmit–Receive **tube**
TRA	1. Telecommunications Regulatory Authority (Egypt)
	2. Telecommunications Resellers Association (U.S.A.)

TRAC	Technical Recommendations Application Committee
tracert	trace route (routing)
TRACON	Terminal Radar Approach Control
TRADACOMS	Trading Data Communications
TRAM	1. Transputer RAM (memory)
	2. Transimpedance Amplifier
transceiver	transmitter and receiver
TRANSEC	Transmission Security
transistor	transfer resistor
transponder	transmitter responder (satcom)
transputer	transistor computer
TRAP	1. Tandem Recursive Algorithm Process
	2. Tactical Receive Application Protocal (military)
TRAPATT	Trapped Plasma Avalanche Triggered Transitor (semiconductors)
TRAU	Transcoding and Rate Adaptation Unit (GSM voice coding)
TRC	1. Transit Routing Control Table
	2. Transverse Redundancy Check (data streams)
TRCO	Trouble Reporting Central Office
TRD	Technical Requirement Document
TRE	Telecommunications Research Establishment
TREE	Transient Radiation Effects on Electronics
TREES	Tropical Ecosystems Environment observation by Satellite
TREX	Transmission Expert
TREG	Telecommunications Regulatory E-mail Grapevine
TRF	1. Tracking & Receiving Facility
	2. Tuned Radio Frequency (circuit)
TRFR	transfer
TRG	Technical Review Group
TRI-CWDM	Tri-Color Wavelength-Division Multiplexing (fiber optic lines)
triac	triode AC (semiconductor switch)
TRIB	tributary
TRIBES	Tri-Band Earth Station (CMI)
triode	A three-electrode electron tube
TRIP	1. Telephone Routing over IP (protocol)
	2. Token-Ring Interface Processor
TRIS	Tactical Reconnaissance Intelligence Service
TrL	Transmission Line
TRL	1. Transistor–Resistor Logic (digital electronics)
	2. Through-Reflect-Line (calibration)
TRM	1. Technical Reference Model

2. **tr**ans**mi**tter
3. **term**inal
4. **Ter**minal **M**ode
5. **Ter**minal **M**ultiplexer

TRMM **T**ropical **R**ainfall **M**easuring **M**ission (remote sensing)
TRMS **T**ransmission **R**esource **M**anagement **S**ystem
TRN **T**oken-**R**ing **N**etwork
TRO **T**emporary **R**estraining **O**rder
TRS 1. **T**elecommunication **R**elay **S**ervice
2. **T**elephone **R**elay **S**ervice
3. **T**runked **R**adio **S**ystem
4. **T**ip, **R**ing, **S**leeve (cables)
TRT **T**iming **R**eference **T**ransponder
TRU **T**one **R**eceiver **U**nit (telephone set)
TRX 1. **transceiver** (Transmitter and Receiver)
2. **Transmission** and **Reception** Unit
TRxxx **T**echnical **R**eference **number**
TS 1. **T**ransport **S**tream (ATM)
2. **T**raffic **S**haping (ATM)
3. **T**echnical **S**pecification
4. **T**ime **S**tamp (ATM)
5. **T**ime **S**lot (switching)
6. **T**ime **S**haring
7. **T**ransaction **S**erver
8. **T**ransmission **S**cheme
9. **T**oll **S**witch
TS3 **T**ime **S**tamp version **3** (ATM)
TSA 1. **T**ime **S**lot **A**llocation
2. **T**ime **S**lot **A**ssignment
3. **T**apered-**S**lot **A**ntenna
4. **T**elecommunications **S**ystem **A**rchitecture
TSAC **T**ime-**S**lot **A**ssigner **C**ircuit
TSACC **T**elecommunications **S**tandards **A**dvisory **C**ouncil of **C**anada
TSAG **T**elecommunications **S**tandardization **A**dvisory **G**roup
TSAP **T**ransport **S**ervice **A**ccess **P**oint (OSI model)
TSAPI **T**elephony **S**ervices **A**pplication **P**rogramming **I**nterface (AT&T)
TSAT **T**-1 **S**mall **A**perture **T**erminal (satcom)
TSB 1. **T**elecommunications **S**tandardization **B**ureau (ITU-T)
2. **T**elecommunications **S**ystem **B**ulletin
TSC 1. **T**ransit **S**witching **C**enter
2. **T**raining **S**equence **C**ode (GSM)

	3. **T**ransport **S**tream **C**ombiner (TV broadcast)
	4. **T**DM-**S**witched **C**apable (GMPLS)
	5. **T**wo-**S**ix **C**ode
	6. **T**ransmission **S**ystem **C**onstruction
	7. **T**echnical **S**ervice **C**enters
TSCM	**T**echnical **S**urveillance **C**ounter**m**easures
TSDU	**T**ransport **S**ervice **D**ata **U**nit (OSI model)
TServer	**T**elephony **Server**
TSF	1. **T**hrough **S**upergroup **F**ilter
	2. **T**élécoms **S**ans **F**rontiérs (France)
	3. **T**rail **S**ignal **F**ailure (MUX)
TSG	**T**ime **S**ignal **G**enerator
TSI	1. **T**ime **S**lot **I**nterchange
	2. **T**ransmitting **S**ubscriber **I**nformation
	3. **T**elecommunication **S**ystem **I**ntegration
TSIC	**T**ime **S**lot **I**nterchange **C**ircuit (PCM)
TSIU	**T**ime **S**lot **I**nterchange **U**nit (switching)
TSK	**T**ransmission **S**ecurity **K**ey
TSM	1. **T**DMA **S**ystem **M**onitor
	2. **T**erminal **S**erver **M**anager
	3. **T**elecommunications **S**ystem **M**anagement
	4. **T**opology-**S**pecific **M**odule (networking)
TSO	1. **T**ime **S**haring **O**peration
	2. **T**ime **S**haring **O**ption (IBM)
	3. **T**elecommunications **S**ystem **O**perator
	4. **T**echnical **S**upport **O**perations
TSOP	**T**hin **S**mall-**O**utline **P**ackage (microchips)
TSP	1. **T**elecommunications **S**ervice **P**riority (FCC)
	2. **T**echnical **S**upport **P**lanning
	3. **T**ransport **S**tream **P**rocessor
	4. **T**emperature-**S**ensitive **P**arameter
	5. **T**elematics **S**ervice **P**rovider
TSPS	**T**raffic **S**ervice **P**osition **S**ystem
TSR	1. **T**erminate-and-**S**tay-**R**esident (programs)
	2. **T**elephone **S**ervice **R**epresentative (agent)
	3. **T**elecommunications **S**ervice **R**equest
	4. **T**ag **S**witching **R**outer
TSRM	**T**elecommunication **S**tandards **R**eference **M**anual
TSS	1. **T**elecommunication **S**upport **S**ystem
	2. **T**elecommunication **S**tandard **S**ection (ITU)
	3. **T**raffic **S**cheduling **S**ystem
	4. **T**ime **S**pace **S**witch
TSSI	**T**ime **S**lot **S**equence **I**ntegrity
TSSOP	1. **T**hin-**S**caled **S**mall-**O**utline **P**ackage (microchips)

	2. **T**hin-**S**hrink **S**mall-**O**utline **P**ackage (microchips)
TST	1. **T**ime–**S**pace–**T**ime system (switching)
	2. **test**
TSTN	**T**riple **S**uper-**T**wisted **N**ematic (display technology)
TSTS	**T**ransaction **S**witching and **T**ransport **S**ervice
TSU	**T**one **S**ender **U**nit (telephone set)
TSW	**T**ransmit **Sw**itch
TT	1. **T**raffic **T**erminal
	2. **T**rue **T**ype (Windows fonts)
	3. **T**ouch-**T**one
	4. **T**ransaction **T**ime
	5. **T**unisie **T**elecom (Tunisia)
	6. **T**ürk **T**elecom (operator)
TT&C	1. **T**elemetry, **T**racking, **and C**ommand (satellites)
	2. **T**elemetry, **T**racking, **and C**ontrol Center (Intelsat)
TT&P	**T**echnical **T**raining **and P**ublications
TTA	**T**elecommunications **T**echnology **A**ssociation
TTAB	**T**ransparent **T**one **A**bove **B**and
TTC	1. **T**ransmit **T**iming **C**ontroller
	2. **T**elecommunications **T**echnology **C**ommittee (Japan)
	3. **T**ime-**T**o-**C**ollection (data acquisition)
TTC&M	**T**elemetry, **T**racking, **C**ommand, **and M**onitoring (satellites)
TTCP	**T**est **TCP**
TTD	**T**rue **T**ime **D**elay (phase shifters)
TTF	**T**ransport **T**ermination **F**unction
TTI	1. **T**ransmit **T**erminal **I**dentification (fax machine)
	2. **T**eam **T**elecom **I**nternational
TTIA	**T**elecommunications **T**echnology **I**nvestment **A**ct
TTIB	**T**ransparent **T**one **I**n **B**and
TTL	1. **T**ransistor–**T**ransistor **L**ogic (digital electronics)
	2. **T**ime-**T**o-**L**ive (packet data transmission)
TTMF	**T**runk **T**ype **M**aster **F**ile (MCI)
TTP	1. **T**rail **T**ermination **P**oint
	2. **T**rusted **T**hird **P**arty (contracts)
TTR	**T**ouch-**T**one **R**eceiver
TTRT	**T**arget **T**oken **R**otation **T**ime (FDDI)
TTS	1. **T**ext-**T**o-**S**peech Synthesis
	2. **T**ransaction **T**racking **S**ystem (networking)
TTTN	**T**andem **T**ie **T**runk **N**etwork
TTX	**T**eletypewriter **Ex**change (service)
TTY	**tele**type (or **tele**typewriter)
TTY OW	**Tele**typewriter **O**rder-**W**ire
TU	1. **T**raffic **U**nit

T

	2. Transmit Unit
	3. Terminal Unit
	4. Tributary Unit (MUX)
	5. Traveling User (SDNS)
TUA	Telecommunications Users Association (U.K.)
TUANZ	Telecommunications Users Association of New Zealand
TUBA	TCP and UDP with Bigger Address (Internet)
TUC	1. Total User Cell (ATM)
	2. Total User Calls
TUG	1. Telecommunication User Group
	2. Tributary Unit Group (MUX)
TUI	Telephone User Interface
TUP	1. Telephone User Part (SS7)
	2. Transponder Uplink system (satcom)
TUR	1. Traffic Usage Recorder
	2. Trunk Utilization Report
TURN	The Utilities Reform Network
TUST	Tunneling and Underground Space Technology
TUV	Technischer Uberwachungs-Verein (Germany)
TV	1. television
	2. Technical Vulnerability
TV/PC	TV with PC smarts (Microsoft)
TVC	Trunk Verification by Customer
TVE	Transversal Equalizer
TVI	Television Interference
TVM	Time-Varying Media (SCSA)
TVOR	Terminal VHF Omni-Range
TVRO	Television Receive-Only (broadcasting)
TVS	1. Trunk Verification by Station
	2. Transient Voltage Suppressor
TVSS	Transient Voltage Surge Suppressor
TWA	1. Traveling-Wave Amplifier (electronics)
	2. Two-Way Alternate (communication system)
TWAIN	Toolkit Without An Interesting Name (scanners standard)
TWAT	Traveling-Wave Amplifier Tube (electronics)
TWC	Two-Wire Circuit
TWS	1. Two-Way Simultaneous (transmission)
	2. Track-While-Scan
TWT	Traveling-Wave Tube (microwave)
TWTA	Traveling-Wave Tube Amplifier
TWX	Teletypewriter Exchange (service)
TX	1. transmit or transmission

	2. **transmitter**
	3. **T**ransit **Ex**change (switching)
	4. **T**erminal **Ex**change
TXD	**Transmit-D**ata (RS232 signal)
Tx/Rx	**Transmitter and Receiver**
.txt	Name extension for ASCII **text** files (computer)

T

U u

U	Symbol for radiation intensity (antennas)
U Frame	Unnumbered **Frame**
U Plane	User **Plane** (ATM)
U-band	Optical Frequency **band** ranging from 1625 to 1675 nm
U-C	User **C**entral (ADSL)
U-NII	**U**nlicensed-**N**ational **I**nformation **I**nfrastructure
U-R	User **R**emote (ADSL)
U/C	Up-**C**onverter (satcom)
U/D	Up/**D**own Chain (Intelsat)
U/L	Up-Link
UA	1. **U**nnumbered **A**cknowledge response (X.25)
	2. **U**ser **A**gent (OSI Application)
UAC	User **A**gent **C**lient (OSI Application)
UADSL	**U**niversal **A**symmetrical **DSL** (access)
UAE	1. **U**ser **A**ccess **E**quipment
	2. **U**ser **A**gent **E**ntity (message systems)
UAEnic	**U**nited **A**rab **E**mirates **n**etwork **i**nformation **c**enter
UAF	**U**niversal **A**ctive **F**ilter (electronics)
UAL	User **A**gent **L**ayer (OSI model)
UAMPT	**U**nion of **A**frican and **M**alagasy **P**ost and **T**elecommunications
UAN	**U**niversal **A**ccess **N**umber (INs)
UAPT	**U**nion of **A**frican **P**ost and **T**elecommunications

U

UARS	Upper Atmospheric Research Satellite
UART	Universal Asynchronous Receiver–Transmitter (computer chips)
UARTO	United Arab Republic Telecommunications Organization
UAS	1. User Agent Server (OSI Application)
	2. Unified Antenna Structure
UASs	Unavailable Seconds
UAT	User Acceptance Testing
UAV	1. Unmanned Aerial Vehicle
	2. Unpiloted Aerial Vehicle
UAWG	Universal ADSL Working Group
UAX	Unit Automatic Exchange
UBR	Unspecified Bit Rate (ATM)
UBS	Uninterruptible Battery System
UC	1. Unified Communications
	2. Universal Controller
	3. Unit Controller
	4. Unit Call
	5. User Channel
UCA	Utility Communications Architecture
UCAID	University Corporation for Advanced Internet Development
UCB	Unit Control Bus
UCC	Uniform Commercial Code
UCD	Uniform Call Distributor
UCF	UNIX Computing Forum
UCITA	Uniform Computer Information Transaction Act
UCM	1. Universal Call Model (SS7 routing)
	2. Universal Controller Module
UCS	1. Uplink Control System (software)
	2. Universal Character Set
UCT	Universal Coordinated Time
UDC	1. Universal Digital Carrier
	2. Universal Device Connector
	3. Universal Data Classification
UDDI	Universal Description Discovery and Integration (e-commerce)
UDF	Universal Disk Format (GIS imaging)
UDI	Unrestricted Digital Information (ISDN)
UDK	Universal Dialing Keyset (GTE)
UDLC	1. Universal Digital Loop Carrier
	2. Universal Data-Link Control (protocol)

UDLR	Uni-Directional Link Routing
UDLT	Universal Digital Loop Transceiver
UDMA	Ultra-DMA
UDOP	Ultimate Dumb, Open Programmable (switches)
UDP	1. User Datagram Protocol (Internet)
	2. Unacknowledged Datagram Protocol (Internet)
UDP/IP	User Datagram Protocol /Internet Protocol
UDPU	Universal Data Patch Unit
UDSL	Unidirectional High bit-rate DSL (access)
UDT	Uniform Data Transfer (OLE technology)
UDTS	Universal Data Transfer Service (Intelsat)
UDTV	Ultra-Definition Television (broadcasting)
UDWDM	Ultra-Dense Wavelength-Division Multiplexing
UECT	Universal Encoding Conversion Technology
UEM	Universal Equipment Module (Northern Telecom)
UEPS	Unit Eruptible Power Supply
UER	Undeleted Error Ratio
UET	Unattended Earth Terminal
UFB	Unfit For Broadcast (Intelsat)
UFMOP	Unintentional Frequency Modulation On Pulse
UFO	UHF Follow-On
UFOs	Unidentified Flying Objects
UFSS	Upstream Failed Signal State
UG	Under-Ground
UGWG	User Glossary Working Group (Internet)
UHF	Ultra-High Frequency (300 MHz to 3 GHz range)
UHTP	Unidirectional Hypertext Transfer Protocol
UHV	1. Ultra-High Voltage (power)
	2. Ultra-High Vacuum
UI	1. User Interface (GSM)
	2. Unnumbered Information
	3. Unix International (consortium)
UICC	Universal Integrated Circuit Card
UID	User ID
UIFN	Universal International Freephone Number
UIL	User Interface Language
UIM	Universal Identity Module
UIR	1. Upper flight Information Region (aviation)
	2. Uniform Impedance Resonator
UIS	Universal Information Services (AT&T)
UIT	Union Internationale Des Telecommunications
UJT	Uni-Junction Transistor (semiconductors)
UKERNA	United Kingdom Education and Research Networking Association (e-learning)

U

UKISC	United **K**ingdom **I**ndustrial **S**pace **C**ommittee (U.K.)
UKNet	University of **K**entucky's campus **Net**work
UKOLN	**U.K** Office for **L**ibrary and Information **N**etworking (ICT)
UL	**U**nderwriters **L**aboratories, Inc. (U.S. standards organization)
ULANA	**U**nified **L**ocal **A**rea **N**etwork **A**rchitecture (U.S. Air Force)
ULF	**U**ltra-**L**ow-**F**requency (300–3000 Hz range)
ULH	**U**ltra-**L**ong-**H**aul (fiber optics)
ULL	**U**ltra-**L**ow-**L**oss (optical fibers))
ULP	**U**pper **L**ayer **P**rotocol (OSI model)
ULS	**U**ser **L**ocation **S**ervice
ULSI	**U**ltra-**L**arge-**S**cale **I**ntegration (microchips)
UM	1. **U**nified **M**essaging
	2. **U**nit **M**anager
	3. **U**nintentional **M**odulation
UMA	1. **U**pper **M**emory **A**rea (computer)
	2. **U**nlicensed **M**obile **A**ccess
UMB	**U**pper **M**emory **B**lock (computer)
UME	**U**NI **M**anagement **E**ntity (ATM)
UMIB	**U**tility **M**essage **I**nterchange **B**us
UMIG	**U**niversal **M**essaging **I**nteroperability **G**roup
UML	**U**nified **M**odeling **L**anguage (programming)
UMS	1. **U**niversal **M**essaging **S**ervice
	2. **U**nified **M**essaging **S**ervice
UMSP	**U**nified **M**emory **S**pace **P**rotocol
UMTS	**U**niversal **M**obile **T**elecommunications **S**ystem (Europe)
UMTSF	**UMTS** **F**orum (wireless)
UN	**U**nited **N**ations
UNA	**U**nified **N**etwork **A**rchitecture
UNC	**U**niversal **N**aming **C**onvention (file naming)
UNCOPUOS	**U**nited **N**ations **C**ommittee **O**n **P**eaceful **U**ses of **O**uter **S**pace
UNE	**U**nbundled **N**etwork **E**lements
UNE-P	**U**nbundled **N**etwork **E**lements—**P**latform
UNI	**U**ser-to-**N**etwork **I**nterface (frame relay)
UNI-RZ	**Uni**polar—**R**eturn-to-**Z**ero (data encoding)
UNIBOL	**Uni**x version of CO**BOL** (programming)
UNII	**U**nlicensed **N**ational **I**nformation **I**nfrastructure (wireless)
UNIVAC	**Univ**ersal **A**utomatic **C**omputer
UNIX	**U**niversal **N**etwork **I**nformation **Ex**change (operating system)

UNMA	Universal Network Management Architecture
UNMR	Universal Network Management Record
UNO	Universal Networked Objects
UNTDI	United Nations Trade Data Interchange
UOI	Utilities Operating Instructions
UPC	1. Uplink Power Control
	2. Usage Parameter Control (ATM)
	3. Universal Product Code (bar code naming)
UPCS	Unlicensed Personal Communications Services (FCC)
UPI	User Protocol Interface
UPM	1. Utility Processor Module
	2. Ultra-Portable Multiplexer
UPN	Universal Personal Number (GSM)
UPnP	Universal Plug **and** Play (computer standard)
UPP	Universal Payment Preamble
UPS	1. Uninterruptible Power Supply (power)
	2. Ultraviolet Photoelectron Spectroscopy
UPSR	Unidirectional Path-Protection Switched Ring (MUX)
UPT	Universal Personal Telecommunications
UPTAA	UPT Access Address
UPTAC	UPT Access Code
UPTAN	UPT Access Number
UR	1. ultrared
	2. your (Morse code transmissions)
URA	Uniform Resource Agent (Internet)
URC	1. Uniform Resource Characteristic (Internet)
	2. Uniform Resource Citation (Internet)
URE	User Range Error
URFS	Up-link RF System (TV satellite broadcast)
URI	Uniform Resource Identifier (Internet)
URL	Universal Resource Locator (Web page address)
URM	User Request Manager
URN	Uniform Resource Name (Internet)
URSEC	Unidad Reguladora de Servicios de Comunicaciones (Uroguay)
US	Unit Separator (character control code)
USAC	Universal Service Administrative Company
USACII	**U.S.A** Standard Code of Information Interchange
USAN	Unites States Advanced Network
USART	Universal Synchronous/Asynchronous Receiver/Transmitter
USAT	Ultra-Small Aperture Terminal (satcom)

U

USB	1. Universal Serial Bus (computer)
	2. Unified S-Band
	3. Upper Side-Band (modulation)
USCM	Universal Service Component Model
USD	Universal Synchronous Data
USDC	United States Digital Cellular
USDLA	United States Distance Learning Association
	(standards)
USDN	United States ISDN
USEDA	Users Society for Electronic Design Automation
USENET	User's Network (networking)
UserID	User Identification
USF	1. Universal Single Frequency
	2. Universal Service telephone Fee (U.S.A.)
USGS	United States Geological Survey
USHR	Unidirectional Self-Healing Ring
USIIA	United States Internet Industry Association
USIM	UMTS Subscriber Identity Module (cellular networks)
USITA	United States Independent Telephone Association
USKA	1. Union Schweizerischer Kurzwellen-Amateure
	2. Union of Swiss Shortwave Amateurs (Switzerland)
USNO	United States Naval Observatory
USO	Universal Service Obligations
USOA	Uniform System Of Accounts (FCC)
USOC	Universal Service Order Code (AT&T)
USOP	User Service Order Profile
USOS	United States On-orbit Segment
USP	1. Universal Service Plan
	2. Usage Sensitive Pricing (tariff)
USPID	User Service Profile Identifier
USPTO	United States Potent and Trademark Office
USRT	Universal Synchronous Receiver-Transmitter
	(computer chip)
USS	1. Unstructured Supplementary Services (GSM)
	2. Unilateral Synchronization System
USSA	United States Suppliers Association
USSB	United States Satellite Broadcasting (company)
USSC	Upper Sideband Suppressed Carrier (modulation)
USSD	Unstructured Supplementary Subscriber Data (GSM)
USSI	Universal Synchronous Serial Interface
USTA	United States Telephone Association (U.S.A.)
USTSA	United States Telephone Suppliers Association (U.S.A.)
USTTI	United States Telecommunications Training Institute
	(standards)

UT	1. Universal **T**ime
	2. User **T**erminal
	3. Upper **T**ester (ATM)
UTA	Utilities **T**elecommunications **A**ssociation (U.S. standards)
UTAM	**U**nlicensed **T**ransition **A**nd **M**anagement (microwave)
UTC	1. Universal **T**ime **C**oordinated
	2. United **T**elecom **C**ouncil
UTDR	Universal **T**runk **D**ata **R**ecord
UTF-8	**U**CS **T**ransformation **F**ormat-**8**
UTOPIA	Universal **T**est and **Op**erations **I**nterface for **A**TM
UTP	**U**nshielded **T**wisted-**P**air (cables)
UTQFP	**U**ltra-**T**hin **Q**uad **F**lat **P**ack (microchips)
UTR	Universal **T**one **R**eceiver
UTRA	**U**MTS **T**errestrial **R**adio **A**ccess
UTRAN	**U**MTS **T**errestrial **R**adio **A**ccess **N**etwork
UTS	Universal **T**elephone **S**ervice
UTSOP	**U**ltra-**T**hin **S**mall-**O**utline **P**ackage (microchips)
UUCP	**U**nix-to-**U**nix **C**opy **P**rotocol
.uud	Name extension for the binary files translated to ASCII format using **UUD**ecode (computer)
UUD	**U**NIX-to-**U**NIX **D**ecoding (computer)
UUDECODE	**U**nix-to-**U**nix **Decod**ing (computer)
.uue	Name extension for the binary files translated to ASCII format using **UUE**ncode (computer)
UUE	**U**NIX-to-**U**NIX **E**ncoding (computer)
UUENCODE	**U**nix-to-**U**nix **Encod**ing
UUI	1. **U**nified **U**ser **I**nterface
	2. **U**nique **U**ser **I**dentifier
	3. **U**ser-to-**U**ser **I**ndicator (ATM)
	4. **U**ser-to-**U**ser **I**nformation (ISDN)
UUID	**U**niversally **U**nique **Id**entifier
UUT	**U**nit **U**nder **T**est
UV	1. **U**ltra-**V**iolet
	2. **U**nified **V**oice
UV-PROM	**U**ltra-**V**iolet erasable **PROM** (memory)
UVA	**U**ltra-**V**iolet **A**lpha
UVB	**U**ltra-**V**iolet **B**eta
UVLO	**U**nder **V**oltage **L**ock-**O**ut (thresholds)
UW	**U**nique **W**ord
UWB	**U**ltra-**W**ide-**B**and
UWC	**U**niversal **W**ireless **C**ommunications
UWCC	**U**niversal **W**ireless **C**ommunications **C**onsortium
UXGA	**U**ltra-**Ex**tended **G**raphics **A**rray (monitors standard)

U

V v

v	Symbol for **v**elocity
V	1. Symbol for **V**olt (unit of voltage or potential difference)
	2. Symbol for **V**ertical polarization (satcom)
	3. Symbol for **V**oltmeter or an electron **V**acuum tube
	4. **V**ocabulary and related subjects (ITU-T)
V.Chip	**V**iolence **Chip** (TV filter)
V.ASVD	**V**ersion **A**nalog **S**imultaneous **V**oice and **D**ata (modems)
V.AVD	**V**ersion **A**lternating **V**oice and **D**ata (modems)
V.DSVD	**V**ersion **D**igital **S**imultaneous **V**oice and **D**ata (modems)
V.fast	**V**ersion **Fast**-speed (modems)
V.FC	**V**ersion **F**ast **C**lass (modems)
V.PCM	**V**ersion **P**ulse-**C**ode **M**odulation (modem)
V&H	**V**ertical **& H**orizontal
V+D	**V**oice **Plus D**ata
V-band	Radio frequency **band** ranging from 40 to 75 GHz (radar)
V-Commerce	**V**oice **Commerce**
V-MOS	**V**-groove **MOS** (semiconductors)
V-pol	**V**ertical **pol**arization
V-root	**V**irtual **root**
V/D	**V**oice **and D**ata
V/F	**V**oltage-**to-F**requency converter (electronics)

V

V/m	Volts **per m**eter (electric field gradient)
V/T	Virtual **T**ime (Internet)
VA	Volt–**A**mpere (electronics)
VAB	1. **V**alue-**A**dded **B**usiness partner (Hewlett-Packard)
	2. **V**oice **A**nswer **B**ack
	3. **V**ehicle **A**ssembly **B**uilding (Kennedy Space Center, Florida)
Vac	Volts **A**lternating **C**urrent (electronics)
VAC	1. **V**oice **A**ctivity **C**ompression
	2. **V**ehicle **A**ccess **C**ontrol
	3. **V**alue-**A**dded **C**arrier
VACC	**V**alue-**A**dded **C**ommon **C**arrier
VAD	1. **V**oice **A**ctivated **D**ialing
	2. **V**oice **A**ctivity **D**etector (GSM)
	3. **V**alue-**A**dded **D**ealer
	4. **V**apor-phase **A**xial **D**eposition process (optical fibers)
VADIS	**V**oice **A**nd **D**ata **I**ntegrated **S**ystem
VADS	**V**alue-**A**dded **D**ata **S**ervice
VADSL	**V**ery-high-speed **A**symmetrical **DSL** (access)
VAFC	**VESA A**dvanced **F**eature **C**onnection
VAIVR	**V**oice **A**ctivated **I**nteractive **V**oice **R**esponse
VAM	1. **V**alue-**A**dded **M**odule (fiber optics)
	2. **V**ersatile **A**ccess **M**ultiplexer
VAN	**V**alue-**A**dded **N**etwork (Intelsat)
VANA	**V**ector **A**utomatic **N**etwork **A**nalyzer
VANS	**V**alue-**A**dded **N**etwork **S**ervice (data transmission)
VAP	1. **V**alue-**A**dded **P**rocess (networking)
	2. **V**ector **A**daptive **P**redictive (coding)
	3. **V**irtual **A**ssembly **P**rogram
VAPD	**V**oice **A**ctivated **P**remier **D**ialing
VAPN	**V**oice **A**ccess to **P**rivate **N**etwork
VAR	1. **V**alue-**A**dded **R**eseller (ICT)
	2. **V**isual-**A**ural **R**ange
	3. **V**olt-**A**mpere-**R**eactive (power generators)
	4. **var**iable
varactor	**var**iable re**act**ance diode (semiconductors)
varicap	**var**iable **cap**acitance diode (semiconductors)
varistor	**var**iable resi**stor** (semiconductors)
VARTI	**V**alue-**A**dded **R**eseller **T**elephone **I**ntegrator (PC telephony)
VAS	**V**alue-**A**dded **S**ervice
VASP	**V**alue **A**dded **S**ervice **P**rovider
VASCAR	**V**isual **A**verage **S**peed **C**omputer **A**nd **R**ecorder

VAX	**V**irtual **A**ddress **Ex**tension (minicomputers)
VB	**V**isual **B**asic (software)
VBA	**V**isual **B**asic for **A**pplications (computer)
VBD	**V**oice-**B**and **D**ata
VBE	**V**ESA **B**IOS **E**xtensions
VBI	**V**ertical **B**linking **I**nterval (TV signal)
VBNS	**V**ery-high-speed **B**ackbone **N**etwork **S**ervice (optics)
VBR	1. **V**ariable **B**it **R**ate (ATM)
	2. **Br**eakdown **V**oltage (FETs)
VBRE	**V**ariable **B**it **R**ate **E**ncoding
VBS	**V**ariable **B**it-rate **S**ervice
VBScript	**V**isual **B**asic **Script**ing edition (software)
VBW	**V**ideo **B**and**w**idth
VBX	**V**isual **B**asic **Ex**tension
VC	1. **V**irtual **C**hannel (SONET and ATM)
	2. **V**irtual **C**ontainer (MUX)
	3. **V**irtual **C**ircuit (X.25)
	4. **V**irtual **C**onnection (ATM)
	5. **V**irtual **C**all (X.25)
	6. **V**ideo **C**ipher
	7. **V**ideo **C**onferencing
VCA	1. **V**oice **C**onnecting **A**greement
	2. **V**oltage-**C**ontrolled **A**mplifier (electronics)
	3. **V**irtual **C**hannel **A**daptation (MUX)
VCACHE	**V**FAT driver **CACHE** (Windows 95)
vCalendar	**v**irtual **Calendar**
VCAPS	**V**ertical **C**avity **A**mplifying **O**ptical **S**witch
vCard	**V**irtual **Card**
VCC	1. **V**irtual **C**hannel **C**onnection (ATM)
	2. **S**upply **V**oltage—**positive**
VCCS	**V**oltage-**C**ontrolled **C**urrent **S**ource (electronics)
VCD	1. **V**ideo **C**ompact **D**isk
	2. **V**ariable **C**apacitance **D**iode (electronics)
	3. **V**ector **C**orrelation **D**etection
VCD IC	**V**ector **C**orrelation **D**etection **I**nterference **C**anceller
VCEP	**V**ideo **C**ompression/**E**xpansion **P**rocessor (chip)
VCF	**V**irtual **C**ard **F**ile
VCFU	**V**ideo **C**rypt **F**eedback **U**nit (TV broadcast)
VCI	**V**irtual **C**hannel **I**dentifier (ATM)
VCL	1. **V**irtual **C**hannel **L**ink (ATM)
	2. **V**isual **C**omponent **L**ibrary (Borland Delphi products)
VCM	1. **V**oice **C**oding **M**odule
	2. **V**irtual **C**hannel **M**emory

V

VCN	1. Virtual Corporate Network
	2. Virtual Circuit Number (X.25)
VCO	Voltage-Controlled Oscillator (electronics)
VCOMM	Virtual Communications (device driver)
VCOS	Visible Caching Operating System
VCPI	Virtual Control Program Interface (MS-DOS programs)
VCPU	Video Compression Processor Unit
VCR	Video-Cassette Recorder
VCS	Virtual Communications System
VCSEL	Vertical-Cavity Surface-Emitting Laser
VCSELD	Vertical-Cavity Surface-Emitting Laser Diode
VCSN	Virtual Channel Sub-Network
VCT	Virtual Channel Termination (MUX)
VCVA	Voltage-Controlled Variable Attenuator
VCVS	Voltage-Controlled Voltage Source (electronics)
VCXC	Virtual Circuit Cross-Connect (MUX)
VCXO	Voltage-Controlled Crystal Oscillator (electronics)
VD	Virtual Destination
VDAP	Voice and Data Access Point
VDB	Visitor Database
VDC	Volts, Direct Current (electronics)
VDD	Virtual Device Driver
V$_{DD}$	Voltage Drain-Drain (electronics)
VDDD	VNET International Direct Distance Dialing (feature)
VDE	1. Visual Development Environment
	2. Video Decompression Engine
VDF	Voice Distribution Frame
VDI	1. Video Device Interface
	2. Virtual Device Interface
VDIF	VESA Display Information Forum
VDISK	Virtual DISK
VDL	Vienna Definition Language (programming)
VDM	1. Voice Data Multiplexer
	2. Video Display Metafile
VDO	Voltage Drop-Out
VDPC	Voice Data Port Card
VDRV	Variable Data Rate Video
VDS	Variable-Depth Sonar
VDSL	Very-high-data-rate DSL (access)
VDT	1. Video Display Terminal (CRT)
	2. Video Dial Tone (service)
VDU	1. Video Display Unit (computer monitor)
	2. Visual Display Unit
VEN	Virtual Ethernet Network

VERONICA	Very Easy Rodent Oriented Netwide Index to Computerized Archives (Internet)
VERT	Vertical (polarization)
VES	Virtual Environment Software
VESA	Video Electronics Standards Association
VESDA	Very Early Smoke Detection Alarm
VF	1. Voice Frequency (signal or channel)
	2. Variance Factor (ATM)
	3. Vacuum Fluorescent (display)
VFAT	Virtual File Allocation Table (computer)
VFC	1. Version Fast Class (modems)
	2. Voltage-to-Frequency Converter (electronics)
VFCTG	Voice Frequency Carrier Telegraph
VFD	Vacuum Fluorescent Display
VFDF	Voice Frequency Distribution Frame
VFDN	Voice Frequency Directory Number (Northern Telecom)
VFET	Vertical FET (semiconductors)
VFG	Virtual Facilities Group
VFIP	Voice File Interchange Protocol
VFO	Variable-Frequency Oscillator (electronics)
VFRAD	Voice Frame Relay Access Device
VFS	1. Virtual File Store
	2. Virtual File System
VFTG	Voice Frequency Telegraph
VG	1. Videotex Gateway
	2. Voice Grade (local loop)
VGA	1. Video Graphics Adapter (640×480 pixel monitors)
	2. Video Graphics Array (monitors standard)
	3. Video Graphic Accelerator (circuit)
	4. Variable Gain Amplifier (electronics)
VGC	1. Voice Grade Channel
	2. Voice Grade Circuit
VGE	1. Voluntary Group of Experts (ITU)
	2. Voice Grade Equivalent
VGF	Voice Grade Facility
VGPL	Voice Grade Private Line
VGrep	Visual Grep (UNIX utility)
VHADSL	Very High-bit-rate Asymmetric DSL (access)
VHD	Very High Density (data storage)
VHDCI	Very High Density Cable Interconnect (PC cable)
VHDL	1. Very High-speed IC hardware Description Language
	2. VHSIC Hardware Description Language (programming)

V

	3. Very High-Density Logic (digital electronics)
VHDSL	Very High-bit-rate **DSL** (access)
VHE	Virtual Home Environment (GSM)
VHF	Very High Frequency (30–300 MHz range)
VHI	Virtual Host Interface
VHLL	Very High-Level Language (programming)
VHS	Video Home Systems (video format)
VHSD	Very High-Speed Data
VHSIC	Very High-Speed Integrated Circuit
VI	Virtual Instrument
VIA	1. Vendors ISDN Association
	2. Virtual Interface Architecture
VIC	Voice Interface Card (Cisco)
VIDA	Voice, Interoperability, Data and Access
VIDF	Vertical Intermediate Distributing Frame
VIEWS	Visualizing Impacts of Earthquakes With Satellite (Japan)
VIL	Vertical Inline
VIM	1. Voice Interface Module
	2. Vendor Independent Messaging
VINES	Virtual Networking Software (networking)
VIP	1. Versatile Interface Processor (Cisco)
	2. Virtual Internet Protocol (Cisco)
VIPA	1. Virtual Image Phased Array
	2. Virtual Internet Protocol Addressing
VIPR	1. Virtual Internet Protocol Routing (Cisco)
	2. Voice over **IP** Router
VIR	Vertical Interval Reference (video)
VIRR	Visible and Infrared Radiometer (remote sensing)
VIRS	Visible and Infrared Scanner (remote sensing)
VISA	1. Virtual Instrument Software Architecture
	2. VXI System Alliance (standard)
VisiCalc	Visible Calculator (historic software)
VISSR	Visible and Infrared Spin Scan Radiometer (remote sensing)
VISTA	VLSI Integrated Set Top Architecture (broadcasting)
Visual J + +	Java Visual programming environment (Microsoft)
VITA	1. Volunteers In Technical Assistance
	2. VME International Trade Association
Vital	VHDL Initiative Toward ASIC Libraries (organization)
VITC	Vertical Interval Time Code (VCR)
VITS	Vertical Interval Test Signal
VIU	Voice Interface Unit
VL Bus	VESA Local **Bus** (computer)

VLAN	Virtual **LAN**
VLB	VESA Local Bus (computer)
VLBI	Very Long Baseline Interferometry (remote sensing)
VLC	Variable Length Coding
VLD	1. Variable Length Decoding
	2. Vehicle Location Detection system
VLDB	Very Large Database
VLF	Very Low Frequency (10 Hz–30 KHz range)
VLIW	Very Long Instruction Word (microprocessor architecture)
VLM	1. Virtual Loadable Module (networking)
	2. Very Large Memory
VLN	Visitor Local Number (GSM)
VLR	Visitors Location Register (cellular networks)
VLRE	Very Low Rate Encoding
VLS	1. Virtual Library System
	2. Virtual LAN Link State protocol (Cabletron Systems)
VLSI	Very Large-Scale Integration (microchips)
VLSIC	Very Large-Scale Integrated Circuit
VLSM	Variable-Length Subnet Mask
VLT	Video Look-up Table
VM	1. Voice Mail
	2. Voice Messaging
	3. Virtual Memory
	4. Virtual Machine
VMC	VESA Media Channel (bus)
VME	1. Virtual Machine Environment (operating system)
	2. Versa Module Eurocard
VMEbus	Versatile Modular **E-bus**
VMEC	Voice Messaging Educational Committee (e-learning)
VMF	Validation Message Fraud
VMI	1. V Series Modem Interface
	2. Vehicle Management Information
VMOS	Vertical **MOS** (semiconductors)
VMR	Validation Monitoring and Removal
VMS	1. Voice Mail Service
	2. Voice Messaging Service (fix and mobile)
	3. Virtual Messaging Service (fix telephony)
	4. Virtual Memory System (operating system)
	5. Vessel Monitoring System (shipping)
VMTP	Versatile Message Transaction Protection
VMUIF	Voice Messaging User Interface Forum
VMWI	Visual Message Waiting Indicator
VMX	Voice Message Exchange

VN	Virtual Network
VNA	1. Vertical Network Analyzer
	2. Vector Network Analyzer
VNET	Virtual private Network (MCI)
VNIR	Visible and Near-Infrared Radiometer (remote sensing)
VNL	Via Net Loss (transmission)
VNLF	Via Net Loss Factor (parameter)
VNN	Voice News Network (broadcaster)
VNS	1. Virtual Network Service
	2. Virtual Network System
VO	Verification Office
VOA	1. Variable Optical Attenuator
	2. Voice Of America (broadcaster)
VoATM	Voice over ATM (service)
VoBB	Voice over Broad-Band
VOC	Virtual Operator's Console
vocoder	voice coder
VoD	Video-on-Demand (service)
VODAS	Voice Operated Device Anti-Sing (telephone circuits)
VODER	Voice Operation Demonstrator
VoDSL	Voice over DSL (service)
VoFR	Voice over Frame-Relay (service)
VOGAD	Voice-Oriented Gain-Adjusting Device
VoHDLC	Voice over HDLC (service)
VoIP	Voice over Internet Protocol (service)
VoIPF	Voice over Internet Protocol Forum
VOL	volume control (circuit diagrams)
VOLCAS	Voice-Operated Loss-Control And Signaling (device)
VOM	Volt-Ohm-Milliameter (test equipment)
VoMBN	Voice-over-Multiservice Broadband Network
VON	Voice-On-the-Net (service)
VoP	Voice-over-Packet (service)
VOR	VHF Omnidirectional Range (navigation system)
VORC	Voice-Operated Relay Circuit
VORDAC	VHF Omnidirectional Range and Distance-measuring equipment for Area Coverage
Vortac	VHF omnidirectional range tacan (air navigation)
VORTAC	VOR collocated and/or combined with TACAN
VOST	VESA Open Set Top (group)
VOTS	VMS OSI Transport Service
VOW	Voice Order Wire
VOX	1. Voice-Operated Transmit
	2. Latin word for Voice meaning Telephone number other than that of Fax number (on business cards)

Voxel	**Vo**lume pi**xel** (3D image)
VoxML	**Voice M**arkup **L**anguage (programming)
VP	1. **V**oice **P**ort
	2. **V**irtual **P**ath (MUX)
	3. **V**ideo **P**rocessor
VP PBX	**V**oice **P**ort **PBX**
VP TEL	**V**oice **P**ort **Tel**ephone
VPA	1. **V**ariable **P**ower **A**ttenuator
	2. **V**irtual **P**ath **A**daptation
	3. **V**oice **P**ort **A**dapter
VPACK	**V**ertical **Pack**age
VPAR	**V**oice **P**ort **A**dapter **R**ack
VPC	1. **V**oice **P**ort **C**luster
	2. **V**irtual **P**ath **C**onnection (ATM)
VPCI	**V**irtual **P**ath **C**onnection **I**dentifier
VPD	1. **V**ariable **P**hase **D**ivider
	2. **V**irtual **P**rinter **D**evice driver
	3. **V**apor-**P**hase **D**issolution (semiconductors)
VPDN	**V**irtual **P**rivate **D**ata **N**etwork
VPDS	**V**irtual **P**rivate **D**ata **S**ervice (MCI)
VPE	**V**apor-**P**hase **E**pitaxy (semiconductors)
VPEC	**V**irginia **P**ower **E**lectronics **C**enter (U.S.A.)
VPI	1. **V**ariable **P**ath **I**dentifier (ATM)
	2. **V**irtual **P**acket **I**dentifier (MPLS)
VPIM	**V**oice **P**rofile for **I**nternet **M**essaging
VPL	**V**irtual **P**ath **L**ink (ATM)
VPM	**V**oice **P**rocessing **M**odule (Sprint)
VPN	**V**irtual **P**rivate **N**etwork (Internet)
VPO	**P**inch-**O**ff **V**oltage (FETs)
VPOTS	**V**ery **P**lain **O**ld **T**elephone **S**ervice
VPP	**V**XI **P**lug-and-**P**lay
VPS	1. **V**ariable **P**hase **S**hifter
	2. **V**apor-**P**hase **S**oldering
VPSN	**V**irtual **P**ath **S**ub-**N**etwork
VPT	1. **V**irtual **P**ath **T**ermination
	2. **V**irtual **P**rivate **T**runking
	3. **V**irtual **P**rinter **T**echnology
VPU	1. **V**irtual **P**hysical **U**nit
	2. **V**oice **P**rocessing **U**nit
VPXC	**V**irtual **P**ath **Cross-C**onnect (MUX)
VQ	**V**ector **Q**uantization
VQFP	**V**ery-fine-pitch **Q**uad **F**lat **P**ack (microchips)
VQL	**V**ariable **Q**uantizing **L**evel

V

VR	1. Virtual Reality (ICT)
	2. Voice Recognition
	3. Voltage Regulator (electronics)
VRAM	Video RAM (memory)
VRC	Vertical Redundancy Check (transmission)
VRD	1. Virtual Ring Down
	2. Virtual Retina Display
VRID	Virtual Router Identifier
VRM	Voice Recognition Module
VRML	1. Virtual Reality Modeling Language (programming)
	2. Virtual Reality Markup Language (programming)
Vrms	Volt root-mean-square (electronics)
VROOMM	Virtual Run-time Object-Oriented Memory Manager
VRPRS	Virtual Route Pacing Response in SNA
VRRP	Virtual Router Redundancy Protocol (IETF)
VRU	Voice Response Unit
vs	versus (technical curves)
VS	1. Virtual Scheduling (ATM)
	2. Virtual Source
	3. Virtual Storage
	4. Vertical Synchronization
VS&F	Voice Store and Forward (service)
VS/VD	Virtual Source/Virtual Destination
VSA	Virtual Scheduling Algorithm
VSAC	Very Small Aperture Check
VSAT	Very Small Aperture Terminal (satcom)
VSB	1. Vestigial Side-Band (modulation)
	2. VME Subsystem Bus
VSC	1. Vertical Service Code (telephony)
	2. Virtual Switch Controller (telephony)
	3. Variable Speech Control (tape recording)
VSE	Voice Services Equipment
VSELP	Vector Sum Excited Linear Predictive coding (voice coding)
VSIM	Virtual Surface Image Memory
VSM	Vestigial-Sideband Modulation
VSNET	Virtual SS7 Network
VSNL	Videsh Sanchar Nigam, Ltd (Indian operator)
VSO	Very Small Outline
VSOP	Very Small Outline Package (Japanese)
VSR	Very Short Reach (fiber optics)
Vss	Voltage Source-Source (electronics)
VSS	1. Voice Server System
	2. VHDL System Simulator

VSW	Voice **Sw**itcher
VSWR	**V**oltage **S**tanding-**W**ave **R**atio (transmission)
VSX	**V**erification **S**uite for **X**/open
VSYNC	**V**ertical **Sync** (video signals)
VT	1. **V**irtual **T**ributary (SONET)
	2. **V**irtual **T**erminal (ISO)
	3. **V**irtual **T**runk
	4. **V**ertical **T**abulation (character)
	5. **V**ideo **T**eleconferencing
VT fuse	**V**ariable-**T**ime **fuse**
VT2	**V**irtual **T**ributary-**2**
VT6	**V**irtual **T**ributary-**6**
VTA	1. **V**ideo **T**erminal **A**dapter (ISDN)
	2. **V**irtual **T**runk **A**gent
VTAC	**V**ermont **T**elecommunications **A**pplications **C**enter
VTAM	**V**irtual **T**elecommunications **A**ccess **M**ethod (IBM PCs)
VTC	**V**ideo **T**ele-**C**onference (U.S. Air Force)
VTCO	**V**oltage–**T**emperature **C**ut-**O**ff (battery charging)
VTD	**V**irtual **T**imer **D**evice driver
VTG	1. **V**oltage **T**iming **G**enerator
	2. **V**irtual **T**ributary **G**roup
VTI	**V**ietnam **T**elecom **I**nternational
VTIR	**V**isible and **T**hermal **I**nfrared **R**adiometer (remote sensing)
VTM	1. **V**oltage-**T**unable **M**agnetron
	2. **V**oltage **T**ransformation **M**odule
	3. **V**essel **T**raffic **M**anagement (shipping)
VTN	**V**endor **T**ype **N**umber
VTNS	**V**irtual **T**elecommunications **N**etwork **S**ervices
VTO	**V**oltage **T**uned **O**scillator
VToA	**V**oice and **T**elephony **o**ver **A**TM (service)
VTOH	**V**irtual **T**ributary **O**ver**h**ead
VTP	1. **V**irtual **T**runking **P**rotocol (Ethernet)
	2. **V**irtual **T**erminal **P**rotocol (ISO)
VTPP	**V**ariable **T**erm **P**ricing **P**lan (AT&T)
VTR	**V**ideo-**T**ape **R**ecorder (equipment)
VTS	1. **V**ehicular **T**echnology **S**ociety
	2. **V**essel **T**raffic **S**ervice (navigation)
	3. **V**irtual **T**erminal **S**ervice
VTTH	**V**ideo **T**o **T**he **H**ome
VTU	1. **V**oucherless **T**op **U**p
	2. **V**ideo **T**eleconferencing **U**nit
VTU-C	**V**DSL **T**ransmission **U**nit—**C**entral Office
VTU-O	**V**DSL **T**ransmission **U**nit—**O**NU

V

VTU-R	VDSL Transmission Unit—Remote
VTVM	Vacuum-Tube Voltmeter (test equipment)
VTX	Videotex
VVA	Voltage-Variable Attenuator
VU	Volume Unit (speech power parameter)
VU Meter	Volume Unit Meter
VUG	Voice User Group
VUI	Video User Interface
VVA	Voltage-Variable Attenuator (microwave)
VWP	Variable Wavelength Path
VxD	Virtual device Driver
VXI	VME Extension for Instrumentation
VXIbus	VME bus Extension for Instrumentation
VXML	Voice XML (programming language)
VXO	Variable Crystal Oscillator (electronics)
VY	Very (Morse code transmissions)

W w

W	1. Symbol for **W**att (unit of electric power)
	2. Symbol for **w**ideband
	3. **W**ait (Hayes compatible modems)
W-band	Radio frequency **band** ranging from 75 to 110 GHz (radar)
W-CDMA	**W**ideband **CDMA** (access)
W-DCS	**W**ideband **D**igital **C**ross-connect **S**ystem
W-OFDM	**W**ideband **OFDM** (transmission)
W/G	**W**ave**g**uide
W/sr	**W**att **per steradian**
W3	**W**orld-**W**ide **W**eb
W3C	**W**orld-**W**ide **W**eb **C**onsortium (standards organization)
W3C2	**W**orld-**W**ide **W**eb **C**onference **C**ommittee
WA DSP	**W**ideband **A**cquisition—**D**igital **S**ignal **P**rocessing
WAA	**W**ireless **A**dvertising **A**ssociation
WAAS	**W**ide-**A**rea **A**ugmentation **S**ystem (navigation)
WABI	**W**indows **A**pplication **B**inary **I**nterface (software)
WAC	**W**ireless **A**ccess **C**ontroller
WACK	**W**ait before transmitting positive **Ack**nowledgement
WACS	**W**hite **A**lice **C**ommunications **S**ystem (AT&T)
WADP	**W**ireless **A**pplication **D**elivery **P**latform
WADS	**W**ide-**A**rea **D**ata **S**ervice
WAE	**W**ireless **A**pplication **E**nvironment

W

WAG	**W**ireless **A**pplication **G**roup (WAP Forum)
WAGPS	**W**ireless **A**ssisted **G**lobal **P**ositioning **S**ystem
WAIP	**W**ireless **A**pplication **I**nfrastructure **P**rovider
WAIS	**W**ide-**A**rea **I**nformation **S**erver (Internet)
WAM	**W**ireless Sp**am**
WAN	**W**ide-**A**rea **N**etwork
WANMC	**WAN M**anagement **C**enter
WAP	1. **W**ireless **A**pplication **P**rotocol
	2. **W**ireless **A**ccess **P**oint
WAPvert	An ad**vert**isement on a **WAP** device
WAR	**W**ireless **A**pplication **R**eader (WAP)
WARC	**W**orld **A**dministrative **R**adio **C**onference (ITU events)
WARC-G	**WARC**—**G**eneral (ITU)
WARC-ST	**WARC** for **S**pace **T**elecommunications
wares	soft**wares** (ICT)
WAS	**W**ireless **A**ccess **S**ystem
WASI	**W**ide-**A**rea **S**ervice **I**dentifier
WASO	**W**estern-**A**rea **S**witch **O**perations
WASP	**W**ireless **A**pplication **S**ervice **P**rovider
WATM	**W**ireless **ATM**
WATS	**W**ide-**A**rea **T**elephone **S**ervices (AT&T)
WATSS	**W**ide-**A**rea **T**elephone **S**ervice **S**ystem
WATTC	**W**orld **A**dministrative **T**elegraph and **T**elephone Conference
.wav	Name extension for Windows **wav**eform audio format files
WAW	**W**aiter-**A**ctor-**W**ebmaster
WAWS	**W**ashington **A**rea **W**ideband **S**ystem
Wb	**W**e**b**er (unit of magnetic flux)
WB	1. **W**ide-**B**and
	2. **W**ire **B**onded
WBA	**W**ireless **B**roadcast **A**ccess
WBC	1. **W**ide-**B**and **C**hannel (FDDI)
	2. **W**ide-**B**and **C**ombiner
WBCS	1. **W**ide-**B**and **C**ommunications **S**ervice
	2. **W**ide-**B**and **C**ommunications **S**ystem
WBD	**W**ide-**B**and **D**ata (channel)
WBEAF	**W**ide-**B**and **E**arth station **A**ntenna **F**eed (Intelsat)
WBEM	**W**eb-**B**ased **E**nterprise **M**anagement (network management)
WBFH	**W**ide-**B**and **F**requency **H**opping (FCC)
WBFM	**W**ide-**B**and **F**requency **M**odulation
WBMCS	**W**ireless **B**roadband **M**ultimedia **C**ommunications **S**ystems

WBPA	Wide-Band Power Amplifier (Intelsat)
WBPM	Wide-Band Phase Modulation
WBT	Web-Based Training (e-learning)
WBTIC	Web-Based Training Information Center (e-learning)
WBVTR	Wide-Band Video Tape Recorder
WC	Wireless Consortium
WCA	1. Web Clipping Application
	2. Wireless Communications Association
WCAPs	Wireless Competitive Access Providers
WCAV	Web Clipping Application Viewer
WCBC	West Coast Billing Center
WCC	Wireline Common Carrier
WCCP	Web-Cache Control Protocol
WCDMA	Wideband CDMA (IMT-2000 access)
WCP	1. Wireless Certificate Profile (WAP Forum)
	2. Wireless Communication Platform/Product
	3. Wireless Communications Protocol
	4. Web Clipping Proxy (server)
WCS	Wireless Communications Services
WCT	WIPO Copyright Treaty
WCV	Weighted Call Value
WD	1. Withdrawn and Deleted
	2. Wavelength Division
WDA	Wireless Digital Assistant
WDCS	Wideband Digital Cross-connect System
WDEF	Windows Definition Function (Macintosh application)
WDF	Wireless Data Forum
WDGPS	Wide-area Differential GPS
WDL	Windows Driver Library
WDM	Wavelength-Division Multiplexing
WDMA	Wavelength-Division Multiple Access
WDMUX	Wavelength Demultiplexer
WDP	1. Workforce Development Program
	2. Wireless Datagram Protocol (WAP Forum)
WDT	Watch-Dog Timer
WDU	Winchester Drive Unit
Web	World-Wide Web
Webcam	Web digital video camera
Webcast	A Broadcast by using World Wide Web
Webinar	Web-based seminar
WebNFS	Web Network File System (client/server protocol)
Webphone	Internet telephone
WebTV	A Web access device for that uses a TV as a display
Webzine	Web-based magazine

W

WEC	Western Electric Company
WECA	Wireless Ethernet Compatibility Alliance
WECC	Web-Enabled Contact Center
WECO	Western Electric Company
WEEE	Waste from Electrical and Electronic Equipment directive
WELL	Whole Earth Lectronic Link (Web site)
WEP	Wired Equivalency Privacy (LANs)
WESTAR	Western-states Air Resources Council
WFC	Winchester Floppy Controller
WFM	Wired For Management Baseline (Intel)
WFQ	Weighted Fair Queuing (data transmission)
WFS	Woodstock File Server (Xerox)
WFW	Windows For Workgroups
WFWG	Windows For Work-Groups
WG	Working Group (IEEE)
WGDTB	Working Group on Digital Television Broadcasting
WGET	Working Group on Emergency Telecommunications (ITU)
WGIG	Working Group on Internet Governance (United Nations)
WGIH	Working Group on Information Highway (TSACC)
WGN	White Gaussian Noise
WGS	Worldwide Geodetic System
Wh	Watt-hour (power)
WH	1. West Hemispheric beam
	2. Western Host
Wi-Fi	Wireless Fidelity (IEEE)
Wi-Fi5	Wireless Fidelity at **5 GHz** range (IEEE)
Wi-LAN	Wireless LAN
WIA	1. World Internetworking Alliance (Internet)
	2. Wireless Institute of Australia
WIC	WAN Interface Card (Cisco)
WID	1. Wireless Integration Device
	2. Wireless Interface Device
WIDS	Wireless Intrusion Detection Systems
WIFS	Wide-Field Sensor (remote sensing)
WIG	Wireless Interoperability Group (WAP Forum)
WiLL	Wireless Local Loop system (Motorola)
WIM	1. WAP Identity Module
	2. Wireless Instant Messaging (cellular networks)
WiMAX	Worldwide interoperability for Microwave **Access** (standard)
WIMP	Windows Icons Menus Pointing device (ICT)

WIN	1. **W**ireless **I**ntelligent **N**etwork
	2. **W**ireless **I**n-building **N**etwork
WIN95	**Win**dows **95**
WIN98	**Win**dows **98**
WIN2000	**Win**dows **2000**
WIN CE	**Win**dows **CE**
Windows ME	**Windows M**illennium **E**dition
Windows NT	**Windows N**ew **T**echnology
WINDS	**W**ideband **I**nter **N**etworking engineering test and **D**emonstration **S**atellite (Japan)
WINF	**W**ireless **I**nformation **N**etworks **F**orum, Inc
WinG	**Win**dows **g**ames
WinISDN	An **ISDN** communications API designed for **Win**dows (interface)
WINS	**W**indows **I**nternet **N**aming **S**ervice
WinSock	**Win**dows **Sock**et API (program)
Wintel	A computer using **Win**dows Operating System and In**tel**' CPU
WIP	1. **W**ashington **I**nternet **P**roject
	2. **W**ork **I**n **P**rocess (equipment)
WIPO	**W**orld **I**ntellectual **P**roperty **O**rganization (Geneva)
WISP	**W**ireless **I**nternet **S**ervice **P**rovider
WIT	**W**ashington **I**nternational **T**eleport
WITS	1. **W**ireless **I**nterface **T**elephone **S**ystem
	2. **W**ashington **I**ntegrated **T**elecommunications **S**ystem (U.S.A.)
WITSA	**W**orld **I**nformation **T**echnology and **S**ervices **A**lliance
WIXC	**W**avelength **I**nterchanging **Cross-C**onnect (fiber optics)
wizywig	Same as **WYSIWYG**
WLAN	**W**ireless **L**ocal-**A**rea **N**etwork
WLANA	**W**ireless **LAN A**lliance (U.S. standards organization)
WLI	**W**ireless **LAN I**nteroperability
WLIF	**W**ireless **LAN I**nteroperability **F**orum
WLL	**W**ireless **L**ocal **L**oop
WLNP	**W**ireless **L**ocal **N**umber **P**ortability
WLRL	**W**ireless **LAN R**esearch **L**aboratory (U.S.A.)
WM	**W**avelength **M**odulation
.wma	Name extension for **W**indows **M**edia **A**udio files (computer)
WMAN	**W**ireless **M**etropolitan-**A**rea **N**etwork
WMATV	**W**ireless **M**aster **A**ntenna **TV** system
.wmf	Name extension for **WMF** files (graphics)
WMF	1. **W**indows **M**etafile **F**ormat (Image format)

W

	2. **W**ireless **M**ultimedia **F**orum
WMI	**W**indows **M**anagement **I**nstrumentation (Microsoft)
WML	1. **W**ireless **M**arkup **L**anguage (programming)
	2. **W**ebsite **M**eta **L**anguage (programming)
WMO	**W**orld **M**eteorological **O**rganization
WMPN	**W**orld's **M**ost **P**owerful **N**etwork (consortium)
WMUX	**W**avelength **M**ultiplex**er**
.wmv	Name extension for **W**indows **M**edia **V**ideo files (computer)
WOL	**W**ake-**O**n-LAN (computer motherboard)
WOM	**W**rite-**O**nly **M**emory (data storage)
WORA	"**W**rite **O**nce, **R**un **A**nywhere" (Java slogan)
WORM	**W**rite **O**nce, **R**ead **M**any times (optical memory disks)
WORN	**W**rite **O**nce, **R**ead **N**ever (as a joke)
WOS	**W**holesale **O**perator **S**ervice
WOSA	**W**indows **O**pen **S**ystem **A**rchitecture
WOTAN	**W**avelength-**A**gile **O**ptical **T**ransport and **A**ccess **N**etwork project
WOTS	**W**ireless **O**ffice **T**elecommunications **S**ystem
.wp	Name extension for **W**ord **P**erfect word processor files (computer)
WP	1. **W**ord **P**rocessing
	2. **W**ire **P**hoto
WP/8B	**W**orking **P**arty **8B** (ITU-R)
WPA	**W**i-Fi **P**rotected **A**ccess
WPABX	**W**ireless **PABX**
WPAD	**W**eb **P**roxy **A**uto-**D**iscovery (protocol)
WPAN	**W**ireless **P**ersonal-**A**rea **N**etwork
WPBX	**W**ireless **P**rivate **B**ranch **E**xchange
.wpg	Name extension for **W**ord**P**erfect **G**raphics format files (graphics)
WPG	**W**ireless **P**rotocols **G**roup (WAP Forum)
WPKI	1. **W**ireless **P**ublic **K**ey **I**nfrastructure
	2. **W**earable **P**ublic **K**ey **I**nfrastructure
WPM	**W**ords **P**er **M**inute (parameter)
WPS	1. **W**ords **P**er **S**econd (parameter)
	2. **W**ireless **P**riority **S**ervices
	3. **W**ork **P**lace **S**hell (OS/2)
WQAM	**W**eighted **Q**audrature **A**mplitude **M**odulation
WR	1. **W**ithdrawn and **R**eplaced
	2. **Wr**ite (logic signal)
WRAM	**W**indows **RAM** (memory)
WRBYV	**W**hite, **R**ed, **B**lue, **Y**ellow, and **V**iolet (telephone cable)
WRC	**W**orld **R**adiocommunications **C**onference (ITU)

WREC	**W**eb **Re**plication and **C**aching
WRED	**W**eighted **R**andom **E**arly **D**etection (QoS)
.wri	Name extension for Microsoft **Wri**te format files (computer)
WRL	**W**ireless **R**ural **L**oop
WRS	**W**orldwide **R**eference **S**ystem (Landsat)
WRSS	**W**estern **R**egion **S**witch **S**upport
WRU	"**W**ho **A**re **Y**ou" (telex control character)
Ws	**W**att**s**econd (power)
WS	**W**ork **S**tation
WS-ANI	**W**ireless **S**ubscriber—**A**utomatic **N**umber **I**dentification
WSA	**W**ireless **S**pecialty **A**pparatus Company
WSC	1. **W**ireless **S**witching **C**enter
	2. **W**ireless **S**erver **C**ertificate
WSDL	**W**eb **S**ervice **D**escription **L**anguage (programming)
WSEAS	**W**orld **S**cientific and **E**ngineering **A**cademy and **S**ociety
WSF	**W**ork **S**tation **F**unction block
WSG	**W**ireless **S**ecurity **G**roup (WAP Forum)
WSI	1. **W**ireless **S**ystems **I**nternational
	2. **W**afer **S**cale **I**ntegration (microchips)
WSIS	**W**orld **S**ummit on the **I**nformation **S**ociety
WSP	1. **W**ireless **S**ervice **P**rovider
	2. **W**ireless **S**ession **P**rotocol
WST	1. **W**AP **S**ecurity **T**oolkit
	2. **W**ireless **S**ubscriber **T**erminal
WSTA	**W**all **S**treet **T**elecommunications **A**ssociation (standards)
WSW	**W**orld **S**pace **W**eek (events)
WSXC	**W**avelength **S**elective **Cross-C**onnect (fiber optics)
WT	**W**ireless **T**elephony
WTA	**W**ireless **T**elephony **A**pplication (WAP Forum)
WTAC	**W**orld **T**elecommunications **A**dvisory **C**ouncil (ITU)
WTB	**W**ireless **T**elecommunications **B**ureau (FCC)
WTDC	**W**orld **T**elecommunications **D**evelopment **C**onference
WTLS	**W**ireless **T**ransport **L**ayer **S**ecurity (WAP Forum)
WTN	**W**orking **T**elephone **N**umber
WTNG	**Wait**ing
WTO	**W**orld **T**rade **O**rganization (Switzerland)
WTP	**W**ireless **T**ransaction **P**rotocol
WTPF	**W**orld **T**elecommunication **P**olicy **F**orum
WTSC	**W**orld **T**elecommunications **S**tandardization Con-ference
WU-ATS	**W**estern **U**nion—**A**dvanced **T**ransmission **S**ystems
WUD	**Wo**ul**d** (Morse code transmissions)

W

WUI	1. Web User Interface (GUI)
	2. Western Union International (MCI)
WUT	Wafer Under Test
WV	Working Voltage (electronics)
WVDC	Working Volts DC (power)
WVI	Web Voice Integration
WVTR	Wideband Video Tape Recorder
WW	1. Wire-Wound (electronics)
	2. Wire-Wrap (electronics)
WWAN	Wireless Wide-Area Network
WWDD	World-Wide Direct Dialing
WWDM	Wide Wavelength-Division Multiplexing (GBE)
WWDSA	World Wide Digital System Architecture
WWMCCS	World Wide Military Command and Control System
WWNWS	World-Wide Navigational Warning Service (IMO)
WWV	World-Wide Vacuum
WWW	1. World-Wide Web (Internet)
	2. World Weather Watch (remote sensing)
WWWC	World-Wide Web Consortium
WX	Weather Report (Morse code transmissions)
Wz	West Zone beam
WZ1	World Zone 1

X x

X	1. Symbol for **ex**change or interchange
	2. Symbol for **ex**tension
	3. Symbol for words commencing with letters sounding like 'X' such as DBx which stands for **D**ata **B**ase **Acc**elerator
	4. Sign for **cross** or **cross-connection** or **coupling**
	5. Sign of prefix **trans**- (from one place or state to another)
	6. Symbol for a **Crystal** component
	7. Symbol for an **unknown** quantity or distance
X.25	**ITU protocol standard for WAN communications**
X-Band	Radio Frequency band (8 GHz–12 GHz)
X-MOS	**High-speed MOS** (semiconductors)
X-Off	Control character to **stop** information flow
X-On	Control character to **start** information flow
X-On/X-Off	**Transmitter On and Transmitter Off**
X-strap	**Cross-strap** (Intelsat)
XA	1. **Ex**tended **A**rchitecture (CD-ROM)
	2. **Ex**change **A**ccess
XA-SMDS	**Ex**change **A**ccess—**SMDS**
XAPIA	**X**.400 **A**pplication **P**rogram **I**nterface **A**ssociation
XAUI	**10GbE A**ttachment **U**nit **I**nterface
Xbar	**Crossbar**

X

XC	**Cross**-**C**onnect (MUX)
XCA	E**x**tended **C**ommunication **A**dapter
XCE	1. **Transponder C**ommand **E**lectronics
	2. **Transponder C**ontrol **E**lectronics
XCM	1. **Cross**-**C**onnect **M**odule (MUX)
	2. **Cross**-**C**onnect **M**ultiplexer (MUX)
XCMD	E**x**ternal **Com**mand (hyper card)
XCOPY	**C**opy E**x**tended (MS-DOS Command)
XCVR	**Transc**ei**v**e**r** (Morse code transmissions)
XDCR	**Transd**u**c**e**r** (electronics)
XDF	E**x**tended **D**istance **F**eature
XDI	**X**RI **D**ata **I**nterchange
XDMA	**X**ing **D**istributed **M**edia **A**rchitecture (Xing Corporation)
XDMCP	**X D**isplay **M**anager **C**ontrol **P**rotocol
XDP	E**x**tra **D**evice **P**ort (Panasonic)
XDR	1. E**x**ternal **D**ata **R**epresentation
	2. E**x**change **D**ata **R**epresentative
xDSL	A **generic** term for **DSL** technologies (access)
xEMS	**x**DSL **E**lement **M**anager **S**ystem
XENIX	Microsoft version of UNIX for microprocessors (OS)
XFCB	E**x**tra-**F**ast **C**omplementary **B**ipolar
XFCN	E**x**ternal **Fun**ctio**n** (hyper card)
XFN	**X**/Open **F**ederated **N**aming (Sun Microsystems)
XFR	Transfer
XGA	1. E**x**tended **G**raphics **A**dapter (1024×768 pixel monitors)
	2. E**x**tended **G**raphics **A**rray (monitors standard)
XGL	E**x**tended **G**raphics **L**ibrary
XHTML	1. **XML**—compatible **V**ersion of **HTML** (programming)
	2. E**x**tensible **HTML** (programming)
XID	E**x**change **Id**entification
XIF	**X**PD **I**mprovement **F**actor
XIP	E**x**ecute-**I**n-**P**lace (graphics)
XIPS	**X**enon **I**on **P**ropulsion **S**ubsystem
XIWT	**Cross**-**I**ndustry **W**orking **T**eam
xLC	**x**DSL **L**ine **C**ard
X_L	Symbol for **Inductive Reactance**
XLib	**X L**ib**r**ary (Windows System)
XLIU	**X.25/X.75 L**ink **I**nterface **U**nit
XLL	E**x**tensible **L**inking **L**anguage (programming)
XLR	**X**-Series **L**ockheed **R**ubber (connectors)
XMA	E**x**tended **M**emory **A**ccess
Xmission	**Transmission** (Ham radio)

Xmit	**Transmit** (Ham radio)
Xmitter	**Transmitter** (Ham radio)
XMF	e**X**tensible **M**usic **F**ormat
XML	E**x**tensible **M**arkup **L**anguage (programming)
XMP	**X**/Open **M**anagement **P**rotocol
XMPP	E**x**tensible **M**essaging and **P**resence **P**rotocol
XMS	E**x**tended **M**emory **S**pecification
XMS API	**X**.400 **M**essage **S**tore **A**pplication **P**rogram **I**nterface
XMSN	**Transmi**ssion
XMT	**Trans**mit (Ham radio)
XMTR	**Transmi**tter (Morse code transmissions)
XMUX	**Trans**mu**l**tiple**x**er
XnLC	**X**DSL **L**ine **C**ard in Access **n**ode Express
XNS	**X**erox **N**etwork **S**ystems (protocol)
XO	**Cr**ystal **O**scillator
XOFF	**Transmitter Off** (control character)
XOMAPI	**X**.400 **O**SI data **M**anipulation **A**pplication **P**rogram **I**nterface
XON	**Transmitter On** (control character)
XON/XOFF	**Transmitter On and Transmitter Off** (protocol)
XOR	E**x**clusive-**OR** gate (digital electronics)
Xover	**Crossover**
XPAD	E**x**ternal **P**acket **A**ssembler/**D**isassembler
XPC	E**x**tended **P**rocessing **C**abinet
XPD	1. **Transp**onder
	2. **Cross-P**olarization **D**iscrimination
	3. **Cross-P**olarization **D**ecoupling
XPDR	**Transp**onde**r**
XPG	**X**/Open **P**ortability **G**uide (standard group)
XPI	1. **Cross-P**olarization **I**solation (Intelsat)
	2. **Cross-P**olarization **I**nterference (Intelsat)
XPIC	**Cross-P**olarization **I**nterference **C**anceller (Intelsat)
XPL	**Cross-P**olarization **L**evel
XPM	1. E**x**tended **P**eripheral **M**odule
	2. **Cross-P**hase **M**odulation
X-POL	1. **Cross-Pol**arized (Intelsat)
	2. **Cross-Pol**arization (antennas)
Xponder	**Transponder**
XPS	**X**-ray **P**hotoelectron **S**pectroscopy
XRB	**Transmit R**eference **B**urst
Xref	**Cross-ref**erence
XRF	1. E**x**tended **R**ecovery **F**acility
	2. **X-R**ay **F**luorescence (spectroscopy)

X

XSG	1. **Ex**tended **S**ervice **G**roup
	2. **X**.25 **S**ervice **G**roup
XSL	**Ex**tensible **S**tyle-sheet **L**anguage (programming)
XSMP	**X S**ession **M**anager **P**rotocol
XT	1. **Crosst**alk
	2. **Ex**tended **T**echnology (IBM PCs)
Xtal	**Crystal**
Xtalk	**Crosstalk**
XTELs	**Tel**eservices protocol (Sun Solaris)
Xterm	A **term**inal for the **X** Windows system
XTI	**X**/Open **T**ransport **I**nterface
XTL	**T**eleservices architecture for Sun Solaris (Sun Microsystems)
XTP	1. **Ex**press **T**ransport **P**rotocol
	2. **Ex**press **T**ransfer **P**rotocol
XUI	**X** Windows **S**ystem **U**ser **I**nterface (toolkit)
XWA	**Ex**perimental **W**ireless **A**pparatus
XWS	**X W**indows **S**ystem
XXX	**Triple X** (protocol)
XXXX	The last **four digits** of a telephone number
XYL	**Former Y**oung **L**ady (morse code transmission)

Y y

y	Symbol for prefix **y**octo-, denoting 10^{-24} or 2^{-80}
Y	1. Symbol for prefix **y**otta-, denoting 10^{24} or 2^{80}
	2. Symbol for **admittance** (electronics)
	3. Symbol for **Y**ttrium
	4. Symbol for **luminosity**
Y cable	A **Y**-shaped **cable**
Y/C	**luminosity and C**olor splitter (color image encoding)
Y2K	**Year 2000** (Millennium Bug)
YAG	**Y**ttrium–**A**luminium **G**arnet crystalline material (laser electronics)
YAHOO	1. **Y**et **A**nother **H**ierarchically **O**fficious **O**racle (Web portal)
	2. **Y**et **A**nother **H**ierarchically **O**rganized **O**racle (Web portal)
YB	**Y**ottabyte, which is 10^{24} or 1024 times zettabyte
YBCO	**Y**ttrium–**B**arium–**C**opper–**O**xide (microwave devices)
yd	**ya**rd (unit of area)
yd²	**square y**ard
YEL	**yel**low
YIG	**Y**ttrium–**I**ron **G**arnet crystalline material (microwave devices)

Y

YL	**Y**oung **L**ady (Morse code transmissions)
YP	**Y**ellow **P**ages (British Telecom)
YTF	**YIG T**uned **F**ilter (microwave)
YTO	**Y**IG-**T**uned **O**scillator

Z z

z	Symbol for prefix **z**epto-, which is 10^{-21} or 2^{-70}
.z	Name extension for UNIX files compressed with the g**z**ip
Z	1. Symbol for prefix **z**etta-, which is 10^{21} or 2^{70}
	2. Symbol for **impedance** (electric circuits)
	3. **Z**ulu time
Z time	**Z**ebra **time**
Z-CAL	**Z**ero Range **Cal**ibration (Intelsat)
Z-marker	**Z**one **marker**
Z$_0$	**Characteristic Impedance** (transmission line)
Z80	An **8**-bit microprocessor developed by **Z**ilog Company
ZAK	**Z**ero **A**dministration **K**it (software)
ZAW	**Z**ero **A**dministration **W**indows (operating system)
ZB	**Z**etta**b**yte, which is 10^{21} or 10^{24} times Exabytes
ZBTSI	**Z**ero **B**yte **T**ime **S**lot Interchange (control technique)
ZCS	1. **Z**ero **C**ode **S**uppression
	2. **Z**ero **C**urrent **S**witching
ZD	**Z**ero **D**efects
ZDNet	**Z**iff **D**avis **Net**work
ZDSF	**Z**ero **D**ispersion **S**hifted **F**iber
ZDW	**Z**ero **D**ispersion **W**avelength
ZIF socket	**Z**ero **I**nsertion **F**orce **socket** (Intel)
ZiffNet	See **ZDNet**

Zin	Input **Impedance**
.zip	Name extension for archive files compressed by PK**ZIP**
Zip	**zip**ped
ZIP	1. **Z**one **I**nformation **P**rotocol
	2. **Z**one **I**nformation **P**lan
	3. **Z**igzag **I**nline **P**ackage
Ziv	Lempel–**Ziv**
Zl	load **Impedance**
ZnO	**Zin**c **O**xide (semiconductors)
.zoo	Name extension for archive files **Zoo** compression utility
Zout	**out**put **Impedance**
ZRL	**Z**ero **R**elative **L**evel
ZRT	**Z**ero **R**ange **T**ransponder (Intelsat)
Z$_s$	source **Impedance**
zsh	**Z** **sh**ell (UNIX)
ZT	**Z**one **T**ime (Intelsat)
ZTLP	**Z**ero **T**ransmission **L**evel **P**oint (telephony)
ZUM	**Z**one **U**sage **M**easured
ZV	**Z**oomed **V**ideo
ZVA	**Z**ero **V**oltage **A**ctivated circuit (electronics)
ZVS	**Z**ero **V**oltage **S**witching

Annex-A (Special Characters)

.	dot or period sign
'	1. apostrophe or quote sign
	2. meaning "**feet**" used when defining length of something
	3. meaning "**minutes**" used when defining latitude and longitude
"	1. Quotation mark (called ditto)
	2. meaning "**inches**" used when defining dimensions of something
	3. meaning "**seconds**" used when defining latitudes and longitudes
	4. the same word as above
"	Double opening quote
"	Double closing quote
'	Single opening quote
'	Single closing quote
—	Em dash
—	1. En dash (also called hyphen)
	2. meaning "**to**" when defining ranges of values
/	1. Oblique stroke or Forward slash (also called diagonal)
	2. Fraction sign meaning "**to**" or "**conversion to**"
\	Backward Slash (also called back slash)
:	Colon
;	Semicolon

=	Equals; is the same as (mathematics)
≠	Does not equal; is different from
#	Sign of "**number**" (also called Hash)
≈	Is approximately equal to (mathematics)
>	Is more than (mathematics)
<	Is less than (mathematics)
°	Symbol for **degree** (plane angle)
°C	Symbol for **degree** Celsius
°F	Symbol for **degree** Fahrenheit
°R	Symbol for **degree** Rankine (absolute temperature scale)
&	Sign of "**and**" (called Ampersand)
@	"**At**" sign (Internet e-mail addresses)
Δ	Greek letter "**delta**" (sign of refractive index contrast)
Δ-β	**Delta–Beta** (switch)
Δ-λ	**Delta–Lambda** (spectral width)
Δ-Σ	**Delta–Sigma** (modulation)
ε	Greek letter "**epsilon**" (sign of electrical permittivity)
λ	Greek letter "**Lambda**" (sign of wavelength)
λ_0	**Zero**-dispersion **wavelength** (fiber optics)
λ_{cc}	cutoff **wavelength** (fiber optics)
μ	1. Greek Letter "**mu**"
	2. Symbol for permeability
	3. Sign of prefix **micro**-, (one-millionth or 10^{-6})
μA	**micro**ampere (electronics)
μbar	**microbar**
μC	**micro**computer
μCi	**micro**curie
μF	**microf**arad (electronics)
μH	**microh**enry (electronics)
μLx	**microl**ux
μm	**microm**eter or micron
μmho	**micromho** (electronics)
μΩ	**microohm** (electronics)
μp	**micro**processor
μrd	**micror**ad
μR	**micror**oentgen
μrem	**microrem**
μrep	**microrep**
μs	**micros**econd
μS	**micros**iemens
μV	**microv**olt (electronics)
μμ	**micromicro** (pico-)
μW	**microw**att (power)

ω	Greek letter "**omega**"; symbol for angular frequency
Ω	1. Greek letter "**omega**"
	2. Sign of **ohm** (unit of resistance and impedance)
Ω/V	**ohm per** Volt
℧	Upside-down capital **omega** (unit of conductivity 'mho")
φ	Greek letter "**phi**" (sign of angle of phase)
ΦM	**Phase** Modulation
π	Greek letter "**pi**", symbol for the number 3.1416 (mathematics)
?	Question mark
ρ	Greek letter "**rho**" (unit of electric charge density)
σ	Greek letter "**sigma**" (unit of electrical conductivity)
*	Symbol for marking an important point (called asterisk, splat, star)
©	**C**opyright
®	**R**egistered
™	**T**rademark
$	U.S. **Dollar** (currency)
₤	**Pound Sterling** (currency)
€	**E**uro (currency)
¥	**Y**en (currency)
%	**P**ercent